Operator Theory: Advances and Applications
Vol. 120

Lie Algebras of Bounded Operators

Daniel Beltiţă
Mihai Şabac

Springer Basel AG

Authors:

Daniel Belțiță
Institute of Mathematics
"Simion Stoilow" of the Romanian Academy
Calea Grivitei No. 21
P.O. Box 1-764
70700 Bucharest
Romania
e-mail: dbeltita@pompeiu.imar.ro

Mihai Şabac
Faculty of Mathematics
University of Bucharest
Str. Academiei 14
70109 Bucharest
Romania
e-mail: msabac@math.math.unibuc.ro

© 2000 Mathematics Subject Classification 47A10, 47A13, 47A25, 47A60, 47B40, 47B47, 17B30, 17B55, 17B65

A CIP catalogue record for this book is available from the
Library of Congress, Washington D.C., USA

Deutsche Bibliothek Cataloging-in-Publication Data

Belțiță, Daniel:
Lie algebras of bounded operators / Daniel Belțiță ; Mihai Şabac. - Basel ; Boston ; Berlin : Birkhäuser, 2001
(Operator theory ; Vol. 120)
ISBN 978-3-7643-6404-5 ISBN 978-3-0348-8332-0 (eBook)
DOI 10.1007/978-3-0348-8332-0

Originally published by Birkhäuser Verlag in 2001

Printed on acid-free paper produced from chlorine-free pulp. TCF ∞
Cover design: Heinz Hiltbrunner, Basel

ISBN 978-3-7643-6404-5

9 8 7 6 5 4 3 2 1

Contents

Introduction

In several proofs from the theory of finite-dimensional Lie algebras, an essential contribution comes from the Jordan canonical structure of linear maps acting on finite-dimensional vector spaces. On the other hand, there exist classical results concerning Lie algebras which advise us to use infinite-dimensional vector spaces as well. For example, the classical Lie Theorem asserts that all *finite-dimensional* irreducible representations of solvable Lie algebras are one-dimensional. Hence, from this point of view, the solvable Lie algebras cannot be distinguished from one another, that is, they cannot be classified. Even this example alone urges the *infinite-dimensional* vector spaces to appear on the stage. But the structure of linear maps on such a space is too little understood; for these linear maps one cannot speak about something like the Jordan canonical structure of matrices.

Fortunately there exists a large class of linear maps on vector spaces of arbitrary dimension, having some common features with the matrices. We mean the bounded linear operators on a complex Banach space. Certain types of bounded operators (such as the Dunford spectral, Foiaş decomposable, scalar generalized or Colojoară spectral generalized operators) actually even enjoy a kind of Jordan decomposition theorem.

One of the aims of the present book is to expound the most important results obtained until now by using bounded operators in the study of Lie algebras. But *the interaction between operator theory and Lie algebra theory turned out to be in both directions.* That is why another aim of the book is to expound a series of answers which Lie algebra theory offers to some long-standing open questions in operator theory related to the construction of a joint spectral theory for non-commuting tuples of operators.

The present exposition is self-contained. However we assume that readers are acquainted with both Lie algebras and general spectral theory. We now present a short exposition of the contents of our book.

The first chapter contains most of the necessary background for reading this book. We must mention that the proofs which can be found in already published books are generally omitted and we indicate these books whenever necessary.

The aim of the second chapter is to describe some features of a very interesting phenomenon, already presented in 1935 by N. Jacobson, which plays a key role throughout the book. We mean the occurrence of nilpotent and quasinilpotent elements in Lie algebras of operators. The axiomatic approach to this phenomenon (cf. §16 and §18) relates it to certain asymmetric commutativity relations (polynomial, respectively analytic, commutativity).

The third chapter contains variants of the classical Lie and Engel Theorems which hold for Lie algebras of bounded operators on arbitrary Banach spaces. In §20 we prove a result generalizing the Lie Theorem in the form that asserts the existence of weights for finite-dimensional representations of solvable Lie algebras. On the other hand, §21 and §22 contain results extending the Lie Theorem in the form that asserts the existence of non-trivial invariant subspaces for a solvable Lie algebra of linear maps on a finite-dimensional vector space. The theorems from these two paragraphs concern certain classes of infinite-dimensional Lie algebras possessing a Levi-Malçev decomposition (for short LM-decomposable Lie algebras). Finally §24 contains generalizations of the Lie Theorem under another aspect: the associative envelope of a solvable Lie algebra of matrices is commutative modulo its Jacobson radical. We prove such results for solvable Lie algebras of operators and §23 contains generalizations of the Engel Theorem, namely sufficient conditions for the associative envelope of a Lie algebra of quasinilpotent operators to consist only of quasinilpotents.

The fourth chapter expounds non-commutative variants of the Taylor joint spectrum for commuting tuples of operators. They possess the expected properties for joint spectra: nonemptyness, compactness, projection property and, in the case of tuples generating a nilpotent Lie algebra, even the polynomial spectral mapping property.

In the fifth chapter, the first two paragraphs are centered around the notion of a normal element in a Lie algebra with involution. In an appropriate framework, this concept allows extension of several properties of the normal operators from Hilbert to Banach spaces. In §31 we treat a situation which should be specific to infinite-dimensional spaces: semisimple Lie algebras of quasinilpotent operators. (Recall that any Lie algebra consisting only of nilpotent matrices is itself nilpotent.)

Several results given in the text are published here for the first time. These results and other hystorical comments are indicated in the "Notes" at the end of each chapter. The book finishes with an extensive bibliography concerning Lie algebras of bounded operators and related topics.

The suggestion to publish this monograph in the series "OT: Advances and Applications" is due to Professor C. Foiaş and Professor I. Gohberg; we take this opportunity to express our gratitude to them. The editor's remarks have improved the language and some technical aspects from the first version of the manuscript and we have also the pleasure to thank him.

Bucharest, May, 2000 The Authors

Chapter I

Preliminaries

This chapter contains a short review of the main objects of our book: Lie algebras and bounded linear operators on a complex Banach space. We will use standard notation for operations with sets and functions. We will denote by $\mathbb{N}$, $\mathbb{Z}$, $\mathbb{R}$, $\mathbb{C}$ the sets of, respectively, natural, integer, real and complex numbers. The classical facts contained in this chapter will be given without proofs.

A. Lie Algebras

§ 1 Basic facts

Let us consider a field Φ. A general definition of a Lie algebra is the following.

Definition 1. A *Lie algebra over the field* Φ is a vector space $\mathcal{L}$ over Φ endowed with a composition law (named Lie product)

$$\mathcal{L} \times \mathcal{L} \to \mathcal{L}, \ (x, y) \mapsto [x, y]$$

which satisfies the following conditions:

(b) it is bilinear, i.e.

$$[\alpha x_1 + \beta x_2, y] = \alpha[x_1, y] + \beta[x_2, y], \ [x, \alpha y_1 + \beta y_2] = \alpha[x, y_1] + \beta[x, y_2]$$

for every $x_1, x_2, y_1, y_2 \in \mathcal{L}$ and $\alpha, \beta \in \Phi$;

(j) it satisfies the Jacobi identity, i. e.

$$[[x, y], z] + [[y, z], x] + [[z, x], y] = 0$$

for every $x, y, z \in \mathcal{L}$;

(i) the equality

$$[x, x] = 0$$

holds for every $x \in \mathcal{L}$.

Remark 1. A Lie algebra $\mathcal{L}$ has the property

$$(s)\quad [x,y]=-[y,x] \text{ for every } x,y\in\mathcal{L}.$$

If the characteristic of the field Φ is different from 2, then the above property (s) can replace the condition (i) in the above definition of a Lie algebra.

Definition 2. A *Lie subalgebra* of a Lie algebra $\mathcal{L}$ is a subset of $\mathcal{L}$ which is a Lie algebra with respect to the restrictions of the operations of $\mathcal{L}$.

Example 1. A very important example of a Lie algebra structure is given by an arbitrary associative algebra over the field Φ. Indeed, if $\mathcal{A}$ is such an associative algebra, i.e. $\mathcal{A}$ is a vector space over the field Φ endowed with a bilinear and associative composition law

$$\mathcal{A}\times\mathcal{A}\to\mathcal{A},\ (x,y)\mapsto xy,$$

then we can put

$$[\cdot,\cdot]:\mathcal{A}\times\mathcal{A}\to\mathcal{A},\ (x,y)\mapsto[x,y],$$
$$[x,y]:=xy-yx,$$

and we obtain a Lie algebra structure induced by the associative algebra structure of $\mathcal{A}$ over Φ.

Example 2. A particular case of the above example is given by the associative algebra of all linear transformations of some vector space into itself. Indeed, let $\mathcal{V}$ be a vector space over Φ and $End(\mathcal{V})$ be the set of all linear transformations of $\mathcal{V}$ into itself. Then for $A,B\in End(\mathcal{V})$ and $\alpha\in\mathcal{V}$ we can define αA, $A+B$, AB by $(\alpha A)v=\alpha(Av)$, $(A+B)v=Av+Bv$, $(AB)v=A(Bv)$ for every $v\in\mathcal{V}$. It is well known that in this way $End(\mathcal{V})$ becomes an associative algebra and particularly, as above, a Lie algebra over Φ.

Definition 3. If $\mathcal{L}_1$, $\mathcal{L}_2$ are Lie algebras over Φ, then a linear map $\rho:\mathcal{L}_1\to\mathcal{L}_2$ is called a *Lie morphism* if it satisfies

$$\rho([x,y])=[\rho(x),\rho(y)] \text{ for every } x,y\in\mathcal{L}_1,$$

where we used the same notation $[\cdot,\cdot]$ for the Lie product in $\mathcal{L}_1$ and $\mathcal{L}_2$. If moreover $\mathcal{L}_2=End(\mathcal{V})$ for a certain vector space $\mathcal{V}$ over Φ, then a Lie morphism $\rho:\mathcal{L}_1\to End(\mathcal{V})$ is called a *representation* of $\mathcal{L}_1$ in the vector space $\mathcal{V}$.

Concerning the above Example 1, we note that, if $\mathcal{L}$ is a finite-dimensional Lie algebra, then there exists a unital associative algebra $U(\mathcal{L})$ and an injective Lie morphism $i:\mathcal{L}\to U(\mathcal{L})$ with the following property: for every unital associative algebra $\mathcal{A}$ and every Lie morphism $\rho:\mathcal{L}\to\mathcal{A}$ there exists a unique morphism of unital associative algebras $\bar\rho:U(\mathcal{L})\to\mathcal{A}$ such that $\bar\rho\circ i=\rho$. Then $U(\mathcal{L})$ is uniquely determined modulo an isomorphism of unital associative algebras and is called the *enveloping algebra* of $\mathcal{L}$. Usually one identifies $\mathcal{L}$ with the Lie subalgebra $i(\mathcal{L})$ of $U(\mathcal{L})$ and in the above property one denotes simply ρ instead of $\bar\rho$.

It is worth noting that, if $\mathcal{L}$ is a complex Lie algebra and $\{e_1, \dots, e_n\}$ is a basis of $\mathcal{L}$, then we can take

$$U(\mathcal{L}) = \mathbb{C}\langle e_1, \dots, e_n \rangle / J,$$

where $\mathbb{C}\langle e_1, \dots, e_n \rangle$ is the free unital algebra generated by $e_1, \dots, e_n$ (i.e. the tensor algebra of the vector space $\mathcal{L} = \mathbb{C}e_1 \oplus \cdots \oplus \mathbb{C}e_n$) and J is its two-sided ideal generated by the set $\{ef - fe - [e, f] \mid e, f \in \mathcal{L}\}(\subset \mathbb{C}\langle e_1, \dots, e_n \rangle)$. The elements of $\mathbb{C}\langle e_1, \dots, e_n \rangle$ are called *polynomials in the noncommuting indeterminates* $e_1, \dots, e_n$. Concerning these polynomials we remark that, if $f_1, \dots, f_m$ are arbitrary elements of $\mathcal{L}$, then, by universality property of the free algebra, there exists a unique algebra morphism $\mathbb{C}\langle X_1, \dots, X_m \rangle \to U(\mathcal{L})$ taking X_i in f_i for $i = 1, \dots, m$. Then the image of a $p \in \mathbb{C}\langle X_1, \dots, X_m \rangle$ under this morphism is denoted usually by $p(f_1, \dots, f_m)(\in U(\mathcal{L}))$ and is called the *image of p in the envelopping algebra* $U(\mathcal{L})$. (Of course, this image depends on the specified system $f_1, \dots, f_m \in U(\mathcal{L})$.)

The following basic result holds.

Theorem 1 (Poincaré-Birkhoff-Witt). *If $\{e_1, \dots, e_n\}$ is a basis of the Lie algebra $\mathcal{L}$, then $\{e_1^{\nu_1} \cdots e_n^{\nu_n} \mid \nu_1, \dots, \nu_n \in \mathbb{N}\}$ is a basis of $U(\mathcal{L})$.*

Definition 4. A *Lie derivation* of a Lie algebra $\mathcal{L}$ is a linear map $D : \mathcal{L} \to \mathcal{L}$ which has the property

$$D([x, y]) = [Dx, y] + [x, Dy] \text{ for every } x, y \in \mathcal{L}.$$

An *inner derivation* of $\mathcal{L}$ is a derivation defined in the following way by means of an element $x \in \mathcal{L}$:

$$D_x : \mathcal{L} \to \mathcal{L}, \; y \mapsto D_x y := [x, y].$$

It is usual to denote the above derivation by adx, i.e.

$$(adx)y = [x, y].$$

Remark 2. If $\mathcal{L}$ is a Lie algebra, then the map

$$ad : \mathcal{L} \to End(\mathcal{L}), \; x \mapsto adx,$$

is a representation of $\mathcal{L}$ which is called the *adjoint representation* of $\mathcal{L}$.

Remark 3. Let $\mathcal{A}$ be an associative algebra and $D : \mathcal{A} \to \mathcal{A}$ be a derivation of $\mathcal{A}$, i.e. D is a linear map such that

$$D(xy) = (Dx)y + x(Dy) \text{ for every } x, y \in \mathcal{A}.$$

Then D is also a Lie derivation with respect to the Lie algebra structure induced by the associative algebra structure of $\mathcal{A}$.

Remark 4. If $\mathcal{V}$ is a vector space over Φ, then an inner derivation of the Lie algebra $End(\mathcal{V})$ is given by an element $A \in End(\mathcal{V})$ in the following way:

$$(ad\,A)B = AB - BA \text{ for every } B \in End(\mathcal{V}).$$

One of the basic concepts in order to describe the structure of Lie algebras is the concept of (Lie) ideal.

Definition 5. An *ideal* of a Lie algebra $\mathcal{L}$ is a vector subspace $\mathcal{I}$ of $\mathcal{L}$ such that

$$[a, x] \in \mathcal{I} \text{ for every } a \in \mathcal{I} \text{ and } x \in \mathcal{L}.$$

Remark 5. An ideal of the Lie algebra $\mathcal{L}$ is an invariant subspace for all the inner derivations of $\mathcal{L}$, i.e.

$$(ad\,x)\mathcal{I} \subseteq \mathcal{I} \text{ for every } x \in \mathcal{L}.$$

Remark 6. If $\mathcal{A}$ is an associative algebra, then an ideal of $\mathcal{A}$ with respect to the induced Lie algebra structure (see Example 1) will be called a *Lie ideal* of $\mathcal{A}$.

Remark 7. If $\mathcal{I}$ is an ideal of the Lie algebra $\mathcal{L}$, then the quotient vector space $\mathcal{L}/\mathcal{I}$ has a canonical structure of Lie algebra:

$$[\tilde{a}, \tilde{b}] = \widetilde{[a, b]} \text{ for every } a, b \in \mathcal{L},$$

where $\mathcal{L} \longrightarrow \mathcal{L}/\mathcal{I}$, $a \mapsto \tilde{a}$ is the quotient map.

Notation. Let us consider two vector subspaces $\mathcal{S}_1, \mathcal{S}_2$ of a Lie algebra $\mathcal{L}$.

a) $\mathcal{S}_1 \cap \mathcal{S}_2$ is the intersection of $\mathcal{S}_1$ and $\mathcal{S}_2$.

b) $\mathcal{S}_1 + \mathcal{S}_2 := \{x_1 + x_2 \mid x_1 \in \mathcal{S}_1, x_2 \in \mathcal{S}_2\}$ is the vector subspace of $\mathcal{L}$ spanned by $\mathcal{S}_1$ and $\mathcal{S}_2$.

c) $[\mathcal{S}_1, \mathcal{S}_2]$ is the vector subspace of $\mathcal{L}$ spanned by the set $\{[x_1, x_2] \mid x_1 \in \mathcal{S}_1, x_2 \in \mathcal{S}_2\}$.

d) $Z_{\mathcal{L}} := \{x \in \mathcal{L} \mid [x, \mathcal{L}] = \{0\}\}$ denotes the *center* of $\mathcal{L}$.

§ 2 Ideals, solvability, nilpotence, radical and semisimplicity

An ideal of a Lie algebra $\mathcal{L}$ is a subspace $\mathcal{I}$ of $\mathcal{L}$ which satisfies $[\mathcal{L}, \mathcal{I}] \subseteq \mathcal{I}$. It is obvious that $\mathcal{I}_1 \cap \mathcal{I}_2, \mathcal{I}_1 + \mathcal{I}_2$ and $[\mathcal{I}_1, \mathcal{I}_2]$ are ideals of $\mathcal{L}$ if $\mathcal{I}_1, \mathcal{I}_2$ are ideals of $\mathcal{L}$.

For any Lie algebra $\mathcal{L}$ we can consider the following two series of ideals in $\mathcal{L}$.

1) The *derived series*, i.e.

$$\mathcal{L} \supseteq \mathcal{L}^{(1)} := \mathcal{L}' := [\mathcal{L}, \mathcal{L}] \supseteq \cdots \supseteq \mathcal{L}^{(k)} := [\mathcal{L}^{(k-1)}, \mathcal{L}^{(k-1)}] \supseteq \cdots.$$

2) The *lower central series*, i.e.

$$\mathcal{L} =: \mathcal{L}^1 \supseteq \mathcal{L}^2 := \mathcal{L}' = [\mathcal{L}, \mathcal{L}^1] \supseteq \cdots \supseteq \mathcal{L}^k := [\mathcal{L}, \mathcal{L}^{k-1}] \supseteq \cdots.$$

The finiteness of these two series defines two important classes of Lie algebras in order to compare them with the *commutative* case (i.e. $[x, y] = 0$ for every $x, y \in \mathcal{L}$).

Definition 1. A Lie algebra $\mathcal{L}$ is *solvable* if $\mathcal{L}^{(n)} = \{0\}$ for some positive integer n.

Definition 2. A Lie algebra $\mathcal{L}$ is *nilpotent* if $\mathcal{L}^n = \{0\}$ for some positive integer n.

Obviously any nilpotent Lie algebra is solvable. Any commutative Lie algebra is nilpotent hence solvable. Now we state other basic properties of the solvable and nilpotent Lie algebras.

Proposition 1.

a) *Every Lie subalgebra of a solvable (respectively nilpotent) Lie algebra is solvable (respectively nilpotent).*
b) *Every image by a Lie morphism of a solvable (respectively nilpotent) Lie algebra is solvable (respectively nilpotent).*
c) *If $\mathcal{I}$ is a solvable ideal of a Lie algebra $\mathcal{L}$ and $\mathcal{L}/\mathcal{I}$ is also solvable, then $\mathcal{L}$ is solvable.*
d) *In every Lie algebra, the sum of any two solvable (respectively nilpotent) ideals is also solvable (respectively nilpotent).*

Now let $\mathcal{L}$ be a finite-dimensional Lie algebra. Using the property *d)* from Proposition 1 we deduce that there exists a solvable ideal $\mathcal{R}$ of $\mathcal{L}$ which contains every solvable ideal of $\mathcal{L}$. This solvable ideal $\mathcal{R}$ is called the *(solvable) radical* of $\mathcal{L}$. The *nil-radical* $\mathcal{N}$ of $\mathcal{L}$ is defined in the same way: it is the greatest nilpotent ideal of $\mathcal{L}$ (i.e. $\mathcal{N}$ is a nilpotent ideal which contains all the other nilpotent ideals of $\mathcal{L}$). Obviously $\mathcal{N} \subseteq \mathcal{R}$.

Another very useful property is the following.

Proposition 2. *If $\mathcal{L}$ is a solvable Lie algebra over a field Φ of characteristic zero then, with the above notation we have $[\mathcal{L}, \mathcal{R}] \subseteq \mathcal{N}$.*

Definition 3. Let $\mathcal{L}$ be a finite-dimensional Lie algebra.

a) If the radical $\mathcal{R}$ of $\mathcal{L}$ equals $\{0\}$, then the Lie algebra $\mathcal{L}$ is called *semisimple* (i.e. $\mathcal{L}$ is semisimple if it contains no non-zero solvable ideal).
b) The Lie algebra $\mathcal{L}$ is called *simple* if it contains no proper ideals (i.e. distinct from $\{0\}$ and $\mathcal{L}$).

Obviously every simple Lie algebra is semisimple.

Because in the following we will study Lie algebras of bounded operators on some complex Banach space, the class of Lie algebras over the complex field $\mathbb{C}$ will be of interest. In the following we will consider $\Phi = \mathbb{C}$ throughout (excepting only the especially mentioned cases). Hence we shall mainly consider complex Lie algebras and complex vector spaces. In this case there exist the following two famous results. (They hold actually even for Lie algebras over an algebraically closed field of characteristic zero.)

Theorem 1 (Engel's Theorem). *Let $\mathcal{L}$ be a Lie algebra over $\mathbb{C}$ and ρ be a representation of $\mathcal{L}$ on the vector space $\mathcal{V}$ of finite dimension n. If $\rho(x)$ is a nilpotent mapping of $\mathcal{V}$ for each $x \in \mathcal{L}$, then there exists a chain*

$$\{0\} = \mathcal{V}_0 \subset \mathcal{V}_1 \subset \cdots \subset \mathcal{V}_k \subset \mathcal{V}_{k+1} \subset \cdots \subset \mathcal{V}_n = \mathcal{V}$$

such that

$$\dim \mathcal{V}_k = k \text{ for every } k = 0, \ldots, n$$

and

$$\rho(x)\mathcal{V}_k \subseteq \mathcal{V}_{k-1} \text{ for every } k = 1, \ldots, n \text{ and every } x \in \mathcal{L}.$$

Theorem 2 (Lie's Theorem). *Let $\mathcal{L}$ be a solvable Lie algebra over $\mathbb{C}$ and ρ be a representation of $\mathcal{L}$ on the vector space $\mathcal{V}$ of finite dimension n. Then there exists a chain*

$$\{0\} = \mathcal{V}_0 \subset \mathcal{V}_1 \subset \cdots \subset \mathcal{V}_k \subset \mathcal{V}_{k+1} \subset \cdots \subset \mathcal{V}_n = \mathcal{V}$$

such that

$$\dim \mathcal{V}_k = k \text{ for every } k = 0, \ldots, n$$

and

$$\rho(x)\mathcal{V}_k \subseteq \mathcal{V}_k \text{ for every } k = 0, \ldots, n \text{ and every } x \in \mathcal{L}.$$

If we apply the above two theorems to the adjoint representation of a Lie algebra, then we obtain the following characterization of the solvable (respectively nilpotent) finite-dimensional complex Lie algebras.

Theorem 3. *Let $\mathcal{L}$ be a complex Lie algebra of finite dimension n. Then $\mathcal{L}$ is solvable iff there exists a chain*

$$\{0\} = \mathcal{I}_0 \subset \mathcal{I}_1 \subset \cdots \subset \mathcal{I}_k \subset \mathcal{I}_{k+1} \subset \cdots \subset \mathcal{I}_n = \mathcal{L}$$

consisting of ideals of $\mathcal{L}$ such that

$$\dim \mathcal{I}_k = k \text{ for every } k = 1, \ldots, n.$$

The Lie algebra $\mathcal{L}$ is nilpotent iff it has a chain of ideals as above and such that

$$(ad\, x)\mathcal{I}_k \subseteq \mathcal{I}_{k-1} \text{ for every } k = 1, \ldots, n \text{ and every } x \in \mathcal{L}.$$

This result suggests the following extension to infinite-dimensional Lie algebras of the concept of solvability and nilpotence from the finite-dimensional case.

Definition 4. A Lie algebra $\mathcal{L}$ is *countably solvable* (or ∞-*solvable*) if there exist $N \leq \aleph_0$ and a sequence $\{\mathcal{I}_k\}_{1 \leq k < N}$ such that

1) $\mathcal{I}_k$ is an ideal of $\mathcal{L}$ such that $\dim \mathcal{I}_k = k$ for every $k \in \mathbb{N}$ with $k < N$;
2) $\mathcal{L}$ is spanned as a vector space by $\bigcup_{1 \leq k < N} \mathcal{I}_k$.

Remark 1. A Lie algebra $\mathcal{L}$ is countably solvable iff there exist $m \leq \infty$ and a chain $\{\mathcal{J}_k\}_{1\leq k<m}$ (i.e. $\mathcal{J}_1 \subset \cdots \subset \mathcal{J}_k \subset \cdots$) such that

1) $\mathcal{J}_k$ is an ideal of $\mathcal{L}$ such that $\dim \mathcal{J}_k = k$ for every $k \in \mathbb{N}$ with $k < m$;
2) $\mathcal{L} = \bigcup_{1\leq k<m} \mathcal{J}_k$.

This class of countably solvable Lie algebras contains the class of the finite-dimensional solvable Lie algebras and is contained in the class of locally solvable and ideally finite (quasisolvable) Lie algebras given in the following definition.

Definition 5. Let $\mathcal{L}$ be a Lie algebra.

a) If $\mathcal{L}$ is spanned as a vector space by a directed family $\{\mathcal{L}_i\}_{i\in I}$ of finite-dimensional subalgebras (i.e. for every $i, j \in I$ there exists $k \in I$ such that $\mathcal{L}_i + \mathcal{L}_j \subseteq \mathcal{L}_k$), then $\mathcal{L}$ is called a *locally finite Lie algebra.* If moreover the family $\{\mathcal{L}_i\}_{i\in I}$ consists only of solvable (respectively nilpotent Lie algebras), then $\mathcal{L}$ is called *locally solvable* (respectively *locally nilpotent*).

b) The Lie algebra $\mathcal{L}$ is called *ideally finite* if it is spanned as a vector space by a family of finite-dimensional ideals. If $\mathcal{L}$ is ideally finite and locally solvable, then it is called *quasisolvable.* If $\mathcal{L}$ is ideally finite and locally nilpotent, then it is called *quasinilpotent.*

Remark 2. Obviously, if $\mathcal{L}$ is a Lie algebra, then the following assertions are equivalent:

(s_0) $\mathcal{L}$ is quasisolvable.
(s_1) $\mathcal{L} \doteq \sum_{\alpha\in\Lambda} \mathcal{I}_\alpha$ for a certain family $\{\mathcal{I}_\alpha\}_{\alpha\in\Lambda}$ consisting of finite-dimensional solvable ideals of $\mathcal{L}$.
(s_2) $\mathcal{L}$ is ideally finite and every finite-dimensional ideal of $\mathcal{L}$ is solvable.

Remark 3. We have also the following equivalent assertions concerning a Lie algebra $\mathcal{N}$:

(n_0) $\mathcal{N}$ is quasinilpotent.
(n_1) $\mathcal{N} = \sum_{\alpha\in\Lambda} \mathcal{I}_\alpha$ for a certain family $\{\mathcal{I}_\alpha\}_{\alpha\in\Lambda}$ consisting of finite-dimensional nilpotent ideals of $\mathcal{L}$.
(n_2) $\mathcal{N}$ is ideally finite and every finite-dimensional ideal of $\mathcal{N}$ is nilpotent.
(n_3) $\mathcal{N}$ is ideally finite and $adx|_{\mathcal{I}}$ is a nilpotent mapping for every $x \in \mathcal{I}$ and every finite-dimensional ideal $\mathcal{I}$ of $\mathcal{N}$.

Remark 4. A finite-dimensional quasisolvable (respectively quasinilpotent) Lie algebra is solvable (respectively nilpotent). A homomorphic image of a quasisolvable (respectively quasinilpotent) Lie algebra is quasisolvable (respectively quasinilpotent).

Remark 5. If $\mathcal{L}$ is an ideally finite Lie algebra, we can consider the family $\{\mathcal{F}_i\}_{i\in I}$ of all finite-dimensional ideals of $\mathcal{L}$. We can define

$$\mathcal{R}(\mathcal{L}) = \sum_{i\in I} \mathcal{R}(\mathcal{F}_i), \quad \mathcal{N}(\mathcal{L}) = \sum_{i\in I} \mathcal{N}(\mathcal{F}_i),$$

where $\mathcal{R}(\mathcal{F}_i)$ (respectively $\mathcal{N}(\mathcal{F}_i)$) denotes the solvable radical (respectively the nil-radical) of $\mathcal{F}_i$ for every $i \in I$. It is clear that $\mathcal{R}(\mathcal{L})$ (respectively $\mathcal{N}(\mathcal{L})$) is the largest quasisolvable (respectively quasinilpotent) ideal of $\mathcal{L}$. We observe also that $\mathcal{R}(\mathcal{L})$ (respectively $\mathcal{N}(\mathcal{L})$) is a quasisolvable (respectively quasinilpotent) ideal of $\mathcal{L}$.

Theorem 4. *If $\mathcal{L}$ is a quasisolvable Lie algebra and $\mathcal{I}$ is a finite-dimensional ideal of $\mathcal{L}$, then there exists an increasing chain of ideals of $\mathcal{L}$,*

$$\{0\} = \mathcal{I}_0 \subset \cdots \subset \mathcal{I}_k \subset \mathcal{I}_{k+1} \subset \cdots \subset \mathcal{I}_n = \mathcal{I}$$

such that $\dim \mathcal{I}_k = k$ $(0 \leq k \leq n)$. *If $\mathcal{L}$ is moreover quasinilpotent, then the above chain can be chosen such that*

$$[\mathcal{L}, \mathcal{I}_k] \subseteq \mathcal{I}_{k-1} \ (1 \leq k \leq n).$$

Proof. We remark that $x \mapsto ad\, x|_{\mathcal{I}}$ is a representation of the Lie algebra $\mathcal{L}$. Because $\dim \mathcal{I} < \infty$ it follows that $\{ad\, x|_{\mathcal{I}} \mid x \in \mathcal{L}\}$ is a finite-dimensional solvable (respectively nilpotent) Lie algebra and the desired result follows by Lie's (respectively Engel's) Theorem. □

§ 3 S^p classes of solvable Lie algebras

Definition 1. For each positive integer p we define S^p as the class of all the complex Lie algebras $\mathcal{A}$ having the following property:

For every $0 \neq x \in \mathcal{A}$ there exists a chain

$$\mathcal{A}_1(x) \subseteq \mathcal{A}_2(x) \subseteq \cdots \subseteq \mathcal{A}_k(x) \subseteq \cdots \subseteq \mathcal{A}$$

consisting of subalgebras of $\mathcal{A}$ such that

1) $x \in \mathcal{A}_1(x)$ and $\dim \mathcal{A}_1(x) \leq p$,
2) $\mathcal{A}_k(x)$ is a solvable Lie algebra for every k,
3) $\mathcal{A}_k(x)$ is an ideal of $\mathcal{A}_{k+1}(x)$ for every k and $\mathcal{A} = \cup_{k\geq 1}\mathcal{A}_k(x)$.

Concerning the above definition let's observe that the chain $\{\mathcal{A}_k(x)\}_{k\geq 1}$ can be refined because a quotient of a solvable Lie algebra by an ideal is also solvable. So we have the following equivalence.

Proposition 1. *Let $\mathcal{A}$ be a complex Lie algebra. Then the following statements are equivalent.*

(i) $\mathcal{A} \in S^p$.

(ii) *For every $0 \neq x \in \mathcal{A}$ there exists a chain*

$$\mathcal{I}_1(x) \subseteq \mathcal{I}_2(x) \subseteq \cdots \subseteq \mathcal{I}_p(x) \subseteq \cdots \subseteq \mathcal{I}_k(x) \subseteq \cdots \subseteq \mathcal{A}$$

consisting of subalgebras of $\mathcal{A}$ such that

1) $x \in \mathcal{I}_p(x)$,
2) $\dim \mathcal{I}_k(x) = k$ *and $\mathcal{I}_k(x)$ is a solvable ideal of $\mathcal{I}_{k+1}(x)$ for* $1 \leq k < \dim \mathcal{A}$,
3) $\cup_{k\geq 1}\mathcal{I}_k(x) = \mathcal{A}$.

Remark 1. For every positive integer p we have $S^p \subset S^{p+1}$.

Remark 2. If $\mathcal{A} \in S^p$ and $\mathcal{C}$ is a subalgebra of $\mathcal{A}$, then $\mathcal{C} \in S^p$.

Remark 3. If $\mathcal{A}$ is a finite-dimensional Lie algebra and p is a positive integer, then the following properties hold.

a) If $\mathcal{A} \in S^p$, then $\mathcal{A}$ is a solvable Lie algebra.

b) If $\dim \mathcal{A} \leq p$, then $\mathcal{A} \in S^p$ iff $\mathcal{A}$ is a solvable Lie algebra.

c) If $\mathcal{A} \in S^p$, then, with the notation of Definition 1, the chain $\{\mathcal{A}_k(x)\}_{k\geq 1}$ is stationary.

d) If $\mathcal{A} \in S^p$ and $p < \dim \mathcal{A}$, then, with the notations of Proposition 1, we have $\mathcal{I}_k(x) = \mathcal{I}_{k+1}(x)$ for $k \geq \dim \mathcal{A}$.

e) If $\mathcal{A} \in S^p$, $p < \dim \mathcal{A}$ and $0 \neq x \in \mathcal{A}$, then there exists a proper ideal $\mathcal{I}(x)$ of $\mathcal{A}$ such that $x \in \mathcal{I}(x)$.

We can prove the following characterization of the classes S^p.

Proposition 2. *If $\mathcal{A}$ is a finite-dimensional Lie algebra, then the following assertions are equivalent.*

(i) $\mathcal{A} \in S^p$.

(ii) *$\mathcal{A}$ is a solvable Lie algebra and every subalgebra $\mathcal{B}$ of $\mathcal{A}$ with* $\dim \mathcal{B} \geq p+1$ *has the property: for every $x \in \mathcal{B}$ there exists a proper ideal $\mathcal{I}(x)$ of $\mathcal{B}$ such that $x \in \mathcal{I}(x)$.*

Proof. $(i) \Rightarrow (ii)$. If $\mathcal{A} \in S^p$, $\mathcal{B}$ is a subalgebra of $\mathcal{A}$, $\dim \mathcal{B} \geq p+1$ and $0 \neq x \in \mathcal{B}$, then we can consider the chain $\{\mathcal{A}_k(x)\}_{k\geq 1}$ given in Definition 1. We have

$$\mathcal{A}_1(x) \cap \mathcal{B} \subseteq \cdots \subseteq \mathcal{A}_k(x) \cap \mathcal{B} \subseteq \cdots \subseteq \mathcal{A} \cap \mathcal{B}.$$

Let k_0 be the first natural number such that $\mathcal{A}_{k_0}(x) \cap \mathcal{B} = \mathcal{B}$. Obviously we have $\{0\} \neq \mathcal{A}_{k_0-1}(x) \cap \mathcal{B} \neq \mathcal{B}$ and $\mathcal{A}_{k_0-1}(x) \cap \mathcal{B}$ is a proper ideal of $\mathcal{B}$ which contains the element x.

$(ii) \Rightarrow (i)$. Let x be an element of $\mathcal{A}$, $x \neq 0$. By (ii) we can find a proper ideal $\mathcal{I}(x)$ of $\mathcal{A}$ such that $x \in \mathcal{I}(x)$. If $\dim \mathcal{I}(x) \leq p$, then we obtain obviously a chain as needed in Definition 1. On the other hand, if $\dim \mathcal{I}(x) > p$, then by (ii) we can find a proper ideal $\mathcal{J}(x)$ of $\mathcal{I}(x)$ such that $x \in \mathcal{J}(x)$, etc. We obtain a chain as in Definition 1 because $\dim \mathcal{A} < \infty$. The assertion (i) is proved. □

Proposition 3. *Let $\mathcal{A}$ be a finite-dimensional Lie algebra. If $\mathcal{A}$ is a solvable Lie algebra and every subalgebra $\mathcal{B}$ of $\mathcal{A}$ with* $\dim \mathcal{B} \geq p+1$ *has the property* $\dim \mathcal{B}^{(1)} \leq \dim \mathcal{B} - 2$, *then $\mathcal{A} \in S^p$.*

Proof. Let $\mathcal{B}$ be a subalgebra of $\mathcal{A}$ with $\dim \mathcal{B} \geq p+1$. We will verify the condition (ii) from Proposition 2. Let us consider $0 \neq x \in \mathcal{B}$ and

$$D_{\mathcal{B}}(x) := sp\{[\mathcal{B}, x], [\mathcal{B}, [\mathcal{B}, x]], \ldots\}.$$

Obviously we have $D_{\mathcal{B}}(x) \subseteq \mathcal{B}^{(1)} = [\mathcal{B}, \mathcal{B}]$ and by the hypothesis we obtain $\dim D_{\mathcal{B}}(x) \leq \dim \mathcal{B} - 2$. Then $\mathcal{I}(x) :=$*span*$\{x, D_{\mathcal{B}}(x)\}$ is an ideal of $\mathcal{B}$ with the properties needed in Proposition 2, because $\dim \mathcal{I}_{\mathcal{B}}(x) \leq \dim \mathcal{B} - 1$. □

Corollary 1. *For every finite-dimensional nilpotent Lie algebra $\mathcal{A}$ we have $\mathcal{A} \in S^p$, for every positive integer p.*

Proof. Indeed, every Lie subalgebra of a nilpotent Lie algebra is nilpotent and for every nilpotent Lie algebra $\mathcal{B}$ we have $\dim \mathcal{B}^{(1)} \leq \dim \mathcal{B} - 2$. □

Proposition 4. *Each of the classes S^p is stable by Lie morphisms.*

Proof. The solvability is stable by Lie morphisms and we have $\dim \rho(E) \leq \dim E$ for every linear mapping ρ of a vector space E into another vector space. □

Corollary 2. *If $\mathcal{A} \in S^p$ and $\mathcal{I}$ is an ideal of $\mathcal{A}$, then $\mathcal{A}/\mathcal{I} \in S^p$.*

§ 4 Radical splitting theorems

The "radical splitting theorem" of Levi is the following.

Theorem 1 (Levi's Theorem). *For any finite-dimensional complex (or, more generally, over a field of characteristic 0) Lie algebra $\mathcal{L}$ with the solvable radical $\mathcal{R}$, there exists a semisimple subalgebra $\mathcal{S}$ of $\mathcal{L}$ such that $\mathcal{L} = \mathcal{R} \oplus \mathcal{S}$.* ($\mathcal{S}$ is called a *Levi factor* of $\mathcal{L}$.)

This theorem can be extended to an ideally finite Lie algebra $\mathcal{L}$ using the ideal $\mathcal{R}(\mathcal{L})$ defined above (see Remark 4 in § 2).

Then one can prove the following extension of Levi's Theorem.

Theorem 2. *If $\mathcal{L}$ is an ideally finite Lie algebra, then there exists a semisimple ideally finite subalgebra $\mathcal{S}$ (which is a maximal semisimple ideally finite subalgebra) of $\mathcal{L}$ such that*

$$\mathcal{L} = \mathcal{R}(\mathcal{L}) \oplus \mathcal{S}.$$

The subalgebra $\mathcal{S}$ from the above theorem has "good properties". It is semisimple. On the other hand there is the following result.

Theorem 3. *Every semisimple ideally finite Lie algebra is a direct sum of finite-dimensional simple ideals.*

Definition 1. An ideal $\mathcal{P}$ of a Lie algebra $\mathcal{L}$ is *primitive* if the following implication is true:

$$\mathcal{I} \text{ ideal in } \mathcal{L},\ [\mathcal{I},\mathcal{I}] \subseteq \mathcal{P} \Rightarrow \mathcal{I} \subseteq \mathcal{P}.$$

Remark 1. A primitive ideal of a Lie algebra $\mathcal{L}$ contains every solvable ideal of $\mathcal{L}$.

Remark 2. If any ideal of a Lie algebra is primitive, then that Lie algebra is semisimple. (Indeed, by the hypothesis, $\{0\}$ will be a primitive ideal. Hence, by the Remark 1, every solvable ideal is contained in $\{0\}$.)

If $\mathcal{L}$ is a semisimple finite-dimensional Lie algebra, then every ideal $\mathcal{I}$ of $\mathcal{L}$ is semisimple and $\mathcal{I} = [\mathcal{I},\mathcal{I}]$. Hence every ideal of a semisimple finite-dimensional Lie algebra is primitive. We have proved the following result.

Proposition 1. *A finite-dimensional Lie algebra $\mathcal{G}$ is semisimple iff every ideal of $\mathcal{G}$ is primitive.*

We can also prove the following property.

Proposition 2. *Let $\mathcal{G}$ be a Lie algebra and θ be a Lie morphism defined on $\mathcal{G}$. The Lie algebra $\theta(\mathcal{G})$ is semisimple iff $Ker\theta$ is a primitive ideal of $\mathcal{G}$.*

Proof. Indeed, the following implications hold:
$\mathcal{J}$ solvable ideal of $\theta(\mathcal{G}) \Rightarrow \exists n \in \mathbb{N}\ [\theta^{-1}(\mathcal{J})]^{(n)} \subseteq Ker\theta$;
$\mathcal{I}$ ideal of $\mathcal{G}$, $[\mathcal{I},\mathcal{I}] \subseteq Ker\theta \Rightarrow \theta(\mathcal{I})$ is a commutative ideal of $\theta(\mathcal{G})$. □

Definition 2. A Lie morphism θ defined on a Lie algebra $\mathcal{G}$ is called *semisimple* if $\theta(\mathcal{G})$ is a semisimple Lie algebra (equivalently $Ker\theta$ is a primitive ideal of $\mathcal{G}$).

Remark 3. If any ideal of $\mathcal{G}$ is primitive, then any Lie morphism defined on $\mathcal{G}$ is semisimple. Particularly every Lie morphism defined on a finite-dimensional semisimple Lie algebra is semisimple.

Now we return to the semisimple Lie algebra $\mathcal{S}$ occuring in the radical splitting theorem for an ideally finite Lie algebra $\mathcal{L}$ (see Theorem 2 above). As we have remarked, $\mathcal{S}$ is a direct sum of finite-dimensional simple ideals (cf. Theorem 3). So we deduce that every ideal of $\mathcal{S}$ is primitive. (Indeed, for every ideal $\mathcal{I}$ of $\mathcal{S}$ we have $\mathcal{I} = [\mathcal{I},\mathcal{I}]$ because $\mathcal{I}$ is a direct sum of finite-dimensional simple ideals of $\mathcal{S}$.) Then the radical spliting theorem for ideally finite Lie algebras suggests the following definition.

Definition 3. A Lie algebra $\mathcal{L}$ is called *LM-decomposable* if we have $\mathcal{L} = \mathcal{R} + \mathcal{G}$, where $\mathcal{R} = \sum_{\alpha\in\Lambda} \mathcal{I}_\alpha$, $\mathcal{I}_\alpha$ finite-dimensional solvable ideals in $\mathcal{L}$ (i.e. $\mathcal{R}$ is a quasisolvable Lie algebra) and $\mathcal{G}$ is a Lie algebra so that any ideal of $\mathcal{G}$ is primitive.

Remark 4. If $\mathcal{A}$ is LM-decomposable and θ is a Lie morphism defined on $\mathcal{A}$ then $\theta(\mathcal{A})$ is also LM-decomposable.

Remark 5. Obviously every quasisolvable Lie algebra, every ideally finite semisimple Lie algebra, every ideally finite Lie algebra (particularly each finite-dimensional Lie algebra) is LM-decomposable.

§ 5 Cartan subalgebras

Other essential objects we need to describe structure theorems for Lie algebras are the Cartan subalgebras. Recall that if $\mathcal{A}$ is a Lie subalgebra of the Lie algebra $\mathcal{L}$ then the *normalizer* $\mathcal{N}$ of $\mathcal{A}$ (in $\mathcal{L}$) is the largest Lie subalgebra of $\mathcal{L}$ which contains $\mathcal{A}$ as an ideal.

Definition 1. A *Cartan subalgebra* of a Lie algebra $\mathcal{L}$ is a Lie subalgebra $\mathcal{C}$ of $\mathcal{L}$ having the following properties:

1. $\mathcal{C}$ is a nilpotent Lie subalgebra of $\mathcal{L}$;
2. $\mathcal{C}$ is its own normalizer (i.e. there is no subalgebra of $\mathcal{L}$ which is different from $\mathcal{C}$ and contains $\mathcal{C}$ as an ideal).

If $\mathcal{B}$ is a nilpotent Lie subalgebra of a finite-dimensional Lie algebra $\mathcal{L}$, we can consider the adjoint representation of $\mathcal{B}$ in $\mathcal{A}$, i.e.

$$\mathcal{B} \to End(\mathcal{A}), \ x \mapsto ad\, x.$$

It is clear that

$$\mathcal{B} \subseteq \mathcal{L}_0(\mathcal{B})$$

where

$$\mathcal{L}_0(\mathcal{B}) = \{x \in \mathcal{L} \mid (ad\, b)^k x = 0 \text{ for some } k \in \mathbb{N} \text{ and every } b \in \mathcal{B}\}.$$

This fact gives the first characterization of the Cartan subalgebras of a finite-dimensional Lie algebra $\mathcal{L}$.

Proposition 1. *A nilpotent Lie subalgebra $\mathcal{B}$ of $\mathcal{L}$ is a Cartan subalgebra iff $\mathcal{L}_0(\mathcal{B}) = \mathcal{B}$.*

Another characterization of the Cartan subalgebras is given by the concept of a regular element of $\mathcal{L}$.

An element $h \in \mathcal{L}$ is called *regular* if the dimension of

$$\mathcal{L}_0(h) := \mathcal{L}_0(\mathbb{C}h) = \{x \in \mathcal{L} \mid (ad\, h)^k x = 0 \text{ for some } k \in \mathbb{N}\}$$

is minimal. The set of regular elements of $\mathcal{G}$ will be denoted by $r(\mathcal{G})$. We have the second characterization of the Cartan subalgebras. (More generally, the following result holds for finite-dimensional Lie algebras over an algebraically closed field Φ of characteristic zero.)

Proposition 2. *$\mathcal{C}$ is a Cartan subalgebra of the finite-dimensional Lie algebra $\mathcal{L}$ iff $\mathcal{C} = \mathcal{L}_0(h)$ for some regular element h of $\mathcal{L}$.*

A basic property of the Cartan subalgebras of a complex finite-dimensional Lie algebra is the conjugation theorem, which we are going to state now. To this end we need to introduce a certain group of automorphisms of a Lie algebra $\mathcal{G}$. Namely, we define the *group of elementary automorphisms* of $\mathcal{G}$, which is denoted $Aut_e(\mathcal{G})$ and is generated by the set

$$\{\exp(ad\, g) \mid g \in \mathcal{G} \text{ and } ad\, g : \mathcal{G} \to \mathcal{G} \text{ is a nilpotent map}\}.$$

Now we can state the *conjugation theorem:*

Theorem 1. *Any two Cartan subalgebras of a complex finite-dimensional Lie algebra $\mathcal{G}$ are conjugated with respect to $Aut_e(\mathcal{G})$. That is, if $\mathcal{H}$ and $\mathcal{K}$ are arbitrary Cartan subalgebras of $\mathcal{G}$, then there exists $\varphi \in Aut_e(\mathcal{G})$ such that $\varphi(\mathcal{H}) = \mathcal{K}$.*

For later use we collect below a few simple facts related to the above introduced notions.

Lemma 1. *If $\mathcal{G}$ is a Lie algebra, $h \in \mathcal{G}$ and $\varphi \in Aut_e(\mathcal{G})$, then we have $h - \varphi(h) \in [\mathcal{G}, \mathcal{G}]$.*

Proof. Obviously it suffices to verify the assertion for $\varphi = \exp(ad g)$, where $g \in \mathcal{G}$ and $ad g$ is nilpotent. In this case we have

$$(\exp(ad g))h = h + [g,h] + \frac{1}{2!}[g,[g,h]] + \cdots \in h + [\mathcal{G},\mathcal{G}]$$

and the proof ends. □

The following facts concern the interplay between Cartan subalgebras and ideals, an aspect which is useful in the study of solvable Lie algebras.

Lemma 2. *Let $\mathcal{G}$ be a complex finite-dimensional solvable Lie algebra.*

(a) *If $c \in \mathcal{G}$ has the property that $ad c : \mathcal{G} \to \mathcal{G}$ is a nilpotent operator, then $c + r(\mathcal{G}) = r(\mathcal{G})$ (i.e. for $x \in \mathcal{G}$ we have $h \in r(\mathcal{G}) \Leftrightarrow h + c \in r(\mathcal{G})$).*

(b) *The equality $[\mathcal{G},\mathcal{G}] + r(\mathcal{G}) = r(\mathcal{G})$ holds.*

Proof. Choose by Lie's Theorem a basis in $\mathcal{G}$ such that any operator $ad g : \mathcal{G} \to \mathcal{G}$ ($g \in \mathcal{G}$) is represented by an upper triangular matrix. For $g \in \mathcal{G}$, the number of occurences of 0 on the diagonal of $ad g$ equals the multiplicity of 0 as a root of the characteristic polynomial of $ad g$, hence it finally equals the dimension of the kernel of $(ad g)^{\dim \mathcal{G}}$.

Now let's come back to the proof of the desired conclusion. Let $c \in \mathcal{G}$ be such that $ad c : \mathcal{G} \to \mathcal{G}$ is a nilpotent map. Then for any $h \in \mathcal{G}$ the matrices of $ad h$ and $ad(h+c)$ have the same number of occurences of 0 on the diagonal (since the diagonal of $ad c$ consists only of zeros). Hence by the remark from the beginning of the proof and the definition of a regular element, we obtain that $h \in r(\mathcal{G})$ iff $h + c \in r(\mathcal{G})$.

Now, for proving the assertion (b) one applies (a) and the fact that for $c \in [\mathcal{G},\mathcal{G}]$ the map $ad c : \mathcal{G} \to \mathcal{G}$ is nilpotent. □

Proposition 3. *Let $\mathcal{G}_0$ be a nilpotent subalgebra of the complex finite-dimensional solvable Lie algebra $\mathcal{G}$.*

(i) *If $\mathcal{G}_0 \cap r(\mathcal{G}) \neq \emptyset$ then $\mathcal{G}_0$ is contained in a Cartan subalgebra of $\mathcal{G}$.*

(ii) *If $\mathcal{G}_0 \cap r(\mathcal{G}) = \emptyset$, then $\mathcal{G}_0$ is contained in an ideal $\mathcal{I}$ of $\mathcal{G}$ such that $\mathcal{I} \neq \mathcal{G}$ (more exactly $\mathcal{I} \cap r(\mathcal{G}) = \emptyset$).*

Proof. If $\mathcal{G}_0 \cap r(\mathcal{G}) \neq \emptyset$, apply Proposition 2 above. On the other hand, if $\mathcal{G}_0 \cap r(\mathcal{G}) = \emptyset$, then consider $\mathcal{I} := \mathcal{G}_0 + [\mathcal{G},\mathcal{G}]$. Obviously $\mathcal{I} \supseteq \mathcal{G}_0$ and $\mathcal{I}$ is an ideal of $\mathcal{G}$. If $\mathcal{I} \cap r(\mathcal{G}) \neq \emptyset$, then by Lemma 2 (b) it easily follows that $\mathcal{G}_0 \cap r(\mathcal{G}) \neq \emptyset$, a contradiction. □

Proposition 4. *Let $\mathcal{H}$ be a Cartan subalgebra of the complex finite-dimensional Lie algebra $\mathcal{G}$. Let $n = \dim \mathcal{H}$ and R be the set of roots of $\mathcal{G}$ with respect to $\mathcal{H}$, i.e. the linear functionals $\alpha : \mathcal{H} \to \mathbb{C}$ such that the corresponding root space*

$$\mathcal{G}^{\alpha} := \{g \in \mathcal{G} \mid \forall h \in \mathcal{H}\, \exists k \in \mathbb{N} : (ad h - \alpha(h))^k g = 0\}$$

is nonzero. Denote

$$\mathcal{C}_{\mathcal{H}} := \bigoplus_{\alpha \in R \setminus \{0\}} \mathcal{G}^{\alpha}.$$

Then the following assertions hold.

(a) *If* $h \in \mathcal{H} \cap r(\mathcal{G})$ *then* $(ad\,h)^n \mathcal{G} = C_{\mathcal{H}}$.

(b) *Let* $\mathcal{K}$ *be another Cartan subalgebra of* $\mathcal{G}$ *and* $C_{\mathcal{K}}$ *be defined as above (with respect to* $\mathcal{K}$*). If* φ *is an automorphism of* $\mathcal{G}$ *such that* $\varphi(\mathcal{H}) = \mathcal{K}$*, then* $\varphi(C_{\mathcal{H}}) = C_{\mathcal{K}}$.

(c) *We have* $C_{\mathcal{H}} \subseteq [\mathcal{G}, \mathcal{G}]$. *More exactly* $[\mathcal{H}, C_{\mathcal{H}}] = C_{\mathcal{H}}$.

(d) *If* $\mathcal{I}$ *is an ideal of* $\mathcal{G}$ *such that* $\dim(\mathcal{G}/\mathcal{I}) = 1$*, then* $C_{\mathcal{H}} \subseteq \mathcal{I}$ *and* $\mathcal{I} = (\mathcal{H} \cap \mathcal{I}) \oplus C_{\mathcal{H}}$.

Proof. (*a*) It is well known that

$$\mathcal{G} = \mathcal{H} \oplus C_{\mathcal{H}}. \tag{1}$$

Moreover, for $h \in \mathcal{H} \cap r(\mathcal{G})$ and $0 \neq \alpha \in R$ we have $\alpha(h) \neq 0$ and the spectrum of the operator $ad\,h$ restricted to its invariant subspace $\mathcal{G}^\alpha$ is $\{\alpha(h)\}$. So $(ad\,h)(\mathcal{G}^\alpha) = \mathcal{G}^\alpha$ for $\alpha \neq 0$, which implies $(ad\,h)(C_{\mathcal{H}}) = C_{\mathcal{H}}$.

On the other hand, since $\mathcal{H}$ is a nilpotent Lie algebra and $h \in \mathcal{H}$, we have $(ad\,h)^n \mathcal{H} = \{0\}$. Now the desired fact follows by (1).

(*b*) It is well known that the set $r(\mathcal{G})$ is invariant to every automorphism of $\mathcal{G}$. Hence if we choose $h \in \mathcal{H} \cap r(\mathcal{G})$, then $k := \varphi(h) \in \mathcal{K} \cap r(\mathcal{G})$. Now, by (*a*) we get

$$\varphi(C_{\mathcal{H}}) = \varphi((ad\,h)^n \mathcal{G}) = (ad\,\varphi(h))^n(\varphi(\mathcal{G})) = (ad\,k)^n \mathcal{G} = C_{\mathcal{K}}.$$

(*c*) By the definition of the root spaces it follows that $[\mathcal{H}, C_{\mathcal{H}}] \subseteq C_{\mathcal{H}}$. Moreover if we choose $h \in \mathcal{H} \cap r(\mathcal{G})$, then $[h, C_{\mathcal{H}}] = C_{\mathcal{H}}$ (cf. the proof of (*a*)).

(*d*) One easily checks that $[\mathcal{G}, \mathcal{G}] \subseteq \mathcal{I}$. So $C_{\mathcal{H}} \subseteq \mathcal{I}$ by (*c*). Now the desired equality immediately follows by (1). □

Definition 2. In the conditions of Proposition 4, the relation (1) is called the *Cartan decomposition* of $\mathcal{G}$ with respect to $\mathcal{H}$.

Proposition 5. *Let* $\mathcal{G}$ *be a complex finite-dimensional solvable Lie algebra and* $\mathcal{I}$ *be an ideal of* $\mathcal{G}$ *such that* $\dim(\mathcal{G}/\mathcal{I}) = 1$. *If* $\mathcal{H}$ *is a Cartan subalgebra of* $\mathcal{G}$*, then* $\mathcal{H} \cap \mathcal{I}$ *is contained in a certain Cartan subalgebra of* $\mathcal{I}$.

Proof. Obviously $\mathcal{H} \cap \mathcal{I}$ is a nilpotent subalgebra of $\mathcal{I}$. Hence by Proposition 3 it suffices to check that $(\mathcal{H} \cap \mathcal{I}) \cap r(\mathcal{I}) \neq \emptyset$. To this end choose $c \in C_{\mathcal{H}}$ and $x \in \mathcal{H} \cap \mathcal{I}$ such that $c + x \in r(\mathcal{I})$ (see Proposition 4 (*d*)). Since $c \in C_{\mathcal{H}}$ it is well known (see Proposition 4 (*c*)) that the map $ad\,c : \mathcal{G} \to \mathcal{G}$ is nilpotent. Hence its restriction $ad\,c : \mathcal{I} \to \mathcal{I}$ is also nilpotent. (We have $c \in C_{\mathcal{H}} \subseteq \mathcal{I}$ by Proposition 4 (*d*).) Now, since $c + x \in r(\mathcal{I})$, by Lemma 2 (*a*) we get $x \in r(\mathcal{I})$, hence $x \in (\mathcal{H} \cap \mathcal{I}) \cap r(\mathcal{I})$. Consequently $(\mathcal{H} \cap \mathcal{I}) \cap r(\mathcal{I}) \neq \emptyset$ and the proof is finished. □

Next we state for later use two theorems concerning the Cartan decompositions of the semisimple Lie algebras. In their statements $\mathcal{G}$ will be a complex finite-dimensional *semisimple* Lie algebra, $\mathcal{H}$ will be a Cartan subalgebra of $\mathcal{G}$, R will be the corresponding set of roots and $\{\mathcal{G}^\alpha\}_{\alpha \in R}$ will be the corresponding set of nonzero root spaces (see Proposition 4 above). In these conditions we have the following theorems:

Theorem 2. *The Cartan subalgebra $\mathcal{H}$ is abelian and for every $\alpha \in R\backslash\{0\}$ we have $\dim \mathcal{G}^\alpha = 1$ and $-\alpha \in R\backslash\{0\}$. Moreover, we can choose a system of generators $\{h_\alpha, e_\alpha \mid \alpha \in R\backslash\{0\}\}$ of $\mathcal{G}$ with the following properties:*

(i) *for every $\alpha \in R\backslash\{0\}$ we have $\mathcal{G}^\alpha = \mathbb{C} \cdot e_\alpha$;*

(ii) *$\mathcal{H}$ is spanned (as a complex vector space) by the set $\{h_\alpha \mid \alpha \in R\backslash\{0\}\}$;*

(iii) *for every $\alpha \in R\backslash\{0\}$ and $h \in \mathcal{H}$ we have $[h, e_\alpha] = \alpha(h)e_\alpha$ and $[e_\alpha, e_{-\alpha}] = -h_\alpha$.*

Furthermore if $\mathcal{G}_u$ is a compact real form of $\mathcal{G}$ (see § 6 below) and $\mathcal{G}$ is endowed with the corresponding involution (i.e. for $t \in \mathcal{G}$ we have $t^* = -t \Leftrightarrow t \in \mathcal{G}_u$; see § 7 below), then the above system of generators can be chosen such that

(iv) *for every $\alpha \in R\backslash\{0\}$ we have $h_\alpha^* = h_\alpha$ and $e_\alpha^* = -e_{-\alpha}$.*

Theorem 3. *For every $t \in \mathcal{G}$ there exists a total ordering $\leq$ on the set of roots R with the following properties.*

1° *If $\alpha, \beta \in R$ and $\alpha \geq \beta$ then $-\alpha \leq -\beta$ and $\alpha + \gamma \geq \beta + \gamma$ for each $\gamma \in R$.*

2° *There exists $h \in \mathcal{H}$ such that $\alpha(h) > 0$ for every $\alpha \in R$ with $\alpha > 0$.*

3° *We have $t \in \mathcal{H} \oplus (\oplus_{0<\alpha\in R}\mathcal{G}^\alpha)$.*

4° *If $ad\, t : \mathcal{G} \to \mathcal{G}$ is nilpotent then $t \in \oplus_{0<\alpha\in R}\mathcal{G}^\alpha$.*

5° *If we denote $\mathcal{N}_+ = \oplus_{0<\alpha\in R}\mathcal{G}^\alpha$ and $\mathcal{N}_- = \oplus_{0>\alpha\in R}\mathcal{G}^\alpha$, then $\mathcal{N}_+$ and $\mathcal{N}_-$ are nilpotent Lie subalgebras of $\mathcal{G}$ and $\mathcal{G} = \mathcal{N}_- \oplus \mathcal{H} \oplus \mathcal{N}_+$. Moreover, the map $ad\, g : \mathcal{G} \to \mathcal{G}$ is nilpotent for every $g \in \mathcal{N}_- \cup \mathcal{N}_+$.*

§ 6 The Killing form and compact Lie algebras

Definition 1. If $\mathcal{L}$ is a finite-dimensional Lie algebra, then the symmetric bilinear form f on $\mathcal{L}$ defined by

$$f(a,b) = Tr((ad\, a)(ad\, b)) \text{ for every } a, b \in \mathcal{L}$$

is called the *Killing form* of $\mathcal{L}$.

If $\mathcal{L}$ is a finite-dimensional Lie algebra over a field of characteristic zero, then the following criteria are known as Cartan's criteria.

i) The fact that $f(a,a) = 0$ for every $a \in \mathcal{L}'$ characterizes the situation when $\mathcal{L}$ is solvable.

ii) The Killing form f is non-degenerate iff $\mathcal{L}$ is semisimple.

If G is a compact Lie group, then the Killing form of the Lie algebra associated to G is negatively defined. This fact suggests the following definition.

Definition 2. A Lie algebra over the field $\mathbb{R}$ of the real numbers is called *compact* if its Killing form is negatively defined.

Remark 1. A compact Lie algebra is semisimple (because its Killing form is non-degenerated) and Lie groups associated with it are compact groups.

Remark 2. In general there is the following criterion: A finite-dimensional real Lie algebra $\mathcal{S}$ is the Lie algebra of some compact Lie group iff there exists on $\mathcal{S}$ a positively defined invariant bilinear form f (i.e. $f(x,x) > 0$ if $0 \neq x \in \mathcal{S}$ and $f([x,y],z) = f(x,[y,z])$ for every $x,y,z \in \mathcal{S}$).

The compact Lie algebras can describe by complexification any complex semisimple Lie algebra. There is the following result.

Theorem 1. *For any semisimple Lie algebra $\mathcal{L}$ over $\mathbb{C}$ there exists a compact Lie algebra $\mathcal{L}_u$ (over $\mathbb{R}$) such that its complexification $(\mathcal{L}_u)_{\mathbb{C}}$ is isomorphic to $\mathcal{L}$.*

Recall that the *complexification* $\mathcal{U}_{\mathbb{C}}$ of a real Lie algebra $\mathcal{U}$ is defined as follows. We consider $\mathbb{C} \otimes \mathcal{U}$ as a tensor product of real vector spaces. We define $[\sum_i \alpha_i \otimes u_i, \sum_j \beta_j \otimes v_j] = \sum_{i,j} \alpha_i\beta_j \otimes [u_i, v_j]$ and $\alpha \sum_i \alpha_i \otimes u_i = \sum_i (\alpha\alpha_i) \otimes u_i$. Then $\mathbb{C} \otimes \mathcal{U}$ becomes a Lie algebra over $\mathbb{C}$ which is denoted by $\mathcal{U}_{\mathbb{C}}$ and is called *the complexification* of $\mathcal{U}$.

Definition 3. In the conditions of Theorem 1, the Lie algebra $\mathcal{L}_u$ is called a *compact form* of the complex Lie algebra $\mathcal{L}$.

The importance of compact forms is given by the method introduced by H. Weyl, the "unitary trick", for transfering problems on representations of Lie groups and algebras to the case of compact Lie groups. Using this method, Weyl gave the first proof of the completely reducibility of the finite-dimensional representations of semisimple Lie algebras over $\mathbb{C}$ (more generally, over an algebraically closed field of characteristic zero).

§ 7 Lie *-algebras

It is possible to describe a Lie algebra structure which contains as a particular situation the Lie algebra structure of an associative *-algebra.

Lemma 1. *Let $\mathcal{L}$ be a Lie algebra over $\mathbb{C}$. The following assertions are equivalent:*

(j) *There is a mapping $^* : \mathcal{L} \to \mathcal{L}$ such that*

$$(a+b)^* = a^* + b^*,\ (\lambda a)^* = \overline{\lambda} a^*,\ (a^*)^* = a \text{ and } [a,b]^* = -[a^*, b^*]$$

for every $a, b \in \mathcal{L}$ and $\lambda \in \mathbb{C}$.

(jj) *There is a real vector subspace $\mathcal{A} \subset \mathcal{L}$ such that*

$$\mathcal{L} = \mathcal{A} \oplus i\mathcal{A} \text{ and } i[a,b] \in \mathcal{A} \text{ for every } a, b \in \mathcal{A}.$$

Proof. If (j) holds, $\mathcal{A} = \{a \in \mathcal{L} \mid a = a^*\}$ and for $a \in \mathcal{L}$ we may uniquely write

$$a = \frac{a+a^*}{2} + i\frac{a-a^*}{2i} \text{ for every } a \in \mathcal{L}.$$

Obviously, $\mathcal{A}$ is a real vector space, $c = -c^*$ iff $ic \in \mathcal{A}$ and $i[a,b] \in \mathcal{A}$ for every $a, b \in \mathcal{A}$. Hence $(j) \Rightarrow (jj)$.

If (jj) holds we may uniquely write for $a \in \mathcal{L}$

$$a = a_1 + ia_2.$$

We define in a natural way,

$$a^* = a_1 - ia_2.$$

Clearly, $a = a^*$ iff $a \in \mathcal{A}$ and the properties of * are obvious. So $(jj) \Rightarrow (j)$. □

Definition 1. A complex Lie algebra $\mathcal{L}$ which satisfies either (j) or (jj) from Lemma 1 is called a *complex Lie algebra with involution* or a *complex Lie *-algebra.* In this case, an element x of $\mathcal{L}$ is *normal* if $[x, x^*] = 0$, *self-adjoint* if $x = x^*$ and *skew* if $x = -x^*$.

We observe that a structure of Lie *-algebra may be implicitly contained in an ordinary Lie algebra structure. A significant example is the following.

Example 1. We consider the Lie algebra $sl(2,\mathbb{C})$ with the basis $\{\tau, x_+, x_-\}$ and the commutation relations $[\tau, x_+] = 2x_+$, $[\tau, x_-] = -2x_-$, $[x_+, x_-] = \tau$. We have $sl(2,\mathbb{C}) = \mathcal{A} \oplus i\mathcal{A}$, $\mathcal{A} = sp_{\mathbb{R}}\{i\tau, x_+, x_-\}$ (i.e. the real vector subspace spanned by $\{i\tau, x_+, x_-\}$) and $\mathcal{A}$ satisfies (jj) of Lemma 1. The *-operation given by $\mathcal{A}$ is the following:

$$\begin{pmatrix} \lambda & \mu \\ \nu & -\lambda \end{pmatrix}^* = \begin{pmatrix} \tilde{\lambda} & \overline{\mu} \\ \overline{\nu} & -\tilde{\lambda} \end{pmatrix}; \begin{cases} \tilde{\lambda} = \widetilde{\lambda_1 + i\lambda_2} = -\lambda_1 + i\lambda_2, \\ \overline{\mu} = \overline{\mu_1 + i\mu_2} = \mu_1 - i\mu_2. \end{cases}$$

A normal element is given by

$$\begin{pmatrix} \lambda & \mu \\ \nu & -\lambda \end{pmatrix} \text{ with } \frac{\overline{\mu}}{\mu} = \frac{\tilde{\lambda}}{\lambda} = \frac{\overline{\nu}}{\nu}.$$

A skew element is given by

$$\begin{pmatrix} \alpha & i\gamma \\ i\beta & -\alpha \end{pmatrix} \text{ with } \alpha, \beta, \gamma \in \mathbb{R}.$$

The Lie subalgebras consisting of normal elements of a certain Lie *-algebra can be characterized as follows.

Theorem 1. *Let $\mathcal{L}$ be a Lie *-algebra and $\mathcal{G} \subseteq \mathcal{L}$ be a real Lie subalgebra consisting of normal elements.*

1) *Then*

$$\mathcal{G}^+ := \mathcal{G} + \mathcal{G}^* + [\mathcal{G}, \mathcal{G}^*]$$

is a Lie algebra and $\mathcal{I} := [\mathcal{G}, \mathcal{G}^*]$ is an ideal in $\mathcal{G}^+$ consisting of skew elements.

2) *If in addition $\mathcal{G}$ is a semisimple ideally finite Lie algebra, then*

$$\mathcal{G}^+ = \mathcal{G}_0 \oplus \mathcal{I},$$

where $\mathcal{G}_0$ is the centralizer of the ideal $\mathcal{I}$ in $\mathcal{G}^+$. Hence for $x, y \in \mathcal{G}$ we may uniquely write $x = x_0 + x_1$, $y = y_0 + y_1$, where x_0, y_0 are normal elements, x_1, y_1 are skew elements, $[x_0, y_1] = [x_1, y_0] = 0$, $[x_0, y_0^*] = 0$.

Proof. First we prove that $[\mathcal{G}, \mathcal{G}^*]$ consists of skew elements. Because $\mathcal{G}$ consists of normal elements we have for any x, y in $\mathcal{G}$, $[x+y, (x+y)^*] = 0$. Hence,

$$0 = [x, x^*] + [x, y^*] + [y, x^*] + [y, y^*] = [x, y^*] + [y, x^*]$$

and it results that $[x, y^*]$ is skew because $[y, x^*] = [x, y^*]^*$. Therefore every element of $[\mathcal{G}, \mathcal{G}^*]$ is skew as real linear combination of elements $[x, y^*]$ with $x, y \in \mathcal{G}$.

Now we prove that $\mathcal{I} = [\mathcal{G}, \mathcal{G}^*]$ is an ideal in $\mathcal{G}^+$. It suffices to prove that $[\mathcal{G}, \mathcal{I}] \subseteq \mathcal{I}$. Indeed $[\mathcal{G}^*, \mathcal{I}] = [\mathcal{G}, \mathcal{I}]^* \subseteq \mathcal{I}^* = \mathcal{I}$ and

$$[\mathcal{I}, \mathcal{I}] = [[\mathcal{G}, \mathcal{G}^*], \mathcal{I}] \subseteq [[\mathcal{G}, \mathcal{I}], \mathcal{G}^*] + [\mathcal{G}, [\mathcal{G}^*, \mathcal{I}]] \subseteq [\mathcal{G}^*, \mathcal{I}] + [\mathcal{G}, \mathcal{I}] \subseteq \mathcal{I}.$$

To prove that $[\mathcal{G}, \mathcal{I}] \subseteq \mathcal{I}$ we have only to prove that $[x, [y, z^*]] \in \mathcal{I}$ if $x, y, z \in \mathcal{G}$.

First we observe that

$$[z, [y, z^*]] = [[z, y], z^*] + [y, [z, z^*]] = [[z, y], z^*] \in \mathcal{I}.$$

Particularly

$$[x + z, [y, (x+z)^*]] \in \mathcal{I},$$

i.e.

$$[x, [y, x^*]] + [x, [y, z^*]] + [z, [y, x^*]] + [z, [y, z^*]] \in \mathcal{I}.$$

Hence

$$[x, [y, z^*]] + [z, [y, x^*]] \in \mathcal{I}. \tag{1}$$

We remember that

$$[y, x^*] = [x, y^*]^* = -[x, y^*] = [y^*, x].$$

Therefore,

$$[z, [y, x^*]] = [z, [y^*, x]] = [[z, y^*], x] + [y^*, [z, x]]$$

and

$$[z, [y, x^*]] - [[z, y^*], x] \in \mathcal{I}.$$

Hence

$$[z, [y, x^*]] \equiv [[z, y^*], x] \,(\mathrm{mod}\mathcal{I}) \equiv [[z^*, y], x] \,(\mathrm{mod}\mathcal{I}) \equiv [x, [y, z^*]] (\mathrm{mod}\mathcal{I})$$

and using the formula (1) we obtain

$$2[x, [y, z^*]] \equiv ([x, [y, z^*]] + [z, [y, x^*]]) \,(\mathrm{mod}\mathcal{I}) \equiv 0 \,(\mathrm{mod}\mathcal{I}),$$

i.e.

$$[x, [y, z^*]] \in \mathcal{I} \text{ for every } x, y, z \in \mathcal{G}.$$

For proving the second part of the theorem we will use Weyl's theorem asserting that any representation of a semisimple complex Lie algebra in a complex finite-dimensional vector space is completely reducible, i.e. any invariant subspace of the representation has an algebraic complement which is also invariant.

I. First we suppose that $\mathcal{G}$ is a finite-dimensional simple Lie algebra. Let us consider $\mathcal{G}_0$ the centralizer of the ideal $\mathcal{I}$ in $\mathcal{G}^+$, i.e.

$$\mathcal{G}_0 = \{x \in \mathcal{G}^+ \mid [x, \mathcal{I}] = \{0\}\}.$$

1. $\mathcal{G}_0$ is an ideal in $\mathcal{G}^+$ because it is the centralizer of the ideal $\mathcal{I}$.

2. $\mathcal{G}_0^* = \mathcal{G}_0$. Indeed, for $x \in \mathcal{G}_0$ we have

$$[x^*, \mathcal{I}] = [x, \mathcal{I}^*]^* = [x, \mathcal{I}]^* = \{0\}$$

because every element of $\mathcal{I} = [\mathcal{G}, \mathcal{G}^*]$ is skew. So $\mathcal{G}_0^* \subseteq \mathcal{G}_0$ and every $x \in \mathcal{G}_0$ satisfies $x = (x^*)^*$, which proves that $\mathcal{G}_0^* = \mathcal{G}_0$.

3. $\mathcal{G}^+ = \mathcal{G}_0 + \mathcal{I}$. If $x \in \mathcal{G}^+, x = x^*$ and $y \in \mathcal{I}$ we have

$$-[x, y] = [x, y]^* = [y^*, x^*] = [-y, x] = [x, y]$$

because every element of $\mathcal{I}$ is skew, i.e. $y = -y^*$. Hence $[x, y] = 0$ for $x = x^* \in \mathcal{G}^+$ and $y \in \mathcal{I}$ i.e.

$$x \in \mathcal{G}^+, x = x^* \Rightarrow x \in \mathcal{G}_0.$$

Particularly $z + z^* \in \mathcal{G}_0$ for every $z \in \mathcal{G}$. If $y, z \in \mathcal{G}$ we can write

$$[y, z] = [y, z + z^*] - [y, z^*].$$

But $[y, z + z^*] \in \mathcal{G}_0$ because $\mathcal{G}_0$ is an ideal, and $[y, z^*] \in \mathcal{I}$ because $\mathcal{I} = [\mathcal{G}, \mathcal{G}^*]$. So we have proved that

$$[\mathcal{G}, \mathcal{G}] \subseteq \mathcal{G}_0 + \mathcal{I}.$$

But $\mathcal{G}$ is semisimple. Hence

$$\mathcal{G} = [\mathcal{G}, \mathcal{G}] \subseteq \mathcal{G}_0 + \mathcal{I}.$$

On the other hand

$$\mathcal{G}^* \subseteq \mathcal{G}_0^* + \mathcal{I}^* = \mathcal{G}_0 + \mathcal{I}.$$

So we have

$$\mathcal{G}^+ = \mathcal{G} + \mathcal{G}^* + \mathcal{I} \subseteq \mathcal{G}_0 + \mathcal{I}.$$

4. $\mathcal{G}_0 \cap \mathcal{I} = \{0\}$, which will finish the proof of the finite-dimensional case. If we denote $\mathcal{I}_0 = \mathcal{G}_0 \cap \mathcal{I}$, we have obviously

$$[\mathcal{G}_0, \mathcal{I}_0] \subseteq [\mathcal{G}_0, \mathcal{I}] = \{0\},$$

$$[\mathcal{I}, \mathcal{I}_0] \subseteq [\mathcal{I}, \mathcal{G}_0] = \{0\},$$

$$[\mathcal{G}^+,\mathcal{I}_0] \subseteq [\mathcal{G}_0+\mathcal{I},\mathcal{I}_0] = \{0\},$$
$$[\mathcal{G},\mathcal{I}_0] \subseteq [\mathcal{G}^+,\mathcal{I}_0] = \{0\}.$$

If we apply Weyl's theorem for $(ad\mathcal{G})|_{\mathcal{I}}$, we obtain that there exists a linear complement $\mathcal{I}_1$ of $\mathcal{I}_0$ in $\mathcal{I}$ such that $(ad\mathcal{G})\mathcal{I}_1 \subseteq \mathcal{I}_1$, i.e.

$$\mathcal{I} = \mathcal{I}_0 \oplus \mathcal{I}_1,\ [g,\mathcal{I}_0] = \{0\},\ [g,\mathcal{I}_1] \subseteq \mathcal{I}_1 \text{ for every } g \in \mathcal{G}.$$

Therefore

$$[\mathcal{G},\mathcal{I}] \subseteq \mathcal{I}_1.$$

If we prove $[\mathcal{G},\mathcal{I}] = \mathcal{I}$, then we have $\mathcal{I}_0 = \{0\}$ as desired. The equality $[\mathcal{G},\mathcal{I}] = \mathcal{I}$ can be obtained as a consequence of Weyl's theorem as follows. We consider the Lie algebra $(ad\mathcal{G})|_{\mathcal{G}^*+\mathcal{I}}$. Then $\mathcal{G}^*+\mathcal{I}$ is completely irreducible under $ad\mathcal{G}$, so there exists a complement $\mathcal{I}^\perp$ of $\mathcal{I}$ in $\mathcal{G}^*+\mathcal{I}$ such that $[\mathcal{G},\mathcal{I}^\perp] \subseteq \mathcal{I}^\perp$. But $[\mathcal{G},\mathcal{I}^\perp] \subseteq \mathcal{I}$ and $[\mathcal{G},\mathcal{I}^\perp] \subseteq \mathcal{I} \cap \mathcal{I}^\perp = \{0\}$. Finally we have

$$\mathcal{I} = [\mathcal{G},\mathcal{G}^*] \subseteq [\mathcal{G},\mathcal{I}^\perp + \mathcal{I}] = [\mathcal{G},\mathcal{I}] \subseteq \mathcal{I}$$

and

$$[\mathcal{G},\mathcal{I}] = \mathcal{I}.$$

Consequently we have obtained $\mathcal{G}^+$ as the direct sum of two ideals $\mathcal{G}_0$ and $\mathcal{I}$. For $x, y \in \mathcal{G}$ we uniquely write $x = x_0 + x_1, y = y_0 + y_1$, where $x_0, y_0 \in \mathcal{G}_0, x_1, y_1 \in \mathcal{I}$. Obviously $[x_0, y_1] = [x_1, y_0] = 0$ and

$$[x_0, y_0^*] = [x - x_1, (y - y_1)^*] \equiv [x, y^*]\,(\text{mod}\mathcal{I}) \equiv 0\,(\text{mod}\mathcal{I}).$$

Therefore $[x_0, y_0^*] \in \mathcal{G}_0 \cap \mathcal{I} = \{0\}$ and the proof is finished in the case $\mathcal{G}$ of finite dimension.

II. The proof of the general case, i.e. $\mathcal{G}$ is a semisimple ideally finite real Lie subalgebra of $\mathcal{L}$ consisting of normal elements.

Hence we may write

$$\mathcal{G} = \bigoplus_{i\in I} \mathcal{G}_i,$$

where each $\mathcal{G}_i$ is a simple ideal of $\mathcal{G}$, $[\mathcal{G}_i, \mathcal{G}_j] = \{0\}$ for $i \neq j$, $[\mathcal{G}_i, \mathcal{G}_i] = \mathcal{G}_i$ for every i and $[\mathcal{G},\mathcal{G}] = \mathcal{G}$. We consider as before $\mathcal{G}^+ = \mathcal{G} + \mathcal{G}^* + \mathcal{I}$, where $\mathcal{I} = [\mathcal{G},\mathcal{G}^*]$ is an ideal in $\mathcal{G}^+$ and $\mathcal{G}_0 = \{x \in \mathcal{G}^+ \mid [x,\mathcal{I}] = \{0\}\}$ is the centralizer of $\mathcal{I}$ in $\mathcal{G}^+$, which is an ideal in $\mathcal{G}^+$. Obviously we have

$$\mathcal{G}^+ = \bigoplus \mathcal{G}_i + \bigoplus \mathcal{G}_i^* + \sum_{i,j} [\mathcal{G}_i, \mathcal{G}_j^*].$$

We have

$$x^* = -x \text{ for every } x \in \mathcal{I} = [\mathcal{G},\mathcal{G}^*] = \sum_{i,j} [\mathcal{G}_i, \mathcal{G}_j^*]$$

and $\mathcal{I}^* = \mathcal{I}$.

If $x \in \mathcal{G}_0$ then the following equalities hold: $[x^*, \mathcal{I}] = [x, \mathcal{I}^*]^* = [x, \mathcal{I}]^* = \{0\}$. Hence $\mathcal{G}_0^* = \mathcal{G}_0$ because $(x^*)^* = x$. For a self-adjoint $x \in \mathcal{G}^+$ and $y \in \mathcal{I}$ we may write

$$-[x, y] = [x, y]^* = [y^*, x^*] = -[y, x] = [x, y].$$

This proves that $[x, y] = 0$. Hence, $x = x^* \in \mathcal{G}^+$ implies $x \in \mathcal{G}_0$.

Let z, y be arbitrary elements of $\mathcal{G}$. We have $z + z^* \in \mathcal{G}_0$ because $z + z^*$ is a self-adjoint element of $\mathcal{G}^+$. But $\mathcal{G}_0$ is an ideal in $\mathcal{G}^+$. Hence,

$$[y, z] + [y, z^*] = [y, z + z^*] \in \mathcal{G}_0$$

and

$$[y, z] \in \mathcal{G}_0 + \mathcal{I} \text{ for every } y, z \in \mathcal{G}$$

because $[y, z^*] \in \mathcal{I}$. So we have

$$\mathcal{G} = [\mathcal{G}, \mathcal{G}] \subseteq \mathcal{G}_0 + \mathcal{I}$$

and

$$\mathcal{G}^+ \subseteq \mathcal{G}_0 + \mathcal{I}$$

since $(\mathcal{G}_0 + \mathcal{I})^* = \mathcal{G}_0^* + \mathcal{I}^* = \mathcal{G}_0 + \mathcal{I}$.

We denote

$$\mathcal{I}_0 := \mathcal{G}_0 \cap \mathcal{I} = \sum_{j,k} \mathcal{G}_0 \cap [\mathcal{G}_j, \mathcal{G}_k^*].$$

Obviously we have

$$[\mathcal{G}_0, \mathcal{I}_0] \subseteq [\mathcal{G}_0, \mathcal{I}] = \{0\},\ [\mathcal{I}, \mathcal{I}_0] \subseteq [\mathcal{I}, \mathcal{G}_0] = \{0\},$$

$$[\mathcal{G}, \mathcal{I}_0] \subseteq [\mathcal{G}_0 + \mathcal{I}, \mathcal{I}_0] = \{0\}.$$

We will describe the action of $ad\mathcal{G}$ on $\mathcal{I}$. Let $x, y, z \in \mathcal{G}$ (normal elements in $\mathcal{L}$) and compute

$$\begin{aligned}[][x + z, y], (x + z)^*] &= [x + z, [y, (x + z)^*]] \\ &= [x, [y, x^*]] + [x, [y, z^*]] + [z, [y, x^*]] + [z, [y, z^*]] \\ &= [[x, y], x^*] + [x, [y, z^*]] + [z, [y, x^*]] + [[z, y], z^*].\end{aligned}$$

We have

$$[z, [y, x^*]] = [z, [y^*, x]] = [[z, y^*], x] + [y^*, [z, x]]$$

because $[y, x^*]$ is skew (i.e. $[y, x^*] = [y^*, x]$). Hence we may write,

$$[[x + z, y], (x + z)^*] = [[x, y], x^*] + 2[x, [y, z^*]] + [y^*, [z, x]] + [[z, y], z^*]$$

or equivalently,

$$2[x, [y, z^*]] = [[x + z, y], (x + z)^*] - [[x, y], x^*] - [y^*, [z, x]] - [[z, y], z^*]. \quad (2)$$

If we take $x \in \mathcal{G}_i, y \in \mathcal{G}_j, z \in \mathcal{G}_k$, by (2) there are the following possibilities:

$$[\mathcal{G}_i, [\mathcal{G}_j, \mathcal{G}_k^*]] = \{0\}, \quad \text{for } i \neq j \neq k \neq i;$$

$$[\mathcal{G}_k, [\mathcal{G}_j, \mathcal{G}_k^*]] \subseteq [\mathcal{G}_j, \mathcal{G}_k^*] = [\mathcal{G}_k, \mathcal{G}_j^*].$$

Also we have,

$$[\mathcal{G}_j, [\mathcal{G}_j, \mathcal{G}_k^*]] = [\mathcal{G}_j, [\mathcal{G}_k^*, \mathcal{G}_j]] \subseteq [\mathcal{G}_k, \mathcal{G}_j^*] = [\mathcal{G}_j, \mathcal{G}_k^*],$$

$$[\mathcal{G}_i, [\mathcal{G}_k, \mathcal{G}_k^*]] \subseteq [\mathcal{G}_k, \mathcal{G}_i^*], \quad [\mathcal{G}_i, [\mathcal{G}_i, \mathcal{G}_i^*]] \subseteq [\mathcal{G}_i, \mathcal{G}_i^*].$$

We denote $\mathcal{I}_{jk} := [\mathcal{G}_j, \mathcal{G}_k^*] (= [\mathcal{G}_k, \mathcal{G}_j^*] = [\mathcal{G}_j^*, \mathcal{G}_k] = [\mathcal{G}_k^*, \mathcal{G}_j])$. Then the above inclusions give the following:

$$[\mathcal{G}, \mathcal{I}_{jk}] \subseteq \mathcal{I}_{jk} \text{ for } j \neq k, \; [\mathcal{G}, \mathcal{I}_{jj}] \subseteq \sum_k \mathcal{I}_{kj}, \; [\mathcal{G}, \sum_k \mathcal{I}_{kj}] \subseteq \sum_k \mathcal{I}_{kj}.$$

Then for $j \neq k$, $\mathcal{I}_{jk}$ is a finite-dimensional ideal which is completely reducible under $ad\,\mathcal{G}$ because $\mathcal{G}$ is semisimple. Let $\mathcal{I}_1$ be a complement of $\mathcal{I}_0 \cap \mathcal{I}_{jk}$ in $\mathcal{I}_{jk}$. We have

$$[\mathcal{G}, \mathcal{I}_{jk}] = [\mathcal{G}, (\mathcal{I}_0 \cap \mathcal{I}_{jk}) + \mathcal{I}_1] = [\mathcal{G}, \mathcal{I}_1] \subseteq \mathcal{I}_1.$$

But $[\mathcal{G}_j, \mathcal{G}_k^* + \mathcal{I}_{jk}] \subseteq \mathcal{I}_{jk}$. Hence, by the same reason we may find a complement $\mathcal{I}_{jk}^{\perp}$ (of $\mathcal{I}_{jk}$ in $\mathcal{G}_k^* + \mathcal{I}_{jk}$) invariant to $ad\,\mathcal{G}_j$. We have,

$$[\mathcal{G}_j, \mathcal{I}_{jk}^{\perp}] \subseteq \mathcal{I}_{jk}^{\perp} \cap \mathcal{I}_{jk} = \{0\},$$

$$\mathcal{I}_{jk} = [\mathcal{G}_j, \mathcal{G}_k^*] \subseteq [\mathcal{G}_j, \mathcal{G}_k^* + \mathcal{I}_{jk}] = [\mathcal{G}_j, \mathcal{I}_{jk} + \mathcal{I}_{jk}^{\perp}] = [\mathcal{G}_j, \mathcal{I}_{jk}] \subseteq \mathcal{I}_{jk}.$$

Hence $\mathcal{I}_{jk} = [\mathcal{G}_j, \mathcal{I}_{jk}]$ and $[\mathcal{G}, \mathcal{I}_{jk}] \supseteq \mathcal{I}_{jk}$. We obtain,

$$\mathcal{I}_{jk} \subseteq [\mathcal{G}, \mathcal{I}_{jk}] \subseteq [\mathcal{G}, \mathcal{I}_1] \subseteq \mathcal{I}_1.$$

It follows that $\mathcal{I}_{jk} = \mathcal{I}_1$, hence $\mathcal{I}_0 \cap \mathcal{I}_{jk} = \{0\}$ for $j \neq k$ and

$$\mathcal{I}_0 = \sum_{jk} \mathcal{G}_0 \cap \mathcal{I}_{jk} = \sum_j \mathcal{G}_0 \cap \mathcal{I}_{jj}.$$

On the other hand, by stage I of the present proof (i.e. the finite-dimensional situation) we obtain

$$\mathcal{G}_0 \cap \mathcal{I}_{jj} = \{z \in \mathcal{G}_j^+ \mid [z, \mathcal{I}_{jj}] = \{0\}\} = \{0\}$$

because $\mathcal{I}_{jj} = [\mathcal{G}_j, \mathcal{G}_j^*] \subseteq \mathcal{G}_j^+ := \mathcal{G}_j + \mathcal{G}_j^* + [\mathcal{G}_j, \mathcal{G}_j^*]$.

Hence $\mathcal{I}_0 = \mathcal{G}_0 \cap \mathcal{I} = \{0\}$ and $\mathcal{G}^+ = \mathcal{G}_0 \oplus \mathcal{I}$. □

B. Complexes

§ 8 Generalities

In the following four paragraphs, unless it is otherwise specified, we denote by $\mathcal{C}$ one of the following categories:

1) the category of complex vector spaces with linear maps as morphisms;
2) the category of complex Banach spaces with bounded linear maps as morphisms;
3) the category of complex Hilbert spaces with bounded linear maps as morphisms.

If $\mathcal{X}, \mathcal{Y}$ are objects of $\mathcal{C}$, then we denote by $\mathcal{B}(\mathcal{X}, \mathcal{Y})$ (resp. $\mathcal{B}(\mathcal{X})$ if $\mathcal{X} = \mathcal{Y}$) the set of all morphisms from $\mathcal{X}$ to $\mathcal{Y}$ (usually denoted as $Hom_{\mathcal{C}}(\mathcal{X}, \mathcal{Y})$); also we shall denote by $\mathcal{X}^* := \mathcal{B}(\mathcal{X}, \mathbb{C})$ the dual object of $\mathcal{X}$ and by $\alpha^* \in \mathcal{B}(\mathcal{Y}^*, \mathcal{X}^*)$ the morphism dual to $\alpha \in \mathcal{B}(\mathcal{X}, \mathcal{Y})$.

Definition 1. By *complex* (in $\mathcal{C}$) we mean a pair of sequences

$$((\mathcal{X}_i)_{i \in \mathbb{Z}}, (\alpha_i)_{i \in \mathbb{Z}})$$

(denoted also $(\mathcal{X}., \alpha.)$) where, for each integer i, $\mathcal{X}_i$ is an object of $\mathcal{C}$ and the $\alpha_i \in \mathcal{B}(\mathcal{X}_i, \mathcal{X}_{i-1})$ satisfy $\alpha_i \circ \alpha_{i+1} = 0$. We shall represent this complex by the following diagram

$$\dots \xleftarrow{\alpha_{i-1}} \mathcal{X}_{i-1} \xleftarrow{\alpha_i} \mathcal{X}_i \xleftarrow{\alpha_{i+1}} \mathcal{X}_{i+1} \xleftarrow{\alpha_{i+2}} \dots . \tag{1}$$

(Sometimes this complex will be denoted by $\mathcal{X}_{.}$ for the sake of simplicity. If there exist the integers n, m with $n < m$ such that $\mathcal{X}_i = \{0\}$ for $i < n$ and for $i > m$, then the complex is also represented in the following way:

$$0 \leftarrow \mathcal{X}_n \xleftarrow{\alpha_{n+1}} \dots \xleftarrow{\alpha_m} \mathcal{X}_m \leftarrow 0.$$

If moreover $m = n + 2$, then the complex is called a *short sequence.*)

Remark 1. In the conditions of Definition 1, for each integer i we have $Ker\,\alpha_i \supseteq Ran\,\alpha_{i+1}$; the quotient vector space

$$\mathbf{H}_i(\mathcal{X}., \alpha.) := Ker\,\alpha_i / Ran\,\alpha_{i+1}$$

is called the *i-th homology space* of the complex $(\mathcal{X}., \alpha.)$ and is denoted $\mathbf{H}_i(\mathcal{X}., \alpha.)$, or simply $\mathbf{H}_i(\mathcal{X}.)$ when no confusion is possible. If $\mathbf{H}_i(\mathcal{X}., \alpha.) = \{0\}$ we say that the complex $(\mathcal{X}., \alpha.)$ is *exact at its i-th term.* We say that the complex is *exact* if it is exact at each of its terms.

Next let's describe some simple means to construct new complexes.

a) Let $\mathcal{X}. = (\mathcal{X}., \alpha.)$ be a complex (in $\mathcal{C}$). For $\lambda \in \mathbb{C}$ we denote by $\lambda\mathcal{X}.$ the complex defined by the pair of sequences

$$((\mathcal{X}_i)_{i \in \mathbb{Z}}, (\lambda\alpha_i)_{i \in \mathbb{Z}}).$$

Moreover for $k \in \mathbb{Z}$ we denote by $\mathcal{X}._{+k}$ the complex defined by the pair

$$((\mathcal{X}_{i+k})_{i\in\mathbb{Z}}, (\alpha_{i+k})_{i\in\mathbb{Z}}).$$

b) Next let's assume that $\mathcal{X}. = (\mathcal{X}., \alpha.)$ is a complex in $\mathcal{C}$ as above and $\mathcal{Y}$ is a complex vector space such that one of the following hypotheses holds:

1) $\mathcal{C}$ is the category of all complex vector spaces and $\mathcal{Y}$ is an arbitrary complex vector space.

2) $\mathcal{C}$ is one of the three categories we work with and $\mathcal{Y}$ is a finite-dimensional complex vector space.

Then we denote by $\mathcal{Y} \otimes \mathcal{X}.$, respectively $\mathcal{X}. \otimes \mathcal{Y}$, the complex in $\mathcal{C}$ defined by the pair of sequences

$$((\mathcal{Y} \otimes \mathcal{X}_i)_{i\in\mathbb{Z}}, (I_{\mathcal{Y}} \otimes \alpha_i)_{i\in\mathbb{Z}}),$$

respectively

$$((\mathcal{X}_i \otimes \mathcal{Y})_{i\in\mathbb{Z}}, (\alpha_i \otimes \mathrm{I}_{\mathcal{Y}})_{i\in\mathbb{Z}}),$$

where as usual $I_{\mathcal{Y}}$ denotes the identity map on $\mathcal{Y}$.

c) If $\mathcal{X}. = (\mathcal{X}., \alpha.)$ and $\mathcal{Z}. = (\mathcal{Z}., \gamma.)$ are complexes in $\mathcal{C}$, then we denote by $\mathcal{X}. \oplus \mathcal{Z}.$ the complex defined by the pair

$$((\mathcal{X}_i \oplus \mathcal{Z}_i)_{i\in\mathbb{Z}}, (\alpha_i \oplus \gamma_i)_{i\in\mathbb{Z}}).$$

d) To a complex $\mathcal{X}. = (\mathcal{X}., \alpha.)$ (in $\mathcal{C}$) we can associate the *dual complex* denoted by $(\mathcal{X}., \alpha.)^*$ (or simply by $(\mathcal{X}.)^*$) and defined by the pair of sequences

$$((\mathcal{X}^*_{-i})_{i\in\mathbb{Z}}, (\alpha^*_{-i+1})_{i\in\mathbb{Z}}).$$

Hence this complex can be represented by the following diagram

$$\cdots \xleftarrow{\alpha^*_{-i+1}} \mathcal{X}^*_{-i} \xleftarrow{\alpha^*_{-i}} \mathcal{X}^*_{-i-1} \xleftarrow{\alpha^*_{-i-1}} \cdots.$$

Remark 2. It is easy to check that for the above constructed complexes one can compute the homology spaces in the following way:

$$a) \quad \mathbf{H}_i(\lambda\mathcal{X}.) = \begin{cases} \mathbf{H}_i(\mathcal{X}.) & \text{if} \quad \lambda \neq 0 \\ \mathcal{X}_i & \text{if} \quad \lambda = 0 \end{cases}, \; \mathbf{H}_i(\mathcal{X}._{+k}) = \mathbf{H}_{i+k}(\mathcal{X}.);$$

$$b) \quad \mathbf{H}_i(\mathcal{Y} \otimes \mathcal{X}.) = \mathcal{Y} \otimes \mathbf{H}_i(\mathcal{Y}.),\ \mathbf{H}_i(\mathcal{X}. \otimes \mathcal{Y}) = \mathbf{H}_i(\mathcal{X}.) \otimes \mathcal{Y};$$

$$c) \quad \mathbf{H}_i(\mathcal{X}. \oplus \mathcal{Z}.) = \mathbf{H}_i(\mathcal{X}.) \oplus \mathbf{H}_i(\mathcal{Z}.).$$

For a result relating the homology spaces of $(\mathcal{X}., \alpha.)$ and $(\mathcal{X}., \alpha.)^*$ see Theorem 1 from § 9 and also Theorem 1 (and especially its Corollary 1) from § 11 below.

Definition 2. By a *bicomplex* we mean a 3-tuple of doubly-indexed sequences

$$((\mathcal{X}_{i,j})_{i,j\in\mathbb{Z}}, (\alpha_{i,j})_{i,j\in\mathbb{Z}}, (\beta_{i,j})_{i,j\in\mathbb{Z}})$$

(denoted also $(\mathcal{X}_{..}, \alpha_{..}, \beta_{..})$) where, for each $i, j \in \mathbb{Z}$, $\mathcal{X}_{i,j}$ is an object of $\mathcal{C}$ and $\alpha_{i,j} \in \mathcal{B}(\mathcal{X}_{i,j}, \mathcal{X}_{i-1,j})$, $\beta_{i,j} \in \mathcal{B}(\mathcal{X}_{i,j}, \mathcal{X}_{i,j-1})$ satisfy $\alpha_{i,j} \circ \alpha_{i+1,j} = 0$, $\beta_{i,j} \circ \beta_{i,j+1} = 0$ and $\beta_{i,j+1} \circ \alpha_{i+1,j+1} = \alpha_{i+1,j} \circ \beta_{i+1,j+1}$. We shall represent this bicomplex by the following commutative diagram:

$$\begin{array}{ccccccc}
 & & \dots & & \dots & & \\
 & & \downarrow{\scriptstyle \alpha_{i+2,j}} & & \downarrow{\scriptstyle \alpha_{i+2,j+1}} & & \\
\dots & \xleftarrow{\beta_{i+1,j}} & \mathcal{X}_{i+1,j} & \xleftarrow{\beta_{i+1,j+1}} & \mathcal{X}_{i+1,j+1} & \xleftarrow{\beta_{i+1,j+2}} & \dots \\
 & & \downarrow{\scriptstyle \alpha_{i+1,j}} & & \downarrow{\scriptstyle \alpha_{i+1,j+1}} & & \\
\dots & \xleftarrow{\beta_{i,j}} & \mathcal{X}_{i,j} & \xleftarrow{\beta_{i,j+1}} & \mathcal{X}_{i,j+1} & \xleftarrow{\beta_{i,j+2}} & \dots \\
 & & \downarrow{\scriptstyle \alpha_{i,j}} & & \downarrow{\scriptstyle \alpha_{i,j+1}} & & \\
 & & \dots & & \dots & &
\end{array} . \qquad (2)$$

(Sometimes the bicomplex will be denoted by $\mathcal{X}_{..}$ for the sake of simplicity. When "many" of the spaces $\mathcal{X}_{i,j}$ equal $\{0\}$ we shall use conventions similar to those from complexes, see Definition 1. For example if there exists an integer m such that $\mathcal{X}_{i,j} = \{0\}$ for each $i < m$ and every n, then we omit in (2) the spaces $\mathcal{X}_{i,j}$ with $i < m - 1$.)

Remark 3. In the conditions of Definition 2, for each $i, j \in \mathbb{Z}$ we have the complexes $(\mathcal{X}_{i.}, \beta_{i.})$ and $(\mathcal{X}_{.j}, \alpha_{.j})$. They will be called the *i-th row*, respectively the *j-th column* of $(\mathcal{X}_{..}, \alpha_{..}, \beta_{..})$.

Remark 4. In the conditions of Definition 2, if $(\mathcal{X}'_{..}, \alpha'_{..}, \beta'_{..})$ is another bicomplex with $\mathcal{X}'_{i,j} \subseteq \mathcal{X}_{i,j}$, $\alpha'_{i,j} = \alpha_{i,j}|_{\mathcal{X}'_{i,j}}$ and $\beta'_{i,j} = \beta_{i,j}|_{\mathcal{X}'_{i,j}}$ for every $i, j \in \mathbb{Z}$, then we say that $(\mathcal{X}'_{..}, \alpha'_{..}, \beta'_{..})$ is a *subbicomplex* of $(\mathcal{X}_{..}, \alpha_{..}, \beta_{..})$. Similarly, if $(\mathcal{X}'_{.}, \alpha'_{.})$ and $(\mathcal{X}_{.}, \alpha_{.})$ are two complexes such that $\mathcal{X}'_i \subseteq \mathcal{X}_i$ and $\alpha'_i = \alpha_i|_{\mathcal{X}'_i}$ for every $i \in \mathbb{Z}$ then $(\mathcal{X}'_{.}, \alpha'_{.})$ is called a *subcomplex* of $(\mathcal{X}_{.}, \alpha_{.})$.

Definition 3. Let $(\mathcal{X}_{..}, \alpha_{..}, \beta_{..})$ be a bicomplex *with finite diagonals*, in the sense that for any $i \in \mathbb{Z}$ the set $\{k \in \mathbb{Z} \mid \mathcal{X}_{i-k,k} \neq \{0\}\}$ is finite. Then the *totalization* of the bicomplex $(\mathcal{X}_{..}, \alpha_{..}, \beta_{..})$ is a complex $(\mathcal{Y}_{.}, \gamma_{.})$ defined in the following way. For any $i \in \mathbb{Z}$,

$$\mathcal{Y}_i := \bigoplus_{k=-\infty}^{+\infty} \mathcal{X}_{i-k,k}$$

and $\gamma_i : \mathcal{Y}_i \to \mathcal{Y}_{i-1}$ is the linear map defined by

$$\gamma_i(x) := (-1)^k \alpha_{i-k,k}(x) + \beta_{i-k,k}(x) \in \mathcal{X}_{i-k-1,k} \oplus \mathcal{X}_{i-k,k-1} \subset \mathcal{Y}_{i-1}$$

for all $x \in \mathcal{X}_{i-k,k}$ and all $k \in \mathbb{Z}$. It is easy to check that $(\mathcal{Y}_{.}, \gamma_{.})$ is a complex in $\mathcal{C}$; it will be denoted by $Tot(\mathcal{X}_{..}, \alpha_{..}, \beta_{..})$ (or simply by $Tot(\mathcal{X}_{..})$).

Now we describe an important example of totalization.

Example 1. Let $\widetilde{\otimes}$ be either the usual algebraic tensor product $\otimes$ if $\mathcal{C}$ is the category of complex vector spaces, or the completed projective tensor product $\widehat{\otimes}$ if $\mathcal{C}$ is the category of complex Banach spaces, or the completed Hilbert tensor product $\bar{\otimes}$ if $\mathcal{C}$ is the category of complex Hilbert spaces. If $\mathcal{X}. = (\mathcal{X}., \beta.)$ and $\mathcal{Y}. = (\mathcal{Y}., \alpha.)$ are two complexes in $\mathcal{C}$ such that either in $\mathcal{X}.$ or in $\mathcal{Y}.$ all but a finite number of terms equal $\{0\}$, then we define the *tensor product of* $\mathcal{X}.$ *with* $\mathcal{Y}.$ (with respect to $\widetilde{\otimes}$) — and denote by $\mathcal{X}.\widetilde{\otimes}\mathcal{Y}.$ — the totalization of the bicomplex defined by the 3-tuple

$$((\mathcal{X}_j\widetilde{\otimes}\mathcal{Y}_i)_{i,j\in\mathbb{Z}}\,,\ (I_{\mathcal{X}_j}\widetilde{\otimes}\alpha_i)_{i,j\in\mathbb{Z}}\,,\ (\beta_j\widetilde{\otimes}I_{\mathcal{X}_i})_{i,j\in\mathbb{Z}}).$$

This bicomplex has finite diagonals in view of the assumptions on $\mathcal{X}.$ and $\mathcal{Y}.$, and can be represented by the following commutative diagram

$$\begin{array}{ccccccc}
 & & \cdots & & \cdots & & \\
 & & \downarrow I_{\mathcal{X}_j}\widetilde{\otimes}\alpha_{i+2} & & \downarrow I_{\mathcal{X}_{j+1}}\widetilde{\otimes}\alpha_{i+2} & & \\
\cdots & \xleftarrow{\beta_j\widetilde{\otimes}I_{\mathcal{Y}_{i+1}}} & \mathcal{X}_j\widetilde{\otimes}\mathcal{Y}_{i+1} & \xleftarrow{\beta_{j+1}\widetilde{\otimes}I_{\mathcal{Y}_{i+1}}} & \mathcal{X}_{j+1}\widetilde{\otimes}\mathcal{Y}_{i+1} & \xleftarrow{\beta_{j+2}\widetilde{\otimes}I_{\mathcal{Y}_{i+1}}} & \cdots \\
 & & \downarrow I_{\mathcal{X}_j}\widetilde{\otimes}\alpha_{i+1} & & \downarrow I_{\mathcal{X}_{j+1}}\widetilde{\otimes}\alpha_{i+1} & & \\
\cdots & \xleftarrow{\beta_j\widetilde{\otimes}I_{\mathcal{Y}_i}} & \mathcal{X}_j\widetilde{\otimes}\mathcal{Y}_i & \xleftarrow{\beta_{j+1}\widetilde{\otimes}I_{\mathcal{Y}_i}} & \mathcal{X}_{j+1}\widetilde{\otimes}\mathcal{Y}_i & \xleftarrow{\beta_{j+2}\widetilde{\otimes}I_{\mathcal{Y}_i}} & \cdots \\
 & & \downarrow I_{\mathcal{X}_j}\widetilde{\otimes}\alpha_i & & \downarrow I_{\mathcal{X}_{j+1}}\widetilde{\otimes}\alpha_i & & \\
 & & \cdots & & \cdots & &
\end{array}.$$

Consequently, we denote

$$\mathcal{X}.\widetilde{\otimes}\mathcal{Y}. : \quad \cdots \xleftarrow{\gamma_m} \mathcal{Z}_m \xleftarrow{\gamma_{m+1}} \cdots$$

where

$$\mathcal{Z}_m = \bigoplus_{p=-\infty}^{+\infty} (\mathcal{X}_p\widetilde{\otimes}\mathcal{Y}_{m-p}) = \bigoplus_{p+q=m} (\mathcal{X}_p\widetilde{\otimes}\mathcal{Y}_q)$$

and

$$\gamma_{p+q} = \gamma_m = \beta_p\widetilde{\otimes}I_{\mathcal{Y}_p} + (-1)^q \cdot I_{\mathcal{X}_q}\widetilde{\otimes}\alpha_p \text{ on } \mathcal{X}_q\widetilde{\otimes}\mathcal{Y}_p, \text{ if } p+q=m.$$

Definition 4. Let $(\mathcal{X}''., \alpha''.)$ and $(\mathcal{X}'., \alpha'.)$ be complexes in $\mathcal{C}$. A *semi-morphism* $\beta. : \mathcal{X}''. \to \mathcal{X}'.$ of these complexes is a sequence of maps $\beta. = (\beta_i)_{i\in\mathbb{Z}}$ where, for each $i \in \mathbb{Z}$, $\beta_i \in \mathcal{B}(\mathcal{X}''_i, \mathcal{X}'_i)$ and there exists $\varepsilon_{i+1} \in \{-1, 1\}$ such that $\varepsilon_{i+1} \cdot (\alpha'_{i+1} \circ \beta_{i+1}) = \beta_i \circ \alpha''_{i+1}$. This semi-morphism gives rise to the following commutative diagram:

$$\begin{array}{ccc}
\cdots & & \cdots \\
\varepsilon_{i+2}\cdot\alpha'_{i+2}\downarrow & & \downarrow\alpha''_{i+2} \\
\mathcal{X}'_{i+1} & \xleftarrow{\beta_{i+1}} & \mathcal{X}''_{i+1} \\
\varepsilon_{i+1}\cdot\alpha'_{i+1}\downarrow & & \downarrow\alpha''_{i+1} \\
\mathcal{X}'_i & \xleftarrow{\beta_i} & \mathcal{X}''_i \\
\varepsilon_i\cdot\alpha'_i\downarrow & & \downarrow\alpha''_i \\
\cdots & & \cdots
\end{array}. \qquad (3)$$

(If $\varepsilon_i = 1$ for every $i \in \mathbb{Z}$, then $\beta.$ is called a *morphism* of complexes.) The *cone* of the semi-morphism of complexes $\beta.$ is a complex $(\mathcal{Y}., \gamma.)$ defined in the following way. For any $i \in \mathbb{Z}$,

$$\mathcal{Y}_{i+1} := \mathcal{X}'_{i+1} \oplus \mathcal{X}''_i$$

and $\gamma_{i+1} : \mathcal{Y}_{i+1} \to \mathcal{Y}_i$ is defined by

$$\gamma_{i+1} : \begin{matrix} \mathcal{X}'_{i+1} \\ \oplus \\ \mathcal{X}''_i \end{matrix} \longrightarrow \begin{matrix} \mathcal{X}'_i \\ \oplus \\ \mathcal{X}''_{i-1} \end{matrix}, \quad \gamma_{i+1} = \begin{pmatrix} \varepsilon_{i+1} \cdot \alpha'_{i+1} & \beta_i \\ 0 & -\alpha''_i \end{pmatrix},$$

that is,

$$\gamma_{i+1}(x'_{i+1} \oplus x''_i) = (\varepsilon_{i+1} \cdot \alpha'_{i+1}(x'_{i+1}) + \beta_i(x''_i)) \oplus (-\alpha''_i(x''_i))$$

for any $x'_{i+1} \oplus x''_i \in \mathcal{X}'_{i+1} \oplus \mathcal{X}''_i = \mathcal{Y}_{i+1}$. It is easy to check that $(\mathcal{Y}., \gamma.)$ is a complex in $\mathcal{C}$; it will be denoted by $Con(\mathcal{X}''., \mathcal{X}'., \beta.)$ (or by $Con(\mathcal{X}'., \beta.)$ if $(\mathcal{X}''., \alpha''.) = (\mathcal{X}'., \alpha'.)$).

Remark 5. In the conditions of Definition 4 let's remark that $(\mathcal{Y}., \gamma.)$ is the totalization of the bicomplex whose 0-th and 1-st column are those from (3) and the other columns consist only of zeros.

Remark 6. In the conditions of Definition 4, by the commutativity of (3), β_i induces a linear map

$$\mathbf{H}_i(\beta.) : \mathbf{H}_i(\mathcal{X}''., \alpha''.) \to \mathbf{H}_i(\mathcal{X}'., \alpha'.) \qquad (i \in \mathbb{Z}).$$

If moreover $(\mathcal{X}'''., \alpha'''.)$ is another complex in $\mathcal{C}$ and $\delta. : \mathcal{X}'''. \to \mathcal{X}''.$ is another semi-morphism of complexes, then it is easy to check that the sequence of maps $\beta. \circ \delta. = (\beta_i \circ \delta_i)_{i \in \mathbb{Z}}$ defines a semi-morphism of complexes, $\beta. \circ \delta. : \mathcal{X}'''. \to \mathcal{X}'.$. Moreover

$$\mathbf{H}_i(\beta. \circ \delta.) = \mathbf{H}_i(\beta.) \circ \mathbf{H}_i(\delta.)$$

for each $i \in \mathbb{Z}$.

The following result can be proved by means of a straightforward chase on diagram.

Proposition 1. *If $\beta. : \mathcal{X}''. \to \mathcal{X}'.$ is a semi-morphism of complexes, then the complex $Con(\mathcal{X}''., \mathcal{X}'., \beta.)$ is exact if and only if for each $i \in \mathbb{Z}$ the map*

$$\mathbf{H}_i(\beta.) : \mathbf{H}_i(\mathcal{X}''.) \to \mathbf{H}_i(\mathcal{X}'.)$$

is a vector space isomorphism.

Remark 7. If $\beta. : \mathcal{X}''. \to \mathcal{X}'.$ is a semi-morphism of complexes such that $\beta_i : \mathcal{X}''_i \to \mathcal{X}'_i$ is a vector space isomorphism for each $i \in \mathbb{Z}$ then

$$\mathbf{H}_i(\beta.) : \mathbf{H}_i(\mathcal{X}''.) \to \mathbf{H}_i(\mathcal{X}'.)$$

is a vector space isomorphism for each $i \in \mathbb{Z}$; in this case $\beta.$ is called a *semi-isomorphism* and we say that the complexes $\mathcal{X}''.$ and $\mathcal{X}'.$ are *semi-isomorphic*.

Remark 8. Let $u_. : \mathcal{X}_. \to \mathcal{Y}_.$ and $v_. : \mathcal{Y}_. \to \mathcal{Z}_.$ be semi-morphisms of complexes in $\mathcal{C}$ with the property that for each $i \in \mathbb{Z}$ the short sequence

$$0 \leftarrow \mathcal{Z}_i \xleftarrow{v_i} \mathcal{Y}_i \xleftarrow{u_i} \mathcal{X}_i \leftarrow 0$$

is exact. Then for each $i \in \mathbb{Z}$ there exists a linear map

$$\delta_{i+1} : \mathbf{H}_{i+1}(\mathcal{Z}_.) \to \mathbf{H}_i(\mathcal{X}_.)$$

such that we have the following exact complex

$$\dots \xleftarrow{H_{i-1}(u_.)} \mathbf{H}_{i-1}(\mathcal{X}_.) \xleftarrow{\delta_i} \mathbf{H}_i(\mathcal{Z}_.) \xleftarrow{H_i(v_.)} \mathbf{H}_i(\mathcal{Y}_.) \xleftarrow{H_i(u_.)} \mathbf{H}_i(\mathcal{X}_.) \xleftarrow{\delta_{i+1}}$$

$$\mathbf{H}_{i+1}(\mathcal{Z}_.) \xleftarrow{H_{i+1}(v_.)} \dots .$$

This is called *the long exact sequence of the homology spaces, associated to the short exact sequence of (semi-morphisms of) complexes*

$$0 \leftarrow \mathcal{Z}_. \xleftarrow{v_.} \mathcal{Y}_. \xleftarrow{u_.} \mathcal{X}_. \leftarrow 0.$$

Next let $(\mathcal{X}_{..}, \alpha_{..}, \beta_{..})$ be a bicomplex in $\mathcal{C}$. For each $i \in \mathbb{Z}$ we denote by $\mathbf{H}_i(\mathcal{X}_{..}, \alpha_{..})$ the complex defined by the pair of sequences

$$((\mathbf{H}_i(\mathcal{X}_{.,j}\ , \alpha_{.,j}\ ))_{j\in\mathbb{Z}}, (\mathbf{H}_i(\beta_{.,j}\ ))_{j\in\mathbb{Z}}).$$

Hence $\mathbf{H}_i(\mathcal{X}_{..}, \alpha_{..})$ can be written

$$\dots \xleftarrow{H_i(\beta_{.,j})} \mathbf{H}_i(\mathcal{X}_{.,j}\ , \alpha_{.,j}\ ) \xleftarrow{H_i(\beta_{.,j+1})} \mathbf{H}_i(\mathcal{X}_{.,j+1}\ , \alpha_{.,j+1}\ ) \xleftarrow{H_i(\beta_{.,j+2})} \dots . \tag{4}$$

We can define similarly the complex $\mathbf{H}_j(\mathcal{X}_{..}, \beta_{..})$ for each $j \in \mathbb{Z}$. Now we can state the following result which allows the computation of the totalization homology in terms of the homology spaces of rows, respectively columns.

Theorem 1. *Let $(\mathcal{X}_{..}, \alpha_{..}, \beta_{..})$ be a bicomplex with finite diagonals.*

(a) *If there exists $i_0 \in \mathbb{Z}$ such that for every $i \in \mathbb{Z}\backslash\{i_0\}$ and $j \in \mathbb{Z}$ we have*

$$\mathbf{H}_j(\mathbf{H}_i(\mathcal{X}_{..}, \alpha_{..})) = \{0\}$$

(i.e. the complex $\mathbf{H}_i(\mathcal{X}_{..}, \alpha_{..})$ is exact for every $i \in \mathbb{Z}\backslash\{i_0\}$), then for every $p \in \mathbb{Z}$ there exists a vector space isomorphism

$$\mathbf{H}_p(Tot(\mathcal{X}_{..})) \cong \mathbf{H}_{p-i_0}(\mathbf{H}_{i_0}(\mathcal{X}_{..}, \alpha_{..})).$$

(b) *If there exists $j_0 \in \mathbb{Z}$ such that for every $i \in \mathbb{Z}$ and $j \in \mathbb{Z}\backslash\{j_0\}$ we have*

$$\mathbf{H}_i(\mathbf{H}_j(\mathcal{X}_{..}, \beta_{..})) = \{0\}$$

(i.e. the complex $\mathbf{H}_j(\mathcal{X}.., \beta..)$ is exact for every $j \in \mathbb{Z}\setminus\{j_0\}$), then for every $p \in \mathbb{Z}$ there exists a vector space isomorphism

$$\mathbf{H}_p(Tot(\mathcal{X}..)) \cong \mathbf{H}_{p-j_0}(\mathbf{H}_{j_0}(\mathcal{X}.., \beta..)).$$

Actually we shall not need this result in its full generality. We shall need its following consequence which contains some sufficient conditions for the totalization to be exact.

Corollary 1. *Let $(\mathcal{X}.., \alpha.., \beta..)$ be a bicomplex with finite diagonals. If one of the following conditions holds*

(a′) *each column of $\mathcal{X}..$ is an exact complex,*
(b′) *each row of $\mathcal{X}..$ is an exact complex,*
(a) *for every $i, j \in \mathbb{Z}$ we have $\mathbf{H}_j(\mathbf{H}_i(\mathcal{X}.., \alpha..)) = \{0\}$,*
(b) *for every $i, j \in \mathbb{Z}$ we have $\mathbf{H}_i(\mathbf{H}_j(\mathcal{X}.., \beta..)) = \{0\}$,*

then the complex $Tot(\mathcal{X}..)$ is exact.

Next we prove other useful results.

Proposition 2. *Let $(\mathcal{X}.., \alpha.., \beta..)$ be a bicomplex with finite diagonals. If $\beta_{ij} = 0$ for every $i, j \in \mathbb{Z}$, then $Tot(\mathcal{X}..)$ is exact if and only if each column of $\mathcal{X}..$ is exact.*

Proof. One simply applies the definition of $Tot(\mathcal{X}..)$. □

Proposition 3. *Let $(\mathcal{X}.., \alpha.., \beta..)$ be a bicomplex with all but a finite number of columns consisting only of zeros. (Consequently it has finite diagonals.) Let's assume that there exists $i_0 \in \mathbb{Z}$ such that the i-th column is exact for every $i \in \mathbb{Z}\setminus\{i_0\}$. Then $Tot(\mathcal{X}..)$ is exact if and only if the i_0-th column is exact.*

Proof. If the i_0-th column is exact, then $Tot(\mathcal{X}..)$ is exact by Corollary 1(a′).

Next let's assume that $Tot(\mathcal{X}..)$ is exact. Without loss of generality we may assume, in view of the hypothesis, that there exists a non-negative integer N such that $\mathcal{X}_{ij} = \{0\}$ for every $j \in \mathbb{Z}\setminus\{0, \dots, N\}$ and every $i \in \mathbb{Z}$. We proceed by induction on N. If $N = 0$ the conclusion is obvious. Now assume that $N \geq 1$ and that the conclusion was obtained for $N - 1$. Let $\mathcal{X}'..$ be the complex obtained from $\mathcal{X}..$ by replacing the N-th column by a column consisting of zeros. We observe that the sequence of maps $\beta._{,N} = (\beta_{i,N})_{i\in\mathbb{Z}}$ defines a semi-morphism of complexes

$$\beta._{,N} : \mathcal{X}._{,N} \to Tot(\mathcal{X}'..)$$

and

$$Tot(\mathcal{X}..) = Con(\mathcal{X}._{,N}\,,\ Tot(\mathcal{X}'..)\,,\ \beta._{,N}). \tag{5}$$

Consequently, in view of Proposition 1 and of the hypothesis that $Tot(\mathcal{X}..)$ is exact, we get that $\mathcal{X}._{,N}$ is exact if and only if $Tot(\mathcal{X}'..)$ is exact.

On the other hand, if $i_0 = N$ then each column of $\mathcal{X}'..$ is exact by hypothesis, so $Tot(\mathcal{X}'..)$ is exact (by Corollary 1 (a′)). Then $\mathcal{X}._{,N}$ is exact in view of what we have just proved above. But $i_0 = N$ so $\mathcal{X}._{,i_0}$ is exact.

Now, if $i_0 \neq N$, then $\mathcal{X}_{\cdot,N}$ is exact by hypothesis, so $Tot(\mathcal{X}'_{\cdot\cdot})$ is exact. But now $\mathcal{X}_{\cdot,i_0}$ is a column of $\mathcal{X}'_{\cdot\cdot}$ so $\mathcal{X}_{\cdot,i_0}$ is exact by the induction hypothesis. □

Proposition 4. *Let $(\mathcal{X}_{\cdot\cdot}, \alpha_{\cdot\cdot}, \beta_{\cdot\cdot})$ be a bicomplex with the following properties:*
(a) *if either $i < 0$ or $j < 0$ then $\mathcal{X}_{ij} = \{0\}$;*
(b) *for every $j \in \mathbb{Z}$ we have $\mathbf{H}_j(\alpha_{1,\cdot}) = 0$;*
(c) *the 0-th row $(\mathcal{X}_{0,\cdot}, \beta_{0,\cdot})$ is not exact;*
(d) *if $j \in \mathbb{Z}$ and $\mathbf{H}_j(\mathcal{X}_{0,\cdot}) = 0$, then $\mathbf{H}_j(\mathcal{X}_{i,\cdot}) = 0$ for every $i \in \mathbb{Z}$.*
Then at least one column of the given bicomplex is not exact.

Proof. Let's assume that every column of $\mathcal{X}_{\cdot\cdot}$ is exact.

In view of the hypotheses (c) and (a) we can choose a non-negative $k \in \mathbb{Z}$ such that $\mathbf{H}_j(\mathcal{X}_{0,\cdot}) = \{0\}$ for each $j < k$ and $\mathbf{H}_k(\mathcal{X}_{0,\cdot}) \neq \{0\}$ (i.e. $\mathcal{X}_{0,k}$ is the first term where the 0-th row $\mathcal{X}_{0,\cdot}$ is not exact). Then we can take $x \in \mathcal{X}_{0,k}$ such that $x \in (Ker\,\beta_{0,k}) \backslash (Ran\,\beta_{0,k+1})$.

Next, in view of the condition (a), of the assumption that each column of $\mathcal{X}_{\cdot\cdot}$ is exact and of the fact that $x \in Ker\,\beta_{0,k}$ we can choose $x_{j+1,k-j} \in \mathcal{X}_{j+1,k-j}$ step by step for $j = 0, \ldots, k$ such that

$$x = \alpha_{1,k}(x_{1,k}), \tag{6}$$

$$\beta_{1,k}(x_{1,k}) = \alpha_{2,k-1}(x_{2,k-1}), \tag{7}$$

....

$$\beta_{k,1}(x_{k,1}) = \alpha_{k+1,0}(x_{k+1,0}).$$

On the other hand we have $\mathbf{H}_j(\mathcal{X}_{0,\cdot}) = \{0\}$ for $j < k$; hence by the hypothesis (d) it is easy to choose $y_{k-j+1,j+1} \in \mathcal{X}_{k-j+1,j+1}$ step by step for $j = 0, 1, \ldots, k-1$ such that

$$x_{k+1,0} = \beta_{k+1,1}(y_{k+1,1}),$$

$$x_{k,1} - \alpha_{k+1,1}(y_{k+1,1}) = \beta_{k,2}(y_{k,2}),$$

....

$$x_{2,k-1} - \alpha_{3,k-1}(y_{3,k-1}) = \beta_{2,k}(y_{2,k}). \tag{8}$$

Now $\alpha_{2,k-1}(x_{2,k-1}) = (\alpha_{2,k-1} \circ \beta_{2,k})(y_{2,k}) = (\beta_{1,k} \circ \alpha_{2,k})(y_{2,k})$ by (8). Hence by (7) we deduce

$$\beta_{1,k}(x_{1,k} - \alpha_{2,k}(y_{2,k})) = 0.$$

This relation and the hypothesis (b) imply $\alpha_{1,k}(x_{1,k} - \alpha_{2,k}(y_{2,k})) \in Ran\,\beta_{0,k+1}$ that is $\alpha_{1,k}(x_{1,k}) \in Ran\,\beta_{0,k+1}$. Hence (6) implies $x \in Ran\,\beta_{0,k+1}$, which contradicts the choice of x. □

§ 9 Banach space complexes

By a *Banach space complex* we mean a complex in the category of complex Banach spaces with bounded linear maps as morphisms. In § 9 we assume that $\mathcal{C}$ is either the category of complex Banach spaces or the category of complex Hilbert spaces.

Theorem 1. *A Banach space complex is exact if and only if its dual complex is exact.*

Theorem 2. *Let T be a topological space and $(\mathcal{X}_i)_{i\in\mathbb{Z}}$ be a sequence of complex Banach spaces with all but a finite number of terms equal zero. For each $i \in \mathbb{Z}$ let*

$$\alpha_i : T \to \mathcal{B}(\mathcal{X}_i, \mathcal{X}_{i-1}),\ t \mapsto \alpha_i(t),$$

be a continuous map (with respect to the norm operator topology on $\mathcal{B}(\mathcal{X}_i, \mathcal{X}_{i-1})$). We assume that for each $t \in T$ the pair of sequences $((\mathcal{X}_i)_{i\in\mathbb{Z}}, (\alpha_i(t))_{i\in\mathbb{Z}})$ defines a Banach space complex, which will be denoted $\mathcal{X}.(t)$. Then the set

$$\{t \in T \mid \textit{The complex } \mathcal{X}.(t) \textit{ is not exact}\}$$

is a closed subset of T.

Another useful result is the following.

Theorem 3 (Słodkowski's Lemma). *Let $\mathcal{X}.$ be a Banach space complex and $\beta. : \mathcal{X}. \to \mathcal{X}.$ be a morphism of complexes. If p is an integer such that $\mathbf{H}_p(\mathcal{X}.) \neq \{0\}$, then there exists $\lambda \in \mathbb{C}$ such that*

$$\mathbf{H}_p(Con(\mathcal{X}., \beta. - \lambda)) \neq \{0\},$$

where $\beta. - \lambda : \mathcal{X}. \to \mathcal{X}.$ is the morphism of complexes defined by the sequence of maps $\beta. - \lambda = (\beta_i - \lambda I_{\mathcal{X}_i})_{i\in\mathbb{Z}}$.

For the proof of this theorem we need the following two lemmas.

Lemma 1. *Let $\mathcal{X}$ be a complex Banach space, $A \in \mathcal{B}(\mathcal{X})$ and $\mathcal{A}$ be the associative unital subalgebra of $\mathcal{B}(\mathcal{X})$ generated by A. If $p : \mathcal{A} \to \mathbb{R}_+$ is a semi-norm with the following properties:*
(i) *for every $B, C \in \mathcal{A}$ we have $p(BC) \leq p(B)\|C\|$,*
(ii) *for every $c \in \mathbb{C}$ there exists $\varepsilon_c > 0$ such that $p((A-c)B) \geq \varepsilon_c \cdot p(A)$ for every $B \in \mathcal{A}$,*
then $p \equiv 0$.

Proof. Let's assume that $p \neq 0$. Then $\mathcal{Y} \neq \{0\}$, where $\mathcal{Y}$ is the Banach space obtained by completion of $\mathcal{A}/Kerp$ with respect to the norm induced by p. By the hypothesis (i), the map $\mathcal{A} \to \mathcal{A}, B \mapsto AB$, determines a bounded linear operator $T \in \mathcal{B}(\mathcal{Y})$. Since $\mathcal{A}/Kerp$ is dense in $\mathcal{Y}$, we deduce by (ii) that for every $c \in \mathbb{C}$ there exists $\varepsilon_c > 0$ such that $\|(T-c)y\| \geq \varepsilon_c\|y\|$ for every $y \in \mathcal{Y}$. But this is not possible since $\mathcal{Y} \neq \{0\}$, so $\sigma(T) \neq \emptyset$ and for c in the topological boundary of $\sigma(T)$ we can find a sequence $\{y_n\}_{n\geq 1}$ of unit vectors from $\mathcal{Y}$ with the property $\lim_{n\to\infty} \|(T-c)y_n\| = 0$. □

Lemma 2. *Let $\mathcal{Z}_0, \mathcal{Z}$ be complex Banach spaces and $D \in \mathcal{B}(\mathcal{Z}_0, \mathcal{Z})$, $A_0 \in \mathcal{B}(\mathcal{Z}_0)$, $A \in \mathcal{B}(\mathcal{Z})$ satisfying $AD = DA_0$. If $D(\mathcal{Z}_0) \neq \mathcal{Z}$, then there exists $c \in \mathbb{C}$ such that $D(\mathcal{Z}_0) + (A-c)\mathcal{Z} \neq \mathcal{Z}$.*

Proof. Let's assume that the conclusion fails. Let $c \in \mathbb{C}$ be arbitrary but fixed for the moment. Then the map

$$\mathcal{Z}_0 \oplus \mathcal{Z} \to \mathcal{Z},\ (z, z_0) \mapsto Dz_0 + (A-c)z$$

is onto, hence its dual map

$$\mathcal{Z}^* \to \mathcal{Z}^* \oplus \mathcal{Z}_0^*,\ z^* \mapsto ((A^*-c)z^*, D^*z^*)$$

is one-to-one and has closed range. Consequently, since $c \in \mathbb{C}$ was arbitrary we get:

$$\forall c \in \mathbb{C}\, \exists \varepsilon_c > 0 : \forall z^* \in \mathcal{Z}^*,\ \|(A^*-c)z^*\| + \|D^*z^*\| \geq \varepsilon_c \|z^*\|. \tag{1}$$

On the other hand, because $D(\mathcal{Z}_0) \neq \mathcal{Z}$, there exists a sequence $\{z_k^*\}_{k\geq 1}$ of unit vectors from $\mathcal{Z}^*$ such that

$$\lim_{k\to\infty} \|D^*z_k^*\| = 0. \tag{2}$$

Now let $\mathcal{A}$ be the associative unital subalgebra of $\mathcal{B}(\mathcal{Z}^*)$ generated by A^*. We define a semi-norm $p : \mathcal{A} \to \mathbb{R}_+$ by

$$p(B) := \limsup_{k\to\infty} \|Bz_k^*\| \quad \text{for } B \in \mathcal{A}.$$

For $B \in \mathcal{A}$ arbitrary there exists a polynomial $q \in \mathbb{C}[X]$ such that $B = q(A^*)$. Hence by $AD = DA_0$ and by (2) we obtain

$$\lim_{k\to\infty} D^*Bz_k^* = \lim_{k\to\infty} D^*q(A^*)z_k^* = \lim_{k\to\infty} q(A_0^*)D^*z_k^* = 0.$$

Hence by applying (1) for $z^* = Bz_k^*$ and taking the superior limit for $k \to \infty$ we obtain

$$\forall c \in \mathbb{C}\, \exists \varepsilon_c > 0 : \forall B \in \mathcal{A},\ p((A^*-c)B) \geq \varepsilon_c p(B).$$

Hence $p \equiv 0$ by Lemma 1. But this is imposible since

$$p(I_{\mathcal{Z}^*}) = \limsup_{k\to\infty} \|z_k^*\| = 1.$$

So we must have $D(\mathcal{Z}_0) + (A-c)\mathcal{Z} \neq \mathcal{Z}$ for some $c \in \mathbb{C}$. □

Proof of Theorem 3. Let $(\mathcal{X}., \alpha.)$ be the given complex. We are going to apply Lemma 2 to the following diagram

$$\begin{array}{ccc} \mathcal{X}_{p+1} & \overset{\beta_{p+1}}{\longleftarrow} & \mathcal{X}_{p+1} \\ \downarrow \alpha_{p+1} & & \downarrow \alpha_{p+1} \\ Ker\,\alpha_p & \overset{\beta_p}{\longleftarrow} & Ker\,\alpha_p\,, \end{array}$$

i.e. for $\mathcal{Z}_0 = \mathcal{X}_{p+1}$, $\mathcal{Z} = Ker\,\alpha_p$, $D = \alpha_{p+1} : \mathcal{X}_{p+1} \to Ker\,\alpha_p$, $A = \beta_p|_{Ker\,\alpha_p} : Ker\,\alpha_p \to Ker\,\alpha_p$ and $A_0 = \beta_{p+1} : \mathcal{X}_{p+1} \to \mathcal{X}_{p+1}$. Since $\mathbf{H}_p(\mathcal{X}.) \neq \{0\}$ we have $D(\mathcal{Z}_0) \neq \mathcal{Z}$. Hence we can apply Lemma 2 to obtain $\lambda \in \mathbb{C}$ such that

$$Ran\,\alpha_{p+1} + (\beta_p - \lambda)(Ker\,\alpha_p) \neq Ker\,\alpha_p.$$

Choose $x_p \in Ker\,\alpha_p$ such that $x_p \notin (\beta_p - \lambda)(Ker\,\alpha_p) + Ran\,\alpha_{p+1}$. It is straightforward to check that for $x_p \oplus 0 \in \mathcal{X}_p \oplus \mathcal{X}_{p-1} = \mathcal{Y}_p$ we have $x_p \oplus 0 \in Ker\,\gamma_p$ and $x_p \oplus 0 \notin Ran\,\gamma_{p+1}$, where $(\mathcal{Y}., \gamma_p)$ denotes the complex $Con(\mathcal{X}., \beta. - \lambda)$. So $\mathbf{H}_p(Con(\mathcal{X}., \beta. - \lambda)) \neq \{0\}$. □

Finally we state a theorem concerning tensor products of complexes in the category of Hilbert spaces (see Example 1 in § 8). Recall that we denote the completed Hilbert tensor product by $\bar{\otimes}$.

Theorem 4. *Let $\mathcal{X}.$ and $\mathcal{Y}.$ be Hilbert space complexes. We assume that each of these complexes has only a finite number of non-zero terms. Then the Hilbert tensor product of complexes $\mathcal{X}.\bar{\otimes}\mathcal{Y}.$ is not exact if and only if neither $\mathcal{X}.$ nor $\mathcal{Y}.$ is exact.*

§ 10 Koszul complexes

In § 10 and § 11 we shall study an important class of complexes, namely the class of Koszul complexes. The present paragraph presents some important general features of these complexes. We will be working in the framework of the category $\mathcal{C}$ (see the beginning of § 8).

As preparation for the definition of a Koszul complex, we recall some notation and notions of exterior algebra. If $\mathcal{E}$ is a complex vector space, we denote by $\Lambda^p\mathcal{E}$ the p-th exterior power of $\mathcal{E}$ for each $p \in \mathbb{N}$. (As usual $\Lambda^0\mathcal{E} = \mathbb{C}$.) We denote by $\Lambda\mathcal{E}$ the exterior algebra of $\mathcal{E}$, that is

$$\Lambda\mathcal{E} := \bigoplus_{p=0}^{\infty} \Lambda^p\mathcal{E}.$$

If p is a positive integer, $u_1, \ldots, u_p \in \mathcal{E}$ and $\underline{u} := u_1 \wedge \cdots \wedge u_p \in \Lambda^p\mathcal{E}$, then for $i, j, \ldots \in \{1, \ldots, p\}$ we denote the omission of factors as usual, that is

$$\overset{i}{\widehat{\underline{u}}} = u_1 \wedge \cdots \wedge u_{i-1} \wedge u_{i+1} \wedge \cdots \wedge u_p \in \Lambda^{p-1}\mathcal{E},$$

$$\overset{i,j}{\widehat{\underline{u}}} = u_1 \wedge \cdots \wedge u_{i-1} \wedge u_{i+1} \wedge \cdots \wedge u_{j-1} \wedge u_{j+1} \wedge \cdots \wedge u_p \in \Lambda^{p-2}\mathcal{E}, \text{ etc.}$$

Definition 1. Let $\mathcal{E}$ be a complex finite-dimensional Lie algebra and $\mathcal{X}$ be an object of $\mathcal{C}$. Let $\rho : \mathcal{E} \to \mathcal{B}(\mathcal{X})$ be a representation of $\mathcal{E}$. The *Koszul complex* of ρ, denoted

$Kos(\rho)$, is a complex $(\mathcal{X}., \alpha.)$ defined in the following way. For $p < 0$ we set $\mathcal{X}_p = \{0\}$; for $p \geq 0$ we set $\mathcal{X}_p := \mathcal{X} \otimes \Lambda^p \mathcal{E}$ and

$$\alpha_p : \mathcal{X} \otimes \Lambda^p \mathcal{E} \to \mathcal{X} \otimes \Lambda^{p-1} \mathcal{E}, \ \alpha_p = \alpha_p^{(1)} + \alpha_p^{(2)},$$

where the maps $\alpha_p^{(1)}, \alpha_p^{(2)} : \mathcal{X} \otimes \Lambda^p \mathcal{E} \to \mathcal{X} \otimes \Lambda^{p-1} \mathcal{E}$ are defined by

$$\alpha_p^{(1)}(x \otimes \underline{u}) = \sum_{i=1}^{p} (-1)^{i-1} \rho(u_i) x \otimes \overset{i}{\widehat{\underline{u}}},$$

$$\alpha_p^{(2)}(x \otimes \underline{u}) = \sum_{1 \leq i < j \leq p} (-1)^{i+j-1} x \otimes [u_i, u_j] \wedge \overset{i,j}{\widehat{\underline{u}}}$$

for every $x \in \mathcal{X}$ and $\underline{u} = u_1 \wedge \cdots \wedge u_p \in \Lambda^p \mathcal{E}$.

Remark 1. A straightforward computation shows that $\alpha_p \circ \alpha_{p+1} = 0$ for every $p \in \mathbb{N}$, so $(\mathcal{X}., \alpha.) = Kos(\rho)$ is indeed a complex (in $\mathcal{C}$). This complex will be often represented by the following diagram:

$$Kos(\rho) : 0 \leftarrow \mathcal{X} \xleftarrow{\alpha_1} \mathcal{X} \otimes \mathcal{E} \xleftarrow{\alpha_2} \cdots \xleftarrow{\alpha_p} \mathcal{X} \otimes \Lambda^p \mathcal{E} \xleftarrow{\alpha_{p+1}} \cdots.$$

Moreover we denote by $\mathbf{H}_p(\rho)$ the p-th homology space of the above complex, that is

$$\mathbf{H}_p(\rho) = \mathbf{H}_p(Kos(\rho)) \quad (p \in \mathbb{N}).$$

Now we are going to describe some concrete situations when the homology spaces are easy to compute.

Example 1. Let's assume that $\dim \mathcal{E} = 1$ and let's choose $e \in \mathcal{E} \backslash \{0\}$. Then it is easy to check that we have

$$Kos(\rho) : \qquad 0 \leftarrow \mathcal{X} \xleftarrow{T} \mathcal{X} \leftarrow 0$$

where $T := \rho(e) \in \mathcal{B}(\mathcal{X})$. So in this case we have $\mathbf{H}_0(\rho) = \mathcal{X}/RanT$ and $\mathbf{H}_1(\rho) = KerT$.

Example 2. Now let $\dim \mathcal{E} = n$. We shall compute the homology spaces at the extreme terms of $Kos(\rho)$, that is the 0-th and the n-th homology spaces. Let $\{e_1, \ldots, e_n\}$ be a basis of $\mathcal{E}$. We have

$$[e_j, e_i] = \sum_{k=1}^{n} c_{ji}^k e_k \quad (1 \leq i < j \leq n).$$

Denote $\underline{e} := e_1 \wedge \cdots \wedge e_n \in \Lambda^n \mathcal{E}$. Using the notation from Definition 1 we get for $x \in \mathcal{X}$

$$\alpha_n(x \otimes \underline{e}) = \sum_{i=1}^{n} (-1)^{i-1} \rho(e_i) x \otimes \overset{i}{\widehat{\underline{e}}} + \sum_{1 \leq i < j \leq n} (-1)^{i+j-1} x \otimes (-\sum_{k=1}^{n} c_{ji}^k e_k) \wedge \overset{i,j}{\widehat{\underline{e}}}.$$

But

$$(\sum_{k=1}^{n} c_{ji}^{k} e_k)\wedge \overset{i,j}{\hat{\underline{e}}} = c_{ji}^{i} e_i \wedge \overset{i,j}{\hat{\underline{e}}} + c_{ji}^{j} e_j \wedge \overset{i,j}{\hat{\underline{e}}} = (-1)^{i-1} c_{ji}^{i} \overset{j}{\hat{\underline{e}}} + (-1)^{j-2} c_{ji}^{j} \overset{i}{\hat{\underline{e}}} .$$

Hence

$$\alpha_n(x\otimes \underline{e}) = \sum_{i=1}^{n}(-1)^{i-1}\rho(e_i)x\otimes \overset{i}{\hat{\underline{e}}} + \sum_{1\le i<j\le n}((-1)^{j-1}c_{ji}^{i}x\otimes \overset{j}{\hat{\underline{e}}} + (-1)^{i}c_{ji}^{j}x\otimes \overset{i}{\hat{\underline{e}}}).$$

Since $c_{ji}^k = -c_{ij}^k$, we obtain finally

$$\alpha_n(x\otimes \underline{e}) = \sum_{i=1}^{n}(-1)^{i-1}(\rho(e_i) + \sum_{\substack{1\le j\le n\\ j\ne i}} c_{ij}^{j})x\otimes \overset{i}{\hat{\underline{e}}} .$$

Hence in view of the identification $\mathcal{X}\otimes(\Lambda^n\mathcal{E})\cong\mathcal{X}$ (since $\Lambda^n\mathcal{E}=\mathbb{C}\cdot\underline{e}$) we obtain

$$\mathbf{H}_n(\rho) = Ker\,\alpha_n = \bigcap_{i=1}^{n} Ker(\rho(e_i) + \sum_{\substack{1\le j\le n\\ j\ne i}} c_{ij}^{j}).$$

To compute $\mathbf{H}_0(\rho)$ let's remark that

$$\alpha_1(x\otimes u) = \rho(u)x \text{ for every } x\in\mathcal{X} \text{ and } u\in\mathcal{E}.$$

Hence

$$\mathbf{H}_0(\rho) = \mathcal{X}/Ran\,\alpha_1 = \mathcal{X}/(\rho(\mathcal{E})\mathcal{X}).$$

Example 3. Now we suppose that $\mathcal{E}$ is a solvable Lie algebra with $\dim\mathcal{E}=2$. Let $\{e_1,e_2\}$ be a basis of $\mathcal{E}$ such that

$$[e_2,e_1] = \gamma e_1$$

for a certain $\gamma\in\mathbb{C}$. Let $\rho:\mathcal{E}\to\mathcal{B}(\mathcal{X})$ be a representation. We denote $T:=\rho(e_2)$ and $N:=\rho(e_1)$. With the notations from Definition 1 we have

$$Kos(\rho): 0 \leftarrow \mathcal{X} \xleftarrow{\alpha_1} \mathcal{X}\otimes\mathcal{E} \xleftarrow{\alpha_2} \mathcal{X}\otimes\Lambda^2\mathcal{E} \leftarrow 0.$$

For $x\in\mathcal{X}$ we have

$$\alpha_1(x\otimes e_1) = \rho(e_1)x = Nx \;, \alpha_1(x\otimes e_2) = \rho(e_2)x = Tx \;,$$

$$\alpha_2(x\otimes(e_1\wedge e_2)) = \rho(e_1)x\otimes e_2 - \rho(e_2)x\otimes e_1 + x\otimes[e_1,e_2]$$
$$= Nx\otimes e_2 - Tx\otimes e_1 + x\otimes(-\gamma e_1) = (-(T+\gamma)x)\otimes e_1 + Nx\otimes e_2 \,.$$

Consequently, in view of the identifications $\mathcal{X} \otimes \mathcal{E} = (\mathcal{X} \otimes e_1) \oplus (\mathcal{X} \otimes e_2) \cong \mathcal{X} \oplus \mathcal{X}$ and $\mathcal{X} \otimes \Lambda^2 \mathcal{E} = \mathcal{X} \otimes (e_1 \wedge e_2) \cong \mathcal{X}$ we can describe the complex $Kos(\rho)$ in the following way:

$$Kos(\rho) : 0 \leftarrow \mathcal{X} \xleftarrow{\alpha_1} \mathcal{X} \oplus \mathcal{X} \xleftarrow{\alpha_2} \mathcal{X} \leftarrow 0,$$

$$\alpha_1(x_1, x_2) = Nx_1 + Tx_2 \quad \forall (x_1, x_2) \in \mathcal{X} \oplus \mathcal{X},$$

$$\alpha_2(x) = (-(T + \gamma)x, Nx) \quad \forall x \in \mathcal{X}\,.$$

Moreover, if $\mathcal{X}$ is finite-dimensional, a computation of dimensions shows that if $Ran\,\alpha_1 = \mathcal{X}$ and $Ker\,\alpha_2 = \{0\}$, then $Ker\,\alpha_1 = Ran\,\alpha_2$ so the complex $Kos(\rho)$ is exact. Hence, if $\dim \mathcal{X} < \infty$ we can characterize the exactness of $Kos(\rho)$ in the following way:

$$Kos(\rho) \text{ exact } \Leftrightarrow Ran\,N + Ran\,T = \mathcal{X} \text{ and } Ker\,N \cap Ker(T + \gamma) = \{0\}.$$

Now let's come back to the general properties of the Koszul complexes.

Let $\rho' : \mathcal{E} \to \mathcal{B}(\mathcal{X}')$ and $\rho'' : \mathcal{E} \to \mathcal{B}(\mathcal{X}'')$ be representations of the finite-dimensional Lie algebra $\mathcal{E}$ (where $\mathcal{X}'$ and $\mathcal{X}''$ are objects of the category $\mathcal{C}$). Let $f \in \mathcal{B}(\mathcal{X}'', \mathcal{X}')$ be a linear map which intertwines the representations ρ' and ρ'' in the sense that $f \circ \rho''(e) = \rho'(e) \circ f$ for every $e \in \mathcal{E}$. In these conditions there exists a natural morphism of complexes $f. : Kos(\rho'') \to Kos(\rho')$ defined by the sequence of maps $(f_p)_{p \in \mathbb{Z}}$, where $f_p := f \otimes I_{\Lambda^p \mathcal{E}} : \mathcal{X}'' \otimes \Lambda^p \mathcal{E} \to \mathcal{X}' \otimes \Lambda^p \mathcal{E}$ for $p \geq 0$ and $f_p = 0$ for $p < 0$. Using this type of morphisms it is easy to obtain the following variant of the long exact sequence of the homology spaces (see Remark 8 in § 8):

Proposition 1. *Let $\rho : \mathcal{E} \to \mathcal{B}(\mathcal{X})$, $\rho' : \mathcal{E} \to \mathcal{B}(\mathcal{X}')$ and $\rho'' : \mathcal{E} \to \mathcal{B}(\mathcal{X}'')$ be representations of the finite-dimensional Lie algebra $\mathcal{E}$ (where $\mathcal{X}$, $\mathcal{X}'$ and $\mathcal{X}''$ are objects of the category $\mathcal{C}$). Assume that $f \in \mathcal{B}(\mathcal{X}, \mathcal{X}'')$ intertwines the representations ρ'' and ρ, $g \in \mathcal{B}(\mathcal{X}', \mathcal{X})$ intertwines the representations ρ and ρ' and there exists the following short exact sequence*

$$0 \leftarrow \mathcal{X}'' \xleftarrow{f} \mathcal{X} \xleftarrow{g} \mathcal{X}' \leftarrow 0.$$

Then there exists an exact complex

$$0 \leftarrow \mathbf{H}_0(\rho'') \leftarrow \mathbf{H}_0(\rho) \leftarrow \mathbf{H}_0(\rho') \leftarrow \cdots$$

$$\leftarrow \mathbf{H}_p(\rho'') \leftarrow \mathbf{H}_p(\rho) \leftarrow \mathbf{H}_p(\rho') \leftarrow \mathbf{H}_{p+1}(\rho'') \leftarrow \cdots.$$

Particularly, if two of the complexes

$$Kos(\rho'),\ Kos(\rho),\ Kos(\rho'')$$

are exact, then the third one is also exact.

The preceding result describes certain connections between representations of the same Lie algebra in different spaces. We concentrate now on the complementary situation. Namely, we shall describe connections between Koszul complexes associated with representations of several Lie algebras in the same space (see Propositions 2 and 3 below). To this end we need the following preliminary result.

Lemma 1. *Let $\mathcal{E},\mathcal{F}$ be finite-dimensional complex vector spaces, $h\in\mathcal{B}(\mathcal{F},\mathcal{E})$ be an onto map and $\mathcal{G}:=Ker\,h$. Consequently we have the following short exact sequence*

$$0\to\mathcal{G}\hookrightarrow\mathcal{F}\xrightarrow{h}\mathcal{E}\to 0.$$

For $p,q\geq 1$ define a linear map

$$\tilde{h}:(\Lambda^{q-1}\mathcal{G})\Lambda(\Lambda^p\mathcal{F})\to(\Lambda^{q-1}\mathcal{G})\otimes(\Lambda^p\mathcal{E})$$

by $\tilde{h}(\underline{g}\wedge f_1\wedge\cdots\wedge f_p):=\underline{g}\otimes h(f_1)\wedge\cdots\wedge h(f_p)$ for $\underline{g}\in\Lambda^{q-1}\mathcal{G}$ and $f_1,\ldots,f_p\in\mathcal{F}$. Then we have a short exact sequence

$$0\to(\Lambda^q\mathcal{G})\Lambda(\Lambda^{p-1}\mathcal{F})\hookrightarrow(\Lambda^{q-1}\mathcal{G})\Lambda(\Lambda^p\mathcal{F})\xrightarrow{\tilde{h}}(\Lambda^{q-1}\mathcal{G})\otimes(\Lambda^p\mathcal{E})\to 0.\qquad(1)$$

Proof. Obviously the short sequence is exact at the extreme terms. It remains only to prove the equality

$$(\Lambda^q\mathcal{G})\Lambda(\Lambda^{p-1}\mathcal{F})=Ker\,\tilde{h}.\qquad(2)$$

The inclusion "$\subseteq$" in (2) is obvious by the definitions of $\mathcal{G}$ and $\tilde{h}$. For proving the converse inclusion it suffices to check that

$$\dim((\Lambda^{q-1}\mathcal{G})\Lambda(\Lambda^p\mathcal{F}))=\dim((\Lambda^q\mathcal{G})\Lambda(\Lambda^{p-1}\mathcal{F})+\dim((\Lambda^{q-1}\mathcal{G})\otimes(\Lambda^p\mathcal{E})).\qquad(3)$$

To this end let's denote $\dim\mathcal{F}=m$ and $\dim\mathcal{G}=n$. Then $\dim\mathcal{E}=m-n$, so

$$\dim((\Lambda^{q-1}\mathcal{G})\otimes(\Lambda^p\mathcal{E}))=\tbinom{n}{q-1}\cdot\tbinom{m-n}{p}.\qquad(4)$$

Now let $\mathcal{G}_1$ be a vector subspace of $\mathcal{F}$ such that $\mathcal{G}\oplus\mathcal{G}_1=\mathcal{F}$. Then $\Lambda^p\mathcal{F}=\bigoplus_{j=0}^p((\Lambda^j\mathcal{G})\Lambda(\Lambda^{p-j}\mathcal{G}_1))$, hence

$$(\Lambda^q\mathcal{G})\Lambda(\Lambda^p\mathcal{F})=\bigoplus_{j=0}^p((\Lambda^{q+j}\mathcal{G})\Lambda(\Lambda^{p-j}\mathcal{G}_1)).$$

Consequently

$$\dim((\Lambda^q\mathcal{G})\Lambda(\Lambda^p\mathcal{F}))=\sum_{j=0}^p\dim((\Lambda^{q+j}\mathcal{G})\Lambda(\Lambda^{p-j}\mathcal{G}_1))$$

$$=\sum_{j=0}^p\dim(\Lambda^{q+j}\mathcal{G})\cdot\dim(\Lambda^{p-j}\mathcal{G}_1)=\sum_{j=0}^p\tbinom{n}{q+j}\cdot\tbinom{m-n}{p-j}.$$

Now replacing in the last formula p by $p-1$, respectively q by $q-1$, we obtain

$$\dim((\Lambda^q\mathcal{G})\Lambda(\Lambda^{p-1}\mathcal{F})) = \sum_{j=0}^{p-1}\binom{n}{q+j}\binom{m-n}{p-1-j} \tag{5}$$

respectively

$$\dim((\Lambda^{q-1}\mathcal{G})\Lambda(\Lambda^{p}\mathcal{F})) = \sum_{j=0}^{p}\binom{n}{q+j-1}\binom{m-n}{p-j}. \tag{6}$$

Now (3) follows from (4)–(6). □

Now we pass to the results announced before the preceding lemma.

Proposition 2. *Let's consider the following short exact sequence*

$$0 \to \mathcal{G} \hookrightarrow \mathcal{F} \xrightarrow{h} \mathcal{E} \to 0,$$

where $\mathcal{G}, \mathcal{F}, \mathcal{E}$ are Lie algebras and h is a Lie algebra morphism. We assume that $[\mathcal{G}, \mathcal{F}] = \{0\}$ (i.e. $\mathcal{G}$ is contained in the center of $\mathcal{F}$). Then for every representation $\rho : \mathcal{E} \to \mathcal{B}(\mathcal{X})$ (where $\mathcal{X}$ is an object of $\mathcal{C}$) we have

$$Kos\,(\rho) \text{ is exact } \iff Kos\,(\rho \circ h) \text{ is exact }. \tag{7}$$

Proof. Let's denote

$$Kos\,(\rho) : 0 \to \mathcal{X} \xleftarrow{\alpha_1} \mathcal{X} \otimes \mathcal{E} \xleftarrow{\alpha_2} \cdots \xleftarrow{\alpha_p} \mathcal{X} \otimes \Lambda^p \mathcal{E} \xleftarrow{\alpha_{p+1}} \cdots,$$

$$Kos\,(\rho \circ h) : 0 \to \mathcal{X} \xleftarrow{\beta_1} \mathcal{X} \otimes \mathcal{F} \xleftarrow{\beta_2} \cdots \xleftarrow{\beta_p} \mathcal{X} \otimes \Lambda^p \mathcal{F} \xleftarrow{\beta_{p+1}} \cdots.$$

The proof of the desired conclusion has several stages.

I. Fix $p, q \geq 1$ for the moment and consider the maps

$$\tilde{h}_i : (\Lambda^{q-1}\mathcal{G})\Lambda(\Lambda^i\mathcal{F}) \to (\Lambda^{q-1}\mathcal{G}) \otimes (\Lambda^i\mathcal{E}) \quad (i = p, p+1)$$

as in Lemma 1. The aim of the present stage is to verify the relations

$$\beta_{p+q}(\mathcal{X} \otimes (\Lambda^{q-l}\mathcal{G})\Lambda(\Lambda^{p+l}\mathcal{F})) \subseteq \mathcal{X} \otimes (\Lambda^{q-l}\mathcal{G})\Lambda(\Lambda^{p+l-1}\mathcal{F}) \;(l = 0, 1) \tag{8}$$

$$(-1)^{q-1}(I_{\Lambda^{q-1}\mathcal{G}} \otimes \alpha_{p+1}) \circ (I_{\mathcal{X}} \otimes \tilde{h}_{p+1}) = (I_{\mathcal{X}} \otimes \tilde{h}_p) \circ (\beta_{p+q}|_{\mathcal{X} \otimes (\Lambda^{q-1}\mathcal{G})\Lambda(\Lambda^{p+1}\mathcal{F})}). \tag{9}$$

This will allow us to write the following commutative diagrams

$$\begin{array}{ccc}
0 & & 0 \\
\downarrow & & \downarrow \\
\mathcal{X} \otimes (\Lambda^q\mathcal{G})\Lambda(\Lambda^{p-1}\mathcal{F}) & \xleftarrow{\beta_{p+q}} & \mathcal{X} \otimes (\Lambda^q\mathcal{G})\Lambda(\Lambda^{p}\mathcal{F}) \\
\downarrow & & \downarrow \\
\mathcal{X} \otimes (\Lambda^{q-1}\mathcal{G})\Lambda(\Lambda^{p}\mathcal{F}) & \xleftarrow{\beta_{p+q}} & \mathcal{X} \otimes (\Lambda^{q-1}\mathcal{G})\Lambda(\Lambda^{p+1}\mathcal{F}) \\
\downarrow I_{\mathcal{X}} \otimes \tilde{h}_p & & \downarrow I_{\mathcal{X}} \otimes \tilde{h}_{p+1} \\
\mathcal{X} \otimes \Lambda^{q-1}\mathcal{G} \otimes \Lambda^{p+1}\mathcal{E} & \xleftarrow{(-1)^{q-1}(I_{\Lambda^{q-1}\mathcal{G}} \otimes \alpha_{p+1})} & \mathcal{X} \otimes \Lambda^{q-1}\mathcal{G} \otimes \Lambda^{p+1}\mathcal{E} \\
\downarrow & & \downarrow \\
0 & & 0
\end{array} \tag{10}$$

To verify (8) consider $x \in \mathcal{X}$, $\underline{g} = g_1 \wedge \cdots \wedge g_{q-l} \in \Lambda^{q-l}\mathcal{G}$ and $\underline{f} = f_1 \wedge \cdots \wedge f_{p+l} \in \Lambda^{p+l}\mathcal{F}$. Since $[\mathcal{G}, \mathcal{F}] = \{0\}$ and $h(\mathcal{G}) = \{0\}$ we deduce

$$\begin{aligned}
\beta_{p+q}(x \otimes \underline{g} \wedge \underline{f}) &= \sum_{j=1}^{q-l} (-1)^{j-1} \rho(h(g_j)) x \otimes \overset{j}{\widehat{\underline{g}}} \wedge \underline{f} \\
&+ \sum_{j=1}^{p+l} (-1)^{q-l+j-1} \rho(h(f_j)) x \otimes \underline{g} \wedge \overset{j}{\widehat{\underline{f}}} \\
&+ \sum_{1 \le j < k \le q-l} (-1)^{j+k-1} x \otimes [g_j, g_k] \wedge \overset{j,k}{\widehat{\underline{g}}} \wedge \underline{f} \\
&+ \sum_{\substack{1 \le j \le q-l \\ 1 \le k \le p+l}} (-1)^{j+k+q-l-1} x \otimes [g_j, f_k] \wedge \overset{j}{\widehat{\underline{g}}} \wedge \overset{k}{\widehat{\underline{f}}} \\
&+ \sum_{1 \le j < k \le p+l} (-1)^{j+k-1} x \otimes [f_j, f_k] \wedge \underline{g} \wedge \overset{j,k}{\widehat{\underline{f}}} \\
&= \sum_{j=1}^{p+l} (-1)^{q-l+j-1} \rho(h(f_j)) x \otimes \underline{g} \wedge \overset{j}{\widehat{\underline{f}}} \\
&+ \sum_{1 \le j < k \le p+l} (-1)^{j+k-1} x \otimes [f_j, f_k] \wedge \underline{g} \wedge \overset{j,k}{\widehat{\underline{f}}} .
\end{aligned}$$

Hence $\beta_{p+q}(x \otimes \underline{g} \wedge \underline{f}) \in \mathcal{X} \otimes (\Lambda^{q-l}\mathcal{G})\Lambda(\Lambda^{p+l-1}\mathcal{F})$.

Next, to verify (9) we take $x \in \mathcal{X}$, $\underline{g} \in \Lambda^{q-1}\mathcal{G}$ and $\underline{f} = f_1 \wedge \cdots \wedge f_{p+1} \in \Lambda^{p+1}\mathcal{F}$. In view of the preceding computation (for $l = 1$) and of the fact that h is a Lie algebra morphism, we get

$$\begin{aligned}
(I_{\mathcal{X}} \otimes \tilde{h}_p)(\beta_{p+q}(x \otimes \underline{g} \wedge \underline{f})) &= \sum_{j=1}^{p+1} (-1)^{q+j-2} \rho(h(f_j)) x \otimes \underline{g} \otimes \overset{j}{\widehat{h(\underline{f})}} \\
&+ \sum_{1 \le j < k \le p+1} (-1)^{j+k+q-2} x \otimes \underline{g} \otimes [h(f_j), h(f_k)] \wedge \overset{j,k}{\widehat{h(\underline{f})}},
\end{aligned} \tag{11}$$

where $h(\underline{f}) := h(f_1) \wedge \cdots \wedge h(f_{p+1})$. On the other hand we have

$$\begin{aligned}
(I_{\Lambda^{q-1}\mathcal{G}} \otimes \alpha_{p+1})&((I_{\mathcal{X}} \otimes \tilde{h}_{p+1})(x \otimes \underline{g} \wedge \underline{f})) = (I_{\Lambda^{q-1}\mathcal{G}} \otimes \alpha_{p+1})(x \otimes \underline{g} \otimes h(\underline{f})) \\
&= \sum_{j=1}^{p+1} (-1)^{j-1} \rho(h(f_j)) x \otimes \underline{g} \otimes \overset{j}{\widehat{h(\underline{f})}} \\
&+ \sum_{1 \le j < k \le p+1} (-1)^{j+k-1} x \otimes \underline{g} \otimes [h(f_j), h(f_k)] \wedge \overset{j,k}{\widehat{h(\underline{f})}} .
\end{aligned} \tag{12}$$

Now, (9) follows from (11) and (12).

II. At this stage, first we keep $q \geq 1$ fixed and let p be variable; hence we continue the rows of (10) at both sides and obtain a short exact sequence of complexes

$$0 \leftarrow \Lambda^{q-1}\mathcal{G} \otimes ((-1)^{q-1} Kos(\rho)) \leftarrow \mathcal{Y}^{q-1}_{\cdot} \leftarrow \mathcal{Y}^{q}._{-1} \leftarrow 0 \tag{13}$$

(cf. the notation $\mathcal{X}._{+k}$ for $k \in \mathbb{Z}$ introduced after Remark 1 from § 8), where the complex $\mathcal{Y}^q.$ is defined by the pair of sequences

$$((\mathcal{X} \otimes (\Lambda^q\mathcal{G})\Lambda(\Lambda^{i+1}\mathcal{F}))_{i\in\mathbb{Z}}\,, (\beta_{q+i+1}|_{\mathcal{X}\otimes(\Lambda^q\mathcal{G})\Lambda(\Lambda^{i+1}\mathcal{F})})_{i\in\mathbb{Z}})$$

(where $\Lambda^j\mathcal{F} := \{0\}$ for $j < 0$). (Obviously $\mathcal{Y}^0. = Kos(\rho \circ h)$.) For each $i \in \mathbb{Z}$ we denote

$$\mathbf{H}_i := \mathbf{H}_i(Kos(\rho)),\ \mathbf{H}_i^q := \mathbf{H}_i(\mathcal{Y}^q_{\cdot}) \quad (\text{so } \mathbf{H}_i(\mathcal{Y}^q_{\cdot-1}) = \mathbf{H}^q_{i-1}).$$

Moreover we have

$$\mathbf{H}_i(\Lambda^{q-1}\mathcal{G} \otimes ((-1)^{q-1} Kos(\rho))) = \Lambda^{q-1}\mathcal{G} \otimes \mathbf{H}_i.$$

Using these relations let's write the long exact sequence of homology spaces associated to (13) for different values of q:

$$q = 1:\ 0 \leftarrow \mathbf{H}_0 \leftarrow \mathbf{H}_0^0 \leftarrow 0 \leftarrow \mathbf{H}_1 \leftarrow \mathbf{H}_1^0 \leftarrow \mathbf{H}_0^1 \leftarrow \cdots, \tag{14}$$

$$q = 2:\ 0 \leftarrow \mathcal{G} \otimes \mathbf{H}_0 \leftarrow \mathbf{H}_0^1 \leftarrow 0 \leftarrow \mathcal{G} \otimes \mathbf{H}_1 \leftarrow \mathbf{H}_1^1 \leftarrow \mathbf{H}_0^2 \leftarrow \cdots, \tag{15}$$

....

$$q + 1:\ 0 \leftarrow \Lambda^q\mathcal{G} \otimes \mathbf{H}_0 \leftarrow \mathbf{H}_0^q \leftarrow 0 \leftarrow \Lambda^q\mathcal{G} \otimes \mathbf{H}_1 \leftarrow \mathbf{H}_1^q \leftarrow \mathbf{H}_0^{q+1} \leftarrow \cdots, \tag{16}$$

....

III. At this stage we come back to the proof of (7). With the above introduced notation we have to prove the following fact:

$$\mathbf{H}_i = \{0\} \text{ for each } i \geq 0 \Leftrightarrow \mathbf{H}_i^0 = \{0\} \text{ for each } i \geq 0.$$

First assume that $\mathbf{H}_i = \{0\}$ for each $i \geq 0$. Then by (16) we get $\mathbf{H}_0^q = \{0\}$ and $\mathbf{H}_{i+1}^q = \mathbf{H}_i^{q+1}$ for every $i, q \geq 0$. Consequently for every $i \geq 0$ we have

$$\mathbf{H}_i^0 = \mathbf{H}_{i-1}^1 = \mathbf{H}_{i-2}^2 = \cdots = \mathbf{H}_1^{i-1} = \mathbf{H}_0^i = \{0\}.$$

Conversely, assume that $\mathbf{H}_i^0 = \{0\}$ for every $i \geq 0$. Then by (14) we get $\mathbf{H}_0 = \mathbf{H}_1 = \{0\}$ and $\mathbf{H}_i = \mathbf{H}_{i-2}^1$ for $i \geq 2$. From $\mathbf{H}_0 = \{0\}$ we get by (16) that $\mathbf{H}_0^q = \{0\}$ for each $q \geq 0$. Next we prove by induction on $i \geq 0$ that

$$\mathbf{H}_0 = \mathbf{H}_1 = \cdots = \mathbf{H}_i = \mathbf{H}_0^q = \mathbf{H}_1^q = \cdots = \mathbf{H}_i^q = \{0\} \text{ for each } q \geq 0. \tag{17}$$

We have just seen that (17) holds for $i = 0$. Now assume that (17) holds and prove that moreover

$$\mathbf{H}_{i+1} = \mathbf{H}_{i+1}^q = \{0\} \text{ for each } q \geq 0. \tag{18}$$

First, since we have already seen that $\mathbf{H}_j = \mathbf{H}^1_{j-2}$, by (17) we get $\mathbf{H}_{i+2} = \mathbf{H}^1_i = \mathbf{H}_{i+1} = \mathbf{H}^1_{i-1} = \{0\}$. Then from $\mathbf{H}_{i+2} = \mathbf{H}_{i+1} = \{0\}$ and (16) we get $\mathbf{H}^q_{i+1} = \mathbf{H}^{q+1}_i$; hence $\mathbf{H}^q_{i+1} = \{0\}$ by (17). Consequently (18) is completely proved. So (17) holds for each $i \geq 0$; particularly $\mathbf{H}_i = \{0\}$ for each $i \geq 0$. □

Proposition 3. *Let's consider the following short exact sequence*

$$0 \to \mathcal{G} \hookrightarrow \mathcal{F} \xrightarrow{h} \mathcal{E} \to 0,$$

where $\mathcal{G}, \mathcal{F}, \mathcal{E}$ are Lie algebras and h is a Lie algebra morphism. We assume that the Lie algebra $\mathcal{F}$ is nilpotent. Then for every representation $\rho : \mathcal{E} \to \mathcal{B}(\mathcal{X})$ (where $\mathcal{X}$ is an object of $\mathcal{C}$) we have

$$Kos\,(\rho) \text{ is exact} \iff Kos\,(\rho \circ h) \text{ is exact}.$$

Proof. Define $\mathcal{G}_1 := [\mathcal{F}, \mathcal{G}]$ and inductively $\mathcal{G}_{i+1} := [\mathcal{F}, \mathcal{G}_i]$ for each $i \geq 1$. Since the Lie algebra $\mathcal{F}$ is nilpotent, there exists $n \geq 0$ such that $\mathcal{G}_{n+1} := [\mathcal{F}, \mathcal{G}_n] = \{0\}$. We have the following sequence of ideals of $\mathcal{F}$

$$\mathcal{G}_1 \supset \mathcal{G}_2 \supset \cdots \supset \mathcal{G}_n$$

and for each $j \in \{1, \ldots, n\}$ we can define the following Lie algebra morphism

$$\pi_j : \mathcal{F}/\mathcal{G}_{j+1} \to \mathcal{F}/\mathcal{G}_j,\ \pi_j(f + \mathcal{G}_{j+1}) := f + \mathcal{G}_j \text{ for each } f \in \mathcal{F}.$$

Moreover we define

$$\pi_0 : \mathcal{F}/\mathcal{G}_1 \to \mathcal{E},\ \pi_0(f + \mathcal{G}_1) := h(f) \text{ for each } f \in \mathcal{F}.$$

(The map π_0 is well defined since $\mathcal{G}_1 \subset \mathcal{G} = Ker\,h$.) Hence we have the following exact sequences consisting of Lie algebras and Lie algebra morphisms

$$\begin{array}{ccccccccc}
0 & \to & \mathcal{G}/\mathcal{G}_1 & \hookrightarrow & \mathcal{F}/\mathcal{G}_1 & \xrightarrow{\pi_0} & \mathcal{E} & \to & 0 \\
0 & \to & \mathcal{G}_1/\mathcal{G}_2 & \hookrightarrow & \mathcal{F}/\mathcal{G}_2 & \xrightarrow{\pi_1} & \mathcal{F}/\mathcal{G}_1 & \to & 0 \\
\cdots & \cdots & \cdots & \cdots & \cdots & \cdots & \cdots & \cdots & \cdots \\
0 & \to & \mathcal{G}_{n-1}/\mathcal{G}_n & \hookrightarrow & \mathcal{F}/\mathcal{G}_n & \xrightarrow{\pi_{n-1}} & \mathcal{F}/\mathcal{G}_{n-1} & \to & 0 \\
0 & \to & \mathcal{G}_n & \hookrightarrow & \mathcal{F} & \xrightarrow{\pi_n} & \mathcal{F}/\mathcal{G}_n & \to & 0.
\end{array}$$

In view of the definitions of $\mathcal{G}_1, \ldots, \mathcal{G}_n$ we can apply Proposition 2 to each of these short exact sequences. We deduce successively $Kos\,(\rho)$ exact $\Leftrightarrow Kos\,(\rho \circ \pi_0)$ exact $\Leftrightarrow Kos\,(\rho \circ \pi_0 \circ \pi_1)$ exact $\Leftrightarrow \ldots \Leftrightarrow Kos\,(\rho \circ \pi_0 \circ \pi_1 \circ \ldots \circ \pi_n)$ exact. Now the proof finishes with the remark that $\pi_0 \circ \pi_1 \circ \cdots \circ \pi_n = h$. □

Now let $\rho : \mathcal{E} \to \mathcal{B}(\mathcal{X})$ be a representation of the finite-dimensional Lie algebra $\mathcal{E}$ (where as usually $\mathcal{X}$ is an object of the category $\mathcal{C}$). Let $\mathcal{I}$ be an ideal of $\mathcal{E}$. We are going to describe some connections between the complexes $Kos\,(\rho)$ and $Kos\,(\rho|_{\mathcal{I}})$. To this end we define for each $e \in \mathcal{E}$ a linear map

$$\theta_e : \mathcal{X} \otimes \Lambda\mathcal{I} \to \mathcal{X} \otimes \Lambda\mathcal{I}$$

by $\theta_e := \theta_e^p$ on $\mathcal{X} \otimes \Lambda^p\mathcal{I}$ for each $p \in \mathbb{N}$, where θ_e^p is a linear map defined in the following way

$$\theta_e^p : \mathcal{X} \otimes \Lambda^p\mathcal{I} \to \mathcal{X} \otimes \Lambda^p\mathcal{I},$$

$$\theta_e^p(x \otimes \underline{u}) = \rho(e)x \otimes \underline{u} + \sum_{i=1}^{p}(-1)^{i-1}x \otimes [e, u_i] \wedge \overset{i}{\widehat{\underline{u}}}$$

for each $x \in \mathcal{X}, \underline{u} = u_1 \wedge \cdots \wedge u_p \in \Lambda^p\mathcal{I}$. (We shall denote $\theta_{\rho,e}$ instead of θ_e when it will be necessary to indicate that θ_e is associated to the representation ρ.) With these notations we can state the following result.

Proposition 4. *a) The map*

$$\theta : \mathcal{E} \to \mathcal{B}(\mathcal{X} \otimes \Lambda\mathcal{I}), \; e \mapsto \theta_e$$

is a representation of the Lie algebra $\mathcal{E}$.

b) For each $e \in \mathcal{E}$ the sequence of maps $\theta_e^{\cdot} = (\theta_e^p)_{p\in\mathbb{Z}}$ (where $\theta_e^p = 0$ for $p < 0$) defines a morphism of complexes

$$\theta_e^{\cdot} : Kos\,(\rho|_{\mathcal{I}}) \to Kos\,(\rho|_{\mathcal{I}}).$$

If moreover $\mathcal{E} = \mathbb{C}e \oplus \mathcal{I}$, then

$$Kos\,(\rho) = Con\,(Kos\,(\rho|_{\mathcal{I}}), \theta_e^{\cdot}).$$

c) If $\mathcal{E} = \mathcal{I}$ and $e \in \mathcal{E}$, then for the morphism of complexes (cf. b)) $\theta_e^{\cdot} : Kos\,(\rho) \to Kos\,(\rho)$ we have $\mathbf{H}_p(\theta_e^{\cdot}) = 0$ for each $p \geq 0$.

Proof. a) The assertion follows by a straightforward computation.

b) The fact that $\theta_e^{\cdot}$ is a morphism of complexes follows also by a direct computation. Now assume that $\mathcal{E} = \mathbb{C}e \oplus \mathcal{I}$ and denote

$$Kos\,(\rho|_{\mathcal{I}}) : 0 \leftarrow \mathcal{X} \xleftarrow{\alpha_1} \mathcal{X} \otimes \mathcal{I} \xleftarrow{\alpha_2} \cdots \xleftarrow{\alpha_p} \mathcal{X} \otimes \Lambda^p\mathcal{I} \xleftarrow{\alpha_{p+1}} \cdots,$$

$$Kos\,(\rho) : 0 \leftarrow \mathcal{X} \xleftarrow{\beta_1} \mathcal{X} \otimes \mathcal{E} \xleftarrow{\beta_2} \cdots \xleftarrow{\beta_p} \mathcal{X} \otimes \Lambda^p\mathcal{E} \xleftarrow{\beta_{p+1}} \cdots.$$

From $\mathcal{E} = \mathbb{C}e \oplus \mathcal{I}$ we deduce that for each $p \geq 0$

$$\Lambda^p\mathcal{E} = (\Lambda^p\mathcal{I}) \oplus (e \wedge (\Lambda^{p-1}\mathcal{I})) \cong (\Lambda^p\mathcal{I}) \oplus (\Lambda^{p-1}\mathcal{I}),$$

so

$$\mathcal{X} \otimes \Lambda^p\mathcal{E} = (\mathcal{X} \otimes \Lambda^p\mathcal{I}) \oplus (\mathcal{X} \otimes (e \wedge (\Lambda^{p-1}\mathcal{I}))) \cong (\mathcal{X} \otimes \Lambda^p\mathcal{I}) \oplus (\mathcal{X} \otimes \Lambda^{p-1}\mathcal{I}).$$

Moreover we have $\beta_p|_{\mathcal{X}\otimes\Lambda^p\mathcal{I}} = \alpha_p$. On the other hand, for arbitrary $\underline{f} = f_1 \wedge \cdots \wedge f_{p-1} \in \Lambda^{p-1}\mathcal{I}$ and $x \in \mathcal{X}$ we have

$$\beta_p(x \otimes e \wedge \underline{f}) = \rho(e)x \otimes \underline{f} + \sum_{i=1}^{p-1}(-1)^i \rho(f_i)x \otimes e \wedge \overset{i}{\widehat{\underline{f}}}$$

$$+ \sum_{1\le i<j\le p-1} (-1)^{i+j-1} x \otimes [f_i, f_j] \wedge e \wedge \overset{i,j}{\widehat{\underline{f}}} + \sum_{i=1}^{p-1}(-1)^{i-1} x \otimes [e, f_i] \wedge \overset{i}{\widehat{\underline{f}}}$$

$$= -e \wedge \alpha_{p-1}(x \otimes \underline{f}) + \theta_e^{p-1}(x \otimes \underline{f}).$$

Consequently, using the identifications

$$\mathcal{X} \otimes \Lambda^p \mathcal{E} \;\cong\; \begin{array}{c} \mathcal{X} \otimes \Lambda^p \mathcal{I} \\ \oplus \\ \mathcal{X} \otimes \Lambda^{p-1}\mathcal{I} \end{array}, \quad \mathcal{X} \otimes \Lambda^{p-1}\mathcal{E} \;\cong\; \begin{array}{c} \mathcal{X} \otimes \Lambda^{p-1} \mathcal{I} \\ \oplus \\ \mathcal{X} \otimes \Lambda^{p-2}\mathcal{I} \end{array}$$

we can write

$$\beta_p : \mathcal{X} \otimes \Lambda^p \mathcal{E} \to \mathcal{X} \otimes \Lambda^{p-1}\mathcal{E}, \; \beta_p = \begin{pmatrix} \alpha_p & \theta_e^{p-1} \\ 0 & -\alpha_{p-1} \end{pmatrix}.$$

Consequently $Kos\,(\rho) = Con\,(Kos\,(\rho|_{\mathcal{I}}), \theta_e^{\cdot})$.

c) Now assume that $\mathcal{I} = \mathcal{E}$ and $e \in \mathcal{E}$. We define a linear map

$$\iota_e : \mathcal{X} \otimes \Lambda\mathcal{E} \to \mathcal{X} \otimes \Lambda\mathcal{E}, \; \iota_e(x \otimes \underline{v}) := x \otimes e \wedge \underline{v} \;\forall x \in \mathcal{X}\, \forall \underline{v} \in \Lambda\mathcal{E}.$$

One easily checks that

$$\iota_e \circ \alpha_p + \alpha_{p+1} \circ \iota_e = \theta_e^p \text{ on } \mathcal{X} \otimes \Lambda^p \mathcal{E}.$$

This implies $\theta_e^p(Ker\,\alpha_p) \subseteq Ran\,\alpha_{p+1}$, hence $\mathbf{H}_p(\theta_e^{\cdot}) = 0$ for $p \ge 0$. □

As a simple application of Proposition 4 we prove the following fact concerning representations of a nilpotent Lie algebra by scalar multiples of the identity map.

Example 4. Let $\mathcal{E}$ be a (finite-dimensional) nilpotent Lie algebra and $\varphi : \mathcal{E} \to \mathbb{C}$ be a Lie algebra morphism (i.e. a character of $\mathcal{E}$). Let $\mathcal{X}$ be an object of the category $\mathcal{C}$ and consider the representation

$$\rho : \mathcal{E} \to \mathcal{B}(\mathcal{X}), \; e \mapsto \varphi(e) I_{\mathcal{X}}.$$

Then the complex $Kos\,(\rho)$ is exact if and only if either $\mathcal{X} = \{0\}$ or $\varphi(e) \neq 0$ for a certain $e \in \mathcal{E}$.

To prove this assertion we may assume obviously that $\mathcal{X} \neq \{0\}$. If $\varphi(e) = 0$ for each $e \in \mathcal{E}$, then $\mathbf{H}_0(Kos\,(\rho)) = \mathcal{X} \neq \{0\}$ (see Example 2 above). Now assume that $\varphi \not\equiv 0$ and prove the exactness of $Kos(\rho)$ by induction on $n = \dim \mathcal{E}$. Choose $u \in \mathcal{E}$ such that $\varphi(u) \neq 0$. There exists an ideal $\mathcal{I}$ of $\mathcal{E}$ such that $u \in \mathcal{I} \neq \mathcal{E}$ (for example $\mathcal{I} = \mathbb{C}u + [\mathcal{E}, \mathcal{E}]$ since $\mathcal{E}$ is nilpotent). Then we can assume that $\dim(\mathcal{E}/\mathcal{I}) = 1$. Take $e \in \mathcal{E} \backslash \mathcal{I}$, so $\mathcal{E} = \mathbb{C}e \oplus \mathcal{I}$. By Proposition 4 *b*) we get

$$Kos(\rho) = Con(Kos(\rho|_{\mathcal{I}}), \theta^{\cdot}_{e}).$$

But $u \in \mathcal{I}$ and $\varphi(u) \neq 0$, so $\varphi|_{\mathcal{I}} \not\equiv 0$ hence $Kos(\rho|_{\mathcal{I}})$ is exact by the induction hypothesis. But this implies the exactness of $Kos(\rho)$ by Corollary 1 (a') and Remark 5 from § 8.

The following result shows that the operators θ^p_e from Proposition 4 have a particular structure when the Lie algebra $\mathcal{E}$ is solvable. We shall use the fact that, if $\dim \mathcal{I} < \infty$ and $p \in \mathbb{N}$, then every element of $\mathcal{B}(\mathcal{X} \otimes \Lambda^p \mathcal{I})$ can be identified, with respect to a fixed basis in $\Lambda^p \mathcal{I}$, with a square matrix (of size $\dim(\Lambda^p \mathcal{I})$) whose entries are elements of $\mathcal{B}(\mathcal{X})$.

Proposition 5. *Let $\rho : \mathcal{E} \to \mathcal{B}(\mathcal{X})$ be a representation of the finite-dimensional solvable Lie algebra $\mathcal{E}$ (where $\mathcal{X}$ is an object of $\mathcal{C}$). Let $\dim \mathcal{E} = n$, $\mathcal{I}$ be an ideal of $\mathcal{E}$ having the dimension m and $\{e_1, \ldots, e_n\}$ be a basis of $\mathcal{E}$ such that*

$$\{e_1, \ldots, e_m\} \subset \mathcal{I} \text{ and } [e_j, e_i] = \sum_{k=1}^{i} c^k_{ji} e_k \ (1 \leq i < j \leq n).$$

For $p \in \mathbb{N}$ consider the canonical basis $\{e_{i_1} \wedge \cdots \wedge e_{i_p} \mid 1 \leq i_1 < \ldots < i_p \leq m\}$ in $\Lambda^p \mathcal{I}$. Then for $l = m, m+1, \ldots, n$ the map $\theta^p_{e_l} \in \mathcal{B}(\mathcal{X} \otimes \Lambda^p \mathcal{I})$ is represented by a triangular matrix whose diagonal entries are

$$\rho(e_l) + c^{i_1}_{l,i_1} + \cdots + c^{i_p}_{l,i_p} \in \mathcal{B}(\mathcal{X}) \quad (1 \leq i_1 < \cdots < i_p \leq m),$$

the other entries being scalar multiples of $I_{\mathcal{X}}$.

Proof. Endow the set $\{(i_1, \ldots, i_p) \mid 1 \leq i_1 < \cdots < i_p \leq m\}$ with the lexicographic order. Namely, if $1 \leq i_1 < \cdots < i_p \leq m$ and $1 \leq i'_1 < \cdots < i'_p \leq m$ then $(i_1, \ldots, i_p) < (i'_1, \ldots, i'_p)$ if and only if there exists $s \in \{1, \ldots, p\}$ such that $i_s < i'_s$ and $i_j = i'_j$ for each $j < s$. Then for $x \in \mathcal{X}$ and $\underline{e} = e_{i_1} \wedge \cdots \wedge e_{i_p} \in \Lambda^p \mathcal{I}$ with $1 \leq i_1 < \cdots < i_p \leq m$ we have

$$\theta^p_{e_l}(x \otimes \underline{e}) = \rho(e_l)x \otimes \underline{e} + \sum_{j=1}^{p} (-1)^{j-1} x \otimes [e_l, e_{i_j}] \wedge \overset{j}{\hat{\underline{e}}}$$

$$= \rho(e_l)x \otimes \underline{e} + \sum_{j=1}^{p} (-1)^{j-1} x \otimes \Big(\sum_{s=1}^{i_j} c^s_{l,i_j} e_s\Big) \wedge \overset{j}{\hat{\underline{e}}}$$

$$= \rho(e_l)x \otimes \underline{e} + \sum_{j=1}^{p} (-1)^{j-1} c^{i_j}_{l,i_j} x \otimes (e_{i_j} \wedge \overset{j}{\hat{\underline{e}}})$$

$$+\sum_{j=1}^{p}(-1)^{j-1}\sum_{s<i_j} c^{s}_{l,i_j} x\otimes (e_s\wedge \overset{j}{\underline{e}})$$

$$=(\rho(e_l)+(\sum_{j=1}^{p} c^{i_j}_{l,i_j})I_{\mathcal{X}})x\otimes \underline{e}+S,$$

where S is a sum of terms of the form $c_{k_1,\dots,k_p}x\otimes(e_{k_1}\wedge\cdots\wedge e_{k_p})$ with $c_{k_1,\dots,k_p}\in\mathbb{C}$ and $(k_1,\dots,k_p)<(i_1,\dots,i_p)$.

Consequently $\theta^p_{e_l}\in\mathcal{B}(\mathcal{X}\otimes\Lambda^p\mathcal{I})$ is represented (with respect to the basis $\{e_{i_1}\wedge\cdots\wedge e_{i_p}\mid 1\le i_1<\cdots<i_p\le m\}$ of $\Lambda^p\mathcal{I}$) by an upper triangular matrix having on its diagonal the entries $\rho(e_l)+(\sum_{j=1}^{p}c^{i_j}_{l,i_j})I_{\mathcal{X}}\in\mathcal{B}(\mathcal{X})$ $(1\le i_1<\ldots<i_p\le m)$ and whose entries are of the form $cI_{\mathcal{X}}$ $(c\in\mathbb{C})$ above the diagonal. □

In the remainder of the present paragraph we shall describe a bicomplex whose totalization has properties generalizing those expressed in Proposition 4. This bicomplex is denoted $\mathcal{B}_\lambda(\mathcal{I},\mathcal{F},\mathcal{X})$ and is associated to a representation of a finite-dimensional Lie algebra $\mathcal{E}$, an ideal $\mathcal{I}$ of $\mathcal{E}$, a Lie subalgebra $\widetilde{\mathcal{F}}$ of $U(\mathcal{E})$ and a character $\lambda:\mathcal{F}\to\mathbb{C}$, as indicated below.

As usually $\mathcal{X}$ is an object of $\mathcal{C}$, $\rho:\mathcal{E}\to\mathcal{B}(\mathcal{X})$ a representation and denote by

$$\theta:\mathcal{E}\to\mathcal{B}(\mathcal{X}\otimes\Lambda\mathcal{E})$$

the representation given by the correspondence $e\mapsto\theta_e$ (cf. Proposition 4 a)). We use the same notation for the canonical extension $\theta:U(\mathcal{E})\to\mathcal{B}(\mathcal{X}\otimes\Lambda\mathcal{E})$ to the enveloping algebra $U(\mathcal{E})$ of $\mathcal{E}$. Moreover we denote

$$\mathcal{F}:=\theta(\widetilde{\mathcal{F}})\subset\mathcal{B}(\mathcal{X}\otimes\Lambda\mathcal{E}).$$

Let's remark that for each $q\ge 0$ the linear subspace $\mathcal{X}\otimes\Lambda^q\mathcal{I}$ is invariant for the Lie algebra $\mathcal{F}$ because $\mathcal{I}$ is an ideal of $\mathcal{E}$. Hence for each $q\ge 0$ we can define a representation of $\mathcal{F}$ in $\mathcal{X}\otimes\Lambda^q\mathcal{I}$ in the following way

$$\rho_q:\mathcal{F}\to\mathcal{B}(\mathcal{X}\otimes\Lambda^q\mathcal{I}),\ f\mapsto f|_{\mathcal{X}\otimes\Lambda^q\mathcal{I}}.$$

Now we can introduce the (announced) bicomplex $\mathcal{B}_\lambda(\mathcal{I},\mathcal{F},\mathcal{X})$. Namely, it is the bicomplex $(\mathcal{Y}_{\cdot\cdot},\alpha_{\cdot\cdot},\beta_{\cdot\cdot})$ constructed in the following way:

$$\begin{aligned}\mathcal{Y}_{qp}&=\mathcal{X}\otimes\Lambda^q\mathcal{I}\otimes\Lambda^p\mathcal{F} &&\text{if } p,q\ge 0,\\ &=0 &&\text{if either } p<0 \text{ or } q<0;\end{aligned}$$

moreover we define $\alpha_{\cdot\cdot}$ and $\beta_{\cdot\cdot}$ such that for $p,q\ge 0$ the p-th row, respectively the q-th column, of the bicomplex are

$$(\mathcal{Y}_{p,\cdot},\beta_{p,\cdot})=Kos(\rho|_{\mathcal{I}}-\lambda)\otimes\Lambda^p\mathcal{F},$$

respectively

$$(\mathcal{Y}_{\cdot,q},\alpha_{\cdot,q})=Kos(\rho_q).$$

Consequently we can represent $\mathcal{B}_\lambda(\mathcal{I},\mathcal{F},\mathcal{X})$ by the following commutative diagram, whose p-th row is $Kos\,(\rho|_{\mathcal{I}}-\lambda)\otimes\Lambda^p\mathcal{F}$ and whose q-th column is the Koszul complex of the representation $\rho_q : \mathcal{F} \to \mathcal{B}(\mathcal{X}\otimes\Lambda^q\mathcal{I})$:

$$
\begin{array}{ccccccccc}
\cdots & \cdots & \cdots & \cdots & \cdots & \cdots & \cdots & \cdots & \cdots \\
 & & \downarrow \alpha_{p+1,0} & & & & \downarrow \alpha_{p+1,q} & & \\
0 & \leftarrow & \mathcal{X}\otimes\Lambda^p\mathcal{F} & \stackrel{\beta_{p,1}}{\longleftarrow} & \cdots & \stackrel{\beta_{p,q}}{\longleftarrow} & \mathcal{X}\otimes\Lambda^q\mathcal{I}\otimes\Lambda^p\mathcal{F} & \stackrel{\beta_{p,q+1}}{\longleftarrow} & \cdots \\
 & & \downarrow \alpha_{p,0} & & & & \downarrow \alpha_{p,q} & & \\
\cdots & \cdots & \cdots & \cdots & \cdots & \cdots & \cdots & \cdots & \cdots \\
 & & \downarrow \alpha_{2,0} & & & & \downarrow \alpha_{2,q} & & \\
0 & \leftarrow & \mathcal{X}\otimes\mathcal{F} & \stackrel{\beta_{1,1}}{\longleftarrow} & \cdots & \stackrel{\beta_{1,q}}{\longleftarrow} & \mathcal{X}\otimes\Lambda^q\mathcal{I}\otimes\mathcal{F} & \stackrel{\beta_{1,q+1}}{\longleftarrow} & \cdots \\
 & & \downarrow \alpha_{1,0} & & & & \downarrow \alpha_{1,q} & & \\
0 & \leftarrow & \mathcal{X} & \stackrel{\beta_{0,1}}{\longleftarrow} & \cdots & \stackrel{\beta_{0,q}}{\longleftarrow} & \mathcal{X}\otimes\Lambda^q\mathcal{I} & \stackrel{\beta_{0,q+1}}{\longleftarrow} & \cdots \\
 & & \downarrow & & & & \downarrow & & \\
 & & 0 & & & & 0 & &
\end{array}
$$

(Let's remark that the above diagram is indeed commutative, that is $\beta_{p,q+1}\circ\alpha_{p+1,q+1} = \alpha_{p+1,q}\circ\beta_{p+1,q+1}$ for $p,q\geq 0$. Indeed, recall that for each $e\in\mathcal{E}$ we have the morphism of complexes $\theta_e^{\cdot} : Kos\,(\rho|_{\mathcal{I}}) \to Kos\,(\rho|_{\mathcal{I}})$, hence $\beta_{0,q+1}\circ\theta_e^{q+1} = \theta_e^q\circ\beta_{0,q+1}$ for $q\geq 0$. This fact easily implies the desired relations in view of the definition of $\beta_{\cdot\cdot}$ and of the fact that each element of $\mathcal{F}$ is a (non-commutative) polynomial of some maps of the form θ_e with $e\in\mathcal{E}$.)

We denote the totalization of $\mathcal{B}_\lambda(\mathcal{I},\mathcal{F},\mathcal{X})$ by $Tot_\lambda(\mathcal{I},\mathcal{F},\mathcal{X})$. When $\lambda\equiv 0$ we denote simply $\mathcal{B}(\mathcal{I},\mathcal{F},\mathcal{X})$, respectively $Tot(\mathcal{I},\mathcal{F},\mathcal{X})$, instead of $\mathcal{B}_0(\mathcal{I},\mathcal{F},\mathcal{X})$, respectively $Tot_0(\mathcal{I},\mathcal{F},\mathcal{X})$. Next we make a few useful remarks concerning the structure of $\mathcal{B}_\lambda(\mathcal{I},\mathcal{F},\mathcal{X})$ and we shall describe its dependence on the ideal $\mathcal{I}$ of $\mathcal{E}$.

Remark 2. Let's assume that $\lambda = 0$ and that there exists $e\in\mathcal{E}$ such that $\widetilde{\mathcal{F}} = \mathbb{C}e \subset \mathcal{E}\subset U(\mathcal{E})$. Then $\Lambda^p\mathcal{F} = \{0\}$ for $p\geq 2$, hence the above diagram reduces to

$$
\begin{array}{ccccccccc}
 & & 0 & & & & 0 & & \\
 & & \downarrow & & & & \downarrow & & \\
0 & \leftarrow & \mathcal{X} & \stackrel{\beta_{1,1}}{\longleftarrow} & \cdots & \stackrel{\beta_{1,q}}{\longleftarrow} & \mathcal{X}\otimes\Lambda^q\mathcal{I} & \stackrel{\beta_{1,q+1}}{\longleftarrow} & \cdots \\
 & & \downarrow \alpha_{1,0} & & & & \downarrow \alpha_{1,q} & & \\
0 & \leftarrow & \mathcal{X} & \stackrel{\beta_{0,1}}{\longleftarrow} & \cdots & \stackrel{\beta_{0,q}}{\longleftarrow} & \mathcal{X}\otimes\Lambda^q\mathcal{I} & \stackrel{\beta_{0,q+1}}{\longleftarrow} & \cdots \\
 & & \downarrow & & & & \downarrow & & \\
 & & 0 & & & & 0 & &
\end{array}
$$

(Since $\dim\mathcal{F} = 1$ we identified $\mathcal{X}\otimes\Lambda^q\mathcal{I}\otimes\mathcal{F}$ with $\mathcal{X}\otimes\Lambda^q\mathcal{I}$.) By definition, each row of this diagram coincides with $Kos\,(\rho|_{\mathcal{I}})$. Also the q-th column is the Koszul complex of the representation

$$\rho_q : \mathcal{F}\to\mathcal{B}(\mathcal{X}\otimes\Lambda^q\mathcal{I}).$$

But by definition $\mathcal{F} = \theta(\widetilde{\mathcal{F}}) = \theta(\mathbb{C}e) = \mathbb{C} \cdot \theta_e \subset \mathcal{B}(\mathcal{X} \otimes \Lambda\mathcal{I})$, hence we have actually $\alpha_{1,q} = \theta_e^q : \mathcal{X} \otimes \Lambda^q\mathcal{I} \to \mathcal{X} \otimes \Lambda^q\mathcal{I}$ (see Example 1 above). So we obtained the bicomplex whose totalization is $Con\,(Kos\,(\rho|_\mathcal{I}), \theta_e^{\cdot})$. Consequently the present framework can be viewed as a generalization of the one from Proposition 4.

Remark 3. For fixed $q \geq 0$ we can identify the complex

$$\mathbf{H}_q(\mathcal{B}(\mathcal{I}, \mathcal{F}, \mathcal{X}), \beta_{\cdot\cdot})$$

(see the notation introduced before Theorem 1 in § 8) with a certain Koszul complex. Namely, the p-th term of the above complex is

$$\mathbf{H}_q(Kos\,(\rho|_\mathcal{I}) \otimes \Lambda^p\mathcal{F}) = \mathbf{H}_q(Kos\,(\rho|_\mathcal{I})) \otimes \Lambda^p\mathcal{F} \quad (p \geq 0).$$

For our aims it suffices to define a representation τ_q of $\mathcal{F}$ in the vector space $\mathbf{H}_q(Kos\,(\rho|_\mathcal{I}))$, that is $\tau_q : \mathcal{F} \to \mathcal{B}(\mathbf{H}_q(Kos\,(\rho|_\mathcal{I})))$, such that

$$\mathbf{H}_q(\mathcal{B}(\mathcal{I}, \mathcal{F}, \mathcal{X}), \beta_{\cdot\cdot}) = Kos\,(\tau_q). \tag{19}$$

To this end take $f \in \mathcal{F} \subset \mathcal{B}(\mathcal{X} \otimes \Lambda\mathcal{E})$. Then the linear subspace $\mathcal{X} \otimes \Lambda\mathcal{I}$ is invariant for f (since $\mathcal{I}$ is an ideal of $\mathcal{E}$) and the sequence $f|_{\mathcal{X}\otimes\Lambda^{\cdot}\mathcal{I}} = (f|_{\mathcal{X}\otimes\Lambda^i\mathcal{I}})_{i\in\mathbb{N}}$ defines a morphism of complexes

$$f|_{\mathcal{X}\otimes\Lambda^{\cdot}\mathcal{I}} : Kos\,(\rho|_\mathcal{I}) \to Kos\,(\rho|_\mathcal{I}).$$

Then we define

$$\tau_q(f) := \mathbf{H}_q(f|_{\mathcal{X}\otimes\Lambda^{\cdot}\mathcal{I}}) : \mathbf{H}_q(Kos\,(\rho|_\mathcal{I})) \to \mathbf{H}_q(Kos\,(\rho|_\mathcal{I})).$$

In other words, the representation τ_q is the natural action of $\mathcal{F}$ on $\mathbf{H}_q(Kos\,(\rho|_\mathcal{I}))$ and (19) is satisfied.

Remark 4. Let's assume that $\mathcal{I} = \mathcal{E}$ and $\mathcal{E}$ is nilpotent. Each $f \in \mathcal{F}$ has the form $f = p_f(\theta_{e_1}, \dots, \theta_{e_k})$ for some $e_1, \dots, e_k \in \mathcal{E}$ and for certain noncommutative polynomial $p_f \in \mathbb{C}\langle X_1, \dots, X_k\rangle$.(Of course p_f is not uniquely determined by f. Although it is easy to see that its free term $p_f(0)$ is uniquely determined.) With the notations of Remark 3 we get, by Proposition 4 c),

$$\tau_q(f) = p_f(0) \cdot I_{H_q(Kos\,(\rho))} \text{ for each } f \in \mathcal{F} \text{ and } q \geq 0.$$

Hence by (19) and by Example 4 we obtain that for $q \geq 0$ the following assertions are equivalent:

- $\mathbf{H}_q(\mathcal{B}(\mathcal{E}, \mathcal{F}, \mathcal{X}), \alpha_{\cdot\cdot})$ is exact;
- either $Kos\,(\rho)$ is exact or $p_f(0) \neq 0$ for some $f \in \mathcal{F}$.

Remark 5. Here we describe a slight generalization of the above remarked equivalence. So let $\mathcal{E}$ be nilpotent and $\mathcal{I} = \mathcal{E}$. If $\lambda : \mathcal{E} \to \mathbb{C}$ is a character, then we can deduce as in Remark 4 that

either $Kos\,(\rho - \lambda)$ is exact or there exists $f = p_f(\theta_{e_1}, \ldots, \theta_{e_k}) \in \mathcal{F}$ with $p_f(\lambda(\theta_{e_1}), \ldots, \lambda(\theta_{e_k})) \neq 0$

if and only if $\mathbf{H}_q(\mathcal{B}_\lambda(\mathcal{E}, \mathcal{F}, \mathcal{X}), \alpha..)$ is exact (for $q \geq 0$ arbitrary). Particularly the above condition is sufficient for $Tot_\lambda(\mathcal{E}, \mathcal{F}, \mathcal{X})$ to be exact (cf. Corollary 1 (*a*) in § 8).

Remark 6. If $\mathcal{J}$ is another ideal of $\mathcal{E}$ such that $\mathcal{I} \subset \mathcal{J}$ and $\lambda \in \widehat{\mathcal{J}}$ is a character of $\mathcal{J}$, then obviously $Tot_{\lambda|_\mathcal{I}}(\mathcal{I}, \mathcal{F}, \mathcal{X})$ is a subcomplex of $Tot_\lambda(\mathcal{J}, \mathcal{F}, \mathcal{X})$. In such a situation we denote by $\alpha..$, $\beta..$ the maps of both $\mathcal{B}_{\lambda|_\mathcal{I}}(\mathcal{I}, \mathcal{F}, \mathcal{X})$ and $\mathcal{B}_\lambda(\mathcal{J}, \mathcal{F}, \mathcal{X})$.

Proposition 6. *Let $\mathcal{E}$ be a finite-dimensional Lie algebra, $\rho : \mathcal{E} \to \mathcal{B}(\mathcal{X})$ be a representation, $\widetilde{\mathcal{F}}$ be a Lie subalgebra of $U(\mathcal{E})$ and $\mathcal{F} := \theta(\widetilde{\mathcal{F}})$ (cf. the notation following the proof of Proposition 5). Let $\mathcal{I}$, $\mathcal{J}$ be ideals of $\mathcal{E}$ such that $[\mathcal{E}, \mathcal{J}] \subseteq \mathcal{I} \subset \mathcal{J}$ and $\dim(\mathcal{J}/\mathcal{I}) = 1$. Then for each $c \in \mathcal{J} \backslash \mathcal{I}$ there exists a morphism of complexes*

$$\delta_c : Tot_{\lambda|_\mathcal{I}}(\mathcal{I}, \mathcal{F}, \mathcal{X}) \to Tot_{\lambda|_\mathcal{I}}(\mathcal{I}, \mathcal{F}, \mathcal{X})$$

such that for each character $\lambda : \mathcal{J} \to \mathbb{C}$ we have

$$Tot_\lambda(\mathcal{J}, \mathcal{F}, \mathcal{X}) = Con\,(Tot_{\lambda|_\mathcal{I}}(\mathcal{I}, \mathcal{F}, \mathcal{X}), \delta_c - \lambda(c)).$$

Proof. We have $\mathcal{J} = \mathcal{I} \oplus \mathbb{C}c$ hence

$$\Lambda^l \mathcal{J} = \Lambda^l \mathcal{I} \oplus (c \wedge (\Lambda^{l-1}\mathcal{I})) \tag{20}$$

for each $l \geq 0$.

The linear map

$$\mathcal{X} \otimes \Lambda^q \mathcal{J} \otimes \Lambda^p \mathcal{F} \xrightarrow{(-1)^q \alpha_{pq} + \beta_{pq}} (\mathcal{X} \otimes \Lambda^{q-1} \mathcal{J} \otimes \Lambda^p \mathcal{F}) \oplus (\mathcal{X} \otimes \Lambda^q \mathcal{J} \otimes \Lambda^{p-1} \mathcal{F}),$$

which is a part of the complex $Tot_\lambda(\mathcal{J}, \mathcal{F}, \mathcal{X})$, will be represented by the following diagram (where we use the decompositions (20)):

$$\begin{array}{ccc}
\mathcal{X} \otimes \Lambda^q \mathcal{I} \otimes \Lambda^p \mathcal{F} & \xrightarrow{\gamma_1} & \mathcal{X} \otimes \Lambda^{q-1} \mathcal{I} \otimes \Lambda^p \mathcal{F} \\
& \nearrow^{\gamma_4} & \\
\mathcal{X} \otimes (c \wedge (\Lambda^{q-1}\mathcal{I})) \otimes \Lambda^p \mathcal{F} & \xrightarrow{\gamma_3} & \mathcal{X} \otimes \Lambda^q \mathcal{I} \otimes \Lambda^{p-1} \mathcal{F} \\
& \searrow_{\gamma_2} & \\
& & \mathcal{X} \otimes (c \wedge (\Lambda^{q-2}\mathcal{I})) \otimes \Lambda^p \mathcal{F} \\
& & \oplus \\
& & \mathcal{X} \otimes (c \wedge (\Lambda^{q-1}\mathcal{I})) \otimes \Lambda^{p-1} \mathcal{F}
\end{array} \quad . \tag{21}$$

We have to define the maps γ_1, γ_2, γ_3 and γ_4.

Since $Tot_{\lambda|_{\mathcal{I}}}(\mathcal{I}, \mathcal{F}, \mathcal{X})$ is a subcomplex of $Tot_\lambda(\mathcal{J}, \mathcal{F}, \mathcal{X})$, it follows that γ_1 coincides with the map $(-1)^q\alpha_{pq} + \beta_{pq}$ from $Tot(\mathcal{I}, \mathcal{F}, \mathcal{X})$ (see also Remark 6 above).

Next we describe the restriction of $(-1)^q\alpha_{pq}+\beta_{pq}$ to $\mathcal{X}\otimes(c\wedge(\Lambda^{q-1}\mathcal{I}))\otimes\Lambda^p\mathcal{F}$ by writing it in the form $\gamma_1+\gamma_2+\gamma_3$. To this end take $x \in \mathcal{X}$, $\underline{u} = u_1\wedge\cdots\wedge u_{q-1} \in \Lambda^{q-1}\mathcal{I}$ and $\underline{f} = f_1 \wedge \cdots \wedge f_p \in \Lambda^p\mathcal{F}$. We have

$$((-1)^q\alpha_{pq} + \beta_{pq})(x \otimes (c \wedge \underline{u}) \otimes \underline{f}) = (-1)^q\alpha_{pq}(x \otimes (c \wedge \underline{u}) \otimes \underline{f})$$

$$+\beta_{pq}(x \otimes (c \wedge \underline{u}) \otimes \underline{f})$$

$$= (-1)^q \left\{ \sum_{k=1}^{p} (-1)^{k-1} (\rho_q(f_k)(x \otimes (c \wedge \underline{u}))) \otimes \overset{k}{\widehat{\underline{f}}} \right.$$

$$\left. + \sum_{1\le k<l\le p} (-1)^{k+l-1} x \otimes (c \wedge \underline{u}) \otimes ([f_k, f_l] \wedge \overset{k,l}{\widehat{\underline{f}}}) \right\}$$

$$+ \left\{ (\rho(c) - \lambda(c)) x \otimes \underline{u} \otimes \underline{f} + \sum_{i=1}^{p-1} (-1)^i \rho(u_i) x \otimes (c \wedge \overset{i}{\widehat{\underline{u}}}) \otimes \underline{f} \right.$$

$$+ \sum_{i=1}^{p-1} (-1)^{i-1} x \otimes ([c, u_i] \wedge \overset{i}{\widehat{\underline{u}}}) \otimes \underline{f}$$

$$\left. + \sum_{1\le i<j\le p-1} (-1)^{i+j-1} x \otimes ([u_i, u_j] \wedge c \wedge \overset{i,j}{\widehat{\underline{u}}}) \otimes \underline{f} \right\}. \tag{22}$$

(Recall that $\rho_q : \mathcal{F} \to \mathcal{B}(\mathcal{X} \otimes \Lambda^q\mathcal{I})$ denotes a certain representation used in the definition of the bicomplex $\mathcal{B}(\mathcal{I}, \mathcal{F}, \mathcal{X})$.)

Now remark that for each $f \in \mathcal{F}$ there exists $w(f) \in \mathcal{X} \otimes \Lambda^q\mathcal{I}$, depending on c only by means of certain commutators of c, such that

$$\rho_q(f)(x \otimes (c \wedge \underline{u})) = c \wedge \rho_{q-1}(x \otimes \underline{u}) + w(f). \tag{23}$$

(Indeed, if there exists $t \in \mathcal{E}$ such that $f = \theta_t \in \mathcal{B}(\mathcal{X} \otimes \Lambda\mathcal{I})$ then

$$\rho_q(f)(x \otimes (c \wedge \underline{u})) = \theta_t(x \otimes (c \wedge \underline{u})) = \rho(t)x \otimes c \wedge \underline{u} + \sum_{i=1}^{q-1} (-1)^i x \otimes [t, u_i] \wedge c \wedge \overset{i}{\widehat{\underline{u}}}$$

$$= c \wedge (\rho(t)x \otimes \underline{u} + \sum_{i=1}^{q-1} (-1)^{i-1} x \otimes [t, u_i] \wedge \overset{i}{\widehat{\underline{u}}}) + x \otimes [t, c] \wedge \underline{u}$$

$$= c \wedge \theta_t^{q-1}(x \otimes \underline{u}) + x \otimes [t, c] \wedge \underline{u} = c \wedge \rho_{q-1}(f)(x \otimes \underline{u}) + x \otimes [t, c] \wedge \underline{u}.$$

Then we proceed by induction on m if f is of the form $\theta_{t_1}\cdots\theta_{t_m}$; generally, f is a linear combination of such products.) In view of this remark we can write (22) in the following form

$$((-1)^q\alpha_{pq}+\beta_{pq})(x\otimes(c\wedge\underline{u})\otimes\underline{f}) = -c\wedge((-1)^{q-1}\alpha_{p,q-1}+\beta_{p,q-1})(x\otimes\underline{u}\otimes\underline{f})$$
$$+\theta_c(x\otimes\underline{u})\otimes\underline{f}-\lambda(c)\cdot x\otimes\underline{u}\otimes\underline{f}+\omega(x\otimes\underline{u}\otimes\underline{f}),$$

where $\omega(x\otimes\underline{u}\otimes\underline{f}) = (-1)^q\sum_{k=1}^{p}(-1)^{k-1}w(f_k)\otimes\overset{k}{\widehat{\underline{f}}}$.

Consequently $\gamma_2 = -((-1)^{q-1}\alpha_{p,q-1}+\beta_{p,q-1})$, $\gamma_3=\omega$ and $\gamma_4=\theta_c\otimes I_{\Lambda\mathcal{F}}-\lambda(c)\cdot I_{\mathcal{X}\otimes\Lambda\mathcal{I}\otimes\Lambda\mathcal{F}}$. We remark that ω is defined only by means of certain commutators of c and $[\mathcal{E},\mathcal{J}]\subseteq\mathcal{I}$ so we have indeed $\gamma_3(\mathcal{X}\otimes(c\wedge(\Lambda^{q-1}\mathcal{I}))\otimes\Lambda^p\mathcal{F})\subseteq\mathcal{X}\otimes\Lambda^{q-1}\mathcal{I}\otimes\Lambda^{p-1}\mathcal{F}$.

Hence in view of (21) and of the identifications $\Lambda^l\mathcal{I}\cong c\wedge(\Lambda^l\mathcal{I})$ we can write

$$Tot_\lambda(\mathcal{J},\mathcal{F},\mathcal{X}) = Con(Tot_{\lambda|_\mathcal{I}}(\mathcal{I},\mathcal{F},\mathcal{X}),\delta_c-\lambda(c))$$

with the morphism of complexes $\delta_c := \theta_c\otimes I_{\Lambda\mathcal{F}}+\omega$. □

Remark 7. If we particularize Proposition 6 for $\widetilde{\mathcal{F}}=\{0\}$ and $\mathcal{J}=\mathcal{E}$, then we obtain Proposition 4 *b*). Hence we can view Proposition 6 as a partial extension of the Proposition 4.

Remark 8. Consider the notation and hypotheses of Proposition 6 and its proof. We have obviously

$$\omega(\mathcal{X}\otimes\Lambda\mathcal{I}\otimes\Lambda^p\mathcal{F})\subseteq\mathcal{X}\otimes\Lambda\mathcal{I}\otimes\Lambda^{p-1}\mathcal{F}.$$

Hence, by definition, the map

$$\delta_c^k:\bigoplus_{q=0}^{k}(\mathcal{X}\otimes\Lambda^q\mathcal{I}\otimes\Lambda^{k-q}\mathcal{F})\to\bigoplus_{q=0}^{k}(\mathcal{X}\otimes\Lambda^q\mathcal{I}\otimes\Lambda^{k-q}\mathcal{F})$$

is given by a block-triangular matrix whose "diagonal" gives the map $\oplus_{q=0}^k(\theta_c^q\otimes I_{\Lambda^{k-q}\mathcal{F}})$.

For later use we prove now three propositions concerning the above structures. We shall use the notation from the definition of $\mathcal{B}(\mathcal{I},\mathcal{F},\mathcal{X})$ with $\mathcal{I}=\mathcal{E}$, that is $\mathcal{B}(\mathcal{E},\mathcal{F},\mathcal{X})$.

Proposition 7. *If $\mathcal{I}=\mathcal{E}$, then $Kos(\rho_0)$ is exact if and only if all the complexes $Kos(\rho_q)$, $q\geq 0$, are exact.*

Proof. First remark that $\theta_{\widetilde{e}}^p=\rho(e)\in\mathcal{B}(\mathcal{X})$ for $p=0$ and $e\in\mathcal{E}$. So for $\widetilde{f}\in\widetilde{\mathcal{F}}(\subset U(\mathcal{E}))$ we have $\theta(\widetilde{f})|_\mathcal{X}=\rho(\widetilde{f})$, i.e. $\rho_0(\theta(\widetilde{f}))=\rho(\widetilde{f})$. On the other hand Proposition 5 easily implies that

$$\theta(\widetilde{f})|_{\mathcal{X}\otimes\Lambda^q\mathcal{E}} = \begin{pmatrix}\rho(\widetilde{f}) & * & * \\ 0 & \ddots & * \\ 0 & 0 & \rho(\widetilde{f})\end{pmatrix}. \tag{24}$$

Consequently we get

$$\rho_q(f) = \begin{pmatrix} \rho(\widetilde{f}) & * & * \\ 0 & \ddots & * \\ 0 & 0 & \rho(\widetilde{f}) \end{pmatrix} \quad \text{for each } f \in \mathcal{F}(= \theta(\widetilde{\mathcal{F}})).$$

(Here we work with $\binom{n}{q} \times \binom{n}{q}$ matrices with entries from $\mathcal{B}(\mathcal{X})$, where $n = \dim \mathcal{E}$.) Now the conclusion easily follows by induction on q, in view of the last assertion of Proposition 1. □

Proposition 8. *Assume that $\mathcal{E}$ is nilpotent and $\mathcal{I} = \mathcal{E}$. If $\widetilde{\mathcal{F}}(\subset U(\mathcal{E}))$ is finite-dimensional, then it is nilpotent and moreover $\mathcal{F}(\subset \mathcal{B}(\mathcal{X} \otimes \Lambda\mathcal{E}))$ is also finite-dimensional and nilpotent.*

Proof. By definition, $\mathcal{F}$ is the image of $\widetilde{\mathcal{F}}$ by a Lie morphism, so it suffices to prove that $\widetilde{\mathcal{F}}$ is nilpotent. Denoting $\mathcal{A}_1 := U(\mathcal{E})$ and, inductively, $\mathcal{A}_{q+1} := [\mathcal{A}_1, \mathcal{A}_q]$ for $q \geq 1$, we will prove that

$$\bigcap_{q=1}^{\infty} \mathcal{A}_q = \{0\} \tag{25}$$

and this implies that every finite-dimensional Lie subalgebra of $U(\mathcal{E})$ (particularly $\widetilde{\mathcal{F}}$) is nilpotent.

For proving (25), recall that the Lie algebra $\mathcal{E}$ is nilpotent and choose a basis $e_1, \ldots, e_n$ of $\mathcal{E}$ such that

$$[e_i, e_j] = \sum_{k>j} c_{ij}^k e_k \quad (1 \leq i < j \leq n). \tag{26}$$

For every $p \geq 1$ and $i_1, \ldots, i_p \in \{1, \ldots, n\}$, denote

$$d(i_1, \ldots, i_p) := \sum_{l=1}^{p} 2^{i_l - 1}.$$

Moreover, for $q \geq 1$, let J_q be the linear span of the monomials $e_1^{\nu_1} \cdots e_n^{\nu_n}$ with $\sum_{k=1}^{n} 2^{k-1} \nu_k \geq q$, and J'_q be the linear span of all monomials $e_{i_1} \cdots e_{i_p}$ with $p \geq 1$, $i_1, \ldots, i_p \in \{1, \ldots, n\}$ and $d(i_1, \ldots, i_p) \geq q$. Obviously $J_q \subseteq J'_q$ and $J'_q \cdot J'_k \subseteq J'_{q+k}$ for all $q, k \geq 1$. We will prove below that actually

$$J_q = J'_q \text{ for } q \geq 1. \tag{27}$$

Since $\bigcap_{q=1}^{\infty} J_q = \{0\}$ by Poincaré-Birkhoff-Witt Theorem (Theorem 1 in § 1), it then follows that $\bigcap_{q=1}^{\infty} J'_q = \{0\}$. But we have obviously $\mathcal{A}_q \subseteq J'_q$ for $q \geq 2$, so (25) follows.

Now, for proving (27), let $p \geq 1$ and $i_1, \ldots, i_p \in \{1, \ldots, n\}$ arbitrary and prove that

$$e_{i_1} \cdots e_{i_p} \in J_{d(i_1, \ldots, i_p)} \tag{28}$$

by induction on the number $\iota(i_1,\ldots,i_p)$ of all pairs (s,t) such that $s,t \in \{1,\ldots,p\}$ and $i_s > i_t$. If this number equals 0, then we have $i_1 \leq \cdots \leq i_p$ and (28) is clear by definition of J_q. In the general case, use the fact that, by (26), for $i < j$, we have

$$e_j e_i = e_i e_j - \sum_{k>j} c_{ij}^k e_k$$

and

$$2^{i-1} + 2^{j-1} (< 2^j) \leq 2^{k-1} \text{ whenever } k > j.$$

Consequently, if for some $l \in \{1,\ldots,p\}$ we have $i_{l+1} > i_l$, then

$$e_{i_1} \cdots e_{i_l} e_{i_{l+1}} \cdots e_{i_p} = e_{i_1} \cdots e_{i_{l+1}} e_{i_l} \cdots e_{i_p} + w,$$

where w is a linear combination of monomials $e_{j_1} \cdots e_{j_r}$ with $\iota(j_1,\ldots,j_r) < \iota(i_1,\ldots,i_p)$ and $d(j_1,\ldots,j_r) > d(i_1,\ldots,i_p)$. This fact obviously justifies the induction step and (28) is completely proved. □

Proposition 9. *Assume that $\mathcal{E}$ is finite-dimensional nilpotent, $\mathcal{I} = \mathcal{E}$ and $\widetilde{\mathcal{F}}(\subset U(\mathcal{E}))$ is finite-dimensional. Denote by $Kos\,(\mathcal{F}_0, \mathcal{X})$ the Koszul complex of the identical representation*

$$id_{\mathcal{F}_0} : \mathcal{F}_0 \to \mathcal{B}(\mathcal{X}),\ f_0 \mapsto f_0.$$

Then all the complexes $Kos\,(\rho_q)$, $q \geq 0$, are exact if and only if the complex $Kos\,(\mathcal{F}_0, \mathcal{X})$ is exact.

Proof. The Lie algebra $\mathcal{F}$ is finite-dimensional nilpotent by Proposition 8. Moreover we have $\rho_0(\mathcal{F}) = \rho(\widetilde{\mathcal{F}}) = \mathcal{F}_0$ (see e.g. the beginning of the proof of Proposition 7). So we have the following short exact sequence

$$0 \to Ker\,\rho_0 \to \mathcal{F} \xrightarrow{\rho_0} \mathcal{F}_0 \to 0.$$

Hence $Kos\,(\mathcal{F}_0, \mathcal{X})$ is exact if and only if $Kos\,(\rho_0)$ is exact (cf. Proposition 3). Now an application of Proposition 7 finishes the proof. □

§ 11 Operations with the Koszul complexes: duality and tensor products

The first aim of this paragraph is to show that, if $\rho : \mathcal{E} \to \mathcal{B}(\mathcal{X})$ is a representation of a solvable finite-dimensional Lie algebra $\mathcal{E}$ ($\mathcal{X}$ is as usual an object of the category $\mathcal{C}$), then the complex $(Kos\,(\rho))^*$ is semi-isomorphic with a certain Koszul complex (see Theorem 1 below).

To begin with, we recall a few facts of exterior algebra. Let $\mathcal{E}$ be a complex vector space of finite dimension n. We denote by $\mathcal{E}^*$ the dual of $\mathcal{E}$ and by $\langle \cdot,\cdot \rangle : \mathcal{E} \times \mathcal{E}^* \to \mathbb{C}$ the duality map. This map extends to a duality map

$$\langle \cdot,\cdot \rangle : \Lambda\mathcal{E} \times \Lambda\mathcal{E}^* \to \mathbb{C} \tag{1}$$

by

$$\begin{aligned}\langle l_1 \wedge \cdots \wedge l_m, f_1 \wedge \cdots \wedge f_p \rangle &= det(\langle l_i, f_j \rangle)_{1 \le i,j \le m} && \text{if } m = p, \\ &= 0 && \text{if } m \ne p,\end{aligned}$$

for every $m, p \in \{1, \dots, n\}$, $l_1, \dots, l_m \in \mathcal{E}$, $f_1, \dots, f_p \in \mathcal{E}^*$; for $l \in \Lambda^0\mathcal{E} = \mathbb{C}$, $f \in \Lambda^0\mathcal{E}^* = \mathbb{C}$ we define $\langle l, f \rangle = lf$ and $\langle l, \underline{f} \rangle = 0 = \langle \underline{l}, f \rangle$ if $\underline{l} \in \Lambda^m\mathcal{E}$, $\underline{f} \in \Lambda^p\mathcal{E}^*$ with $m, p \ge 1$. Moreover for every $u \in \Lambda\mathcal{E}$ we define

$$\varepsilon(u) : \Lambda\mathcal{E} \to \Lambda\mathcal{E},\ v \mapsto u \wedge v,$$

$$\iota(u) : \Lambda\mathcal{E}^* \to \Lambda\mathcal{E}^*,\ \iota(u) := \varepsilon(u)^*.$$

(In the definition of $\iota(u)$ we use the identification of $\Lambda\mathcal{E}^*$ with $(\Lambda\mathcal{E})^*$ by means of (1).) Obviously for each $u, v \in \Lambda\mathcal{E}$ we have

$$\varepsilon(u \wedge v) = \varepsilon(u)\varepsilon(v),\ \iota(u \wedge v) = \iota(v)\iota(u).$$

Since $\mathcal{E}$ is finite-dimensional we have $(\mathcal{E}^*)^* = \mathcal{E}$ and $(\Lambda\mathcal{E}^*)^* = \Lambda\mathcal{E}$. In view of these identifications we can define (as above) the linear maps $\varepsilon(a) : \Lambda\mathcal{E}^* \to \Lambda\mathcal{E}^*$ and $\iota(a) : \Lambda\mathcal{E} \to \Lambda\mathcal{E}$ for $a \in \Lambda\mathcal{E}^*$.

Using these maps we shall associate to an arbitrary (but fixed) $\underline{w} \in \Lambda^n \backslash \{0\}$ the vector space isomorphism

$$\tau_{\underline{w}}^{\mathcal{E}} : \Lambda\mathcal{E}^* \to \Lambda\mathcal{E},\ a \mapsto \iota(a)\underline{w},$$

which has the property $\tau_{\underline{w}}^{\mathcal{E}}(\Lambda^p\mathcal{E}^*) = \Lambda^{n-p}\mathcal{E}$ for $p = 0, 1, \dots, n$. When no confusion will be possible, we shall denote $\tau^{\mathcal{E}}$, $\tau_{\underline{w}}$ or simply τ instead of $\tau_{\underline{w}}^{\mathcal{E}}$. To describe explicitly the action of this isomorphism choose a basis $\{x_1, \dots, x_n\}$ of $\mathcal{E}$ such that $\underline{w} = x_1 \wedge \cdots \wedge x_n$ and denote by $\{x_1^*, \dots, x_n^*\}$ the dual basis in $\mathcal{E}^*$. Fix $p \in \{1, \dots, n-1\}$ and $1 \le i_1 < \ldots < i_p \le n$. We are going to compute $\tau_{\underline{w}}^{\mathcal{E}}(a)$ for $a := x_{i_1}^* \wedge \cdots \wedge x_{i_p}^* \in \Lambda^p\mathcal{E}^*$. For each $u \in \Lambda\mathcal{E}^*$ we have

$$\langle \tau_{\underline{w}}^{\mathcal{E}}(a), u \rangle = \langle \iota(a)\underline{w}, u \rangle = \langle \underline{w}, \iota(a)^*u \rangle = \langle \underline{w}, \varepsilon(a)u \rangle$$

$$= \langle \underline{w}, a \wedge u \rangle = \langle x_1 \wedge \cdots \wedge x_n, x_{i_1}^* \wedge \cdots \wedge x_{i_p}^* \wedge u \rangle.$$

Consequently we have

$$\langle \tau_{\underline{w}}^{\mathcal{E}}(a), u \rangle \ne 0 \Leftrightarrow 0 \ne u \in \mathbb{C} \cdot x_{j_1}^* \wedge \cdots \wedge x_{j_{n-p}}^*$$

where

$$1 \le j_1 < \cdots < j_{n-p} \le n \text{ and } \{j_1, \dots, j_{n-p}\} = \{1, \dots, n\} \backslash \{i_1, \dots, i_p\}.$$

So by using the duality map we have $\tau_{\underline{w}}^{\mathcal{E}}(a) = \mu \cdot x_{j_1} \wedge \cdots \wedge x_{j_{n-p}}$, since $\langle \tau_{\underline{w}}^{\mathcal{E}}(a), x_{j_1}^* \wedge \ldots \wedge x_{j_{n-p}}^* \rangle = \mu \langle x_{j_1} \wedge \cdots \wedge x_{n-p}, x_{j_1}^* \wedge \ldots \wedge x_{j_{n-p}}^* \rangle$, where $\mu = \langle x_1 \wedge \cdots \wedge x_n, x_{i_1}^* \wedge \cdots \wedge x_{i_p}^* \wedge x_{j_1}^* \wedge \cdots \wedge x_{j_{n-p}}^* \rangle \in \{-1, 1\}$ and verifies

$$x_{i_1}^* \wedge \cdots \wedge x_{i_p}^* \wedge x_{j_1}^* \wedge \cdots \wedge x_{j_{n-p}}^* = \mu \cdot x_1^* \wedge \cdots \wedge x_n^*.$$

Now it is straightforward to deduce that $\mu = (-1)^{(i_p-p)+\ldots+(i_1-1)}$. So the action of $\tau_{\underline{w}}^{\mathcal{E}}$ is given by

$$\tau_{\underline{w}}^{\mathcal{E}}(x_{i_1}^* \wedge \cdots \wedge x_{i_p}^*) = (-1)^{\sum_{k=1}^{p} i_k - \frac{p(p+1)}{2}} \cdot x_{j_1} \wedge \cdots \wedge x_{j_{n-p}}.$$

In the following we assume that $\mathcal{E}$ has moreover a Lie algebra structure $[\cdot,\cdot]$. That is, $\mathcal{E}$ will be a Lie algebra of finite dimension n.

Then for every $v \in \mathcal{E}$ we define a linear map

$$\Delta(v) : \Lambda\mathcal{E} \to \Lambda\mathcal{E}$$

by

$$\Delta(v)(\underline{u}) := \sum_{i=1}^{p} (-1)^{i-1} [v, u_i] \wedge \overset{i}{\widehat{\underline{u}}} = \sum_{i=1}^{p} u_1 \wedge \cdots \wedge u_{i-1} \wedge [v, u_i] \wedge u_{i+1} \wedge \cdots \wedge u_p$$

for $\underline{u} = u_1 \wedge \cdots \wedge u_p \in \Lambda^p\mathcal{E}$, $p \geq 1$, and $\Delta(v) = 0$ on $\Lambda^0\mathcal{E}(= \mathbb{C})$. One easily checks that the linear map $\Delta(v)$ has the properties

$$\Delta(v)(u \wedge a) = (\Delta(v)u) \wedge a + u \wedge (\Delta(v)a) \text{ for } u, a \in \Lambda\mathcal{E} \tag{2}$$

and

$$\Delta(v)(\Lambda^p\mathcal{E}) \subseteq \Lambda^p\mathcal{E} \text{ for } p = 0, \ldots, n.$$

We define also the linear map

$$\Delta^*(v) : \Lambda\mathcal{E}^* \to \Lambda\mathcal{E}^*,\ \Delta^*(v) := -(\Delta(v))^*.$$

This map has properties similar to those of $\Delta(v)$, namely

$$\Delta^*(v)(a \wedge u) = (\Delta^*(v)a) \wedge u + a \wedge (\Delta^*(v)u) \text{ for } u, a \in \Lambda\mathcal{E}^* \tag{3}$$

and

$$\Delta^*(v)(\Lambda^p\mathcal{E}^*) \subseteq \Lambda^p\mathcal{E}^* \text{ for } p = 0, \ldots, n.$$

Now we prove a formula relating the above introduced maps.

Lemma 1. *Let $\underline{w} \in \Lambda^n\mathcal{E}\backslash\{0\}$ and $\tau_{\underline{w}}^{\mathcal{E}} : \Lambda\mathcal{E}^* \to \Lambda\mathcal{E}$ as above. For every $a \in \Lambda\mathcal{E}^*$ and $v \in \mathcal{E}$, the maps $\Delta(v) : \Lambda\mathcal{E} \to \Lambda\mathcal{E}$, $\Delta^*(v) : \Lambda\mathcal{E}^* \to \Lambda\mathcal{E}^*$ and $\iota(a) : \Lambda\mathcal{E} \to \Lambda\mathcal{E}$ are related by the formula*

$$(\tau_{\underline{w}}^{\mathcal{E}} \circ \Delta^*(v))a = (\Delta^*(v) \circ \tau_{\underline{w}}^{\mathcal{E}})a - (\iota(a) \circ \Delta(v))\underline{w}.$$

Proof. By (3) we have $\Delta^*(v) \circ \varepsilon(a) = \varepsilon(\Delta^*(v)a) + \varepsilon(a) \circ \Delta^*(v)$, so

$$\varepsilon(\Delta^*(v)a) = \Delta^*(v) \circ \varepsilon(a) - \varepsilon(a) \circ \Delta^*(v) : \Lambda\mathcal{E}^* \to \Lambda\mathcal{E}^*.$$

By duality we get

$$\iota(\Delta^*(v)a) = -\iota(a) \circ \Delta(v) + \Delta(v) \circ \iota(a) : \Lambda\mathcal{E} \to \Lambda\mathcal{E}.$$

If we evaluate both members of this equality at $\underline{w} \in \Lambda^n\mathcal{E}$ we obtain

$$\iota(\Delta^*(v)a)\underline{w} = -(\iota(a) \circ \Delta(v))\underline{w} + \Delta(v)(\iota(a)\underline{w}),$$

that is

$$\tau^{\mathcal{E}}_{\underline{w}}(\Delta^*(v)a) = -(\iota(a) \circ \Delta(v))\underline{w} + \Delta(v)(\tau^{\mathcal{E}}_{\underline{w}}(a)),$$

which is obviously equivalent to the desired formula. □

Corollary 1. *If* $\underline{w} \in \Lambda^n\mathcal{E}\backslash\{0\}$ *and* $v \in \mathcal{E}$ *then*

$$\tau^{\mathcal{E}}_{\underline{w}} \circ \Delta^*(v) = (\Delta(v) - \lambda) \circ \tau^{\mathcal{E}}_{\underline{w}},$$

where $\lambda := Tr(ad\, v)$.

Proof. We remark that

$$\Delta(v)\underline{w} = Tr(ad\, v) \cdot \underline{w}$$

and then apply Lemma 1. □

In the following we denote by $\mathcal{E}^{op}$ the Lie algebra *opposed* to $\mathcal{E}$. This means that $\mathcal{E}^{op}$ coincides with $\mathcal{E}$ as vector space but its bracket is defined by $(e, f) \mapsto -[e, f]$. We have obviously $\Lambda\mathcal{E}^{op} = \Lambda\mathcal{E}$ and

$$ad^{op}\, v = -ad\, v,\ \Delta^{op}(v) = -\Delta(v) \text{ for each } v \in \mathcal{E}^{op},$$

where $ad^{op}\, v : \mathcal{E}^{op} \to \mathcal{E}^{op}$ and $\Delta^{op}(v) : \Lambda\mathcal{E}^{op} \to \Lambda\mathcal{E}^{op}$ are defined for $\mathcal{E}^{op}$ similarly to the definitions of $ad\, v : \mathcal{E} \to \mathcal{E}$ and $\Delta(v) : \Lambda\mathcal{E} \to \Lambda\mathcal{E}$ for $\mathcal{E}$. Since $\Lambda\mathcal{E}^{op} = \Lambda\mathcal{E}$ we may consider

$$\tau^{\mathcal{E}}_{\underline{w}} : \Lambda\mathcal{E}^* \to \Lambda\mathcal{E}^{op}$$

(for $\underline{w} \in \Lambda^n\mathcal{E}$).

Remark 1. Let $\mathcal{I}$ be an ideal of $\mathcal{E}$ with $\dim(\mathcal{E}/\mathcal{I}) = 1$. Let $\{u_1, \dots, u_{n-1}\}$ be a basis in $\mathcal{I}$, $v \in \mathcal{E}\backslash\mathcal{I}$, $\underline{u} := u_1 \wedge \cdots \wedge u_{n-1} \in \Lambda^{n-1}\mathcal{I}$ and $\underline{w} := \underline{u} \wedge v \in \Lambda^n\mathcal{E}$. Choose $v^* \in \mathcal{E}^*$ such that $Ker\, v^* = \mathcal{I}$ and $v^*(v) = 1$. Then we have

$$\mathcal{E} = \mathcal{I} \oplus \mathbb{C}v \text{ and } \mathcal{E}^* = \mathcal{I}^* \oplus \mathbb{C}v^*.$$

Generally, for $p = 0, \dots, n$ we have

$$\Lambda^p\mathcal{E} = (\Lambda^p\mathcal{I}) \oplus (v \wedge (\Lambda^{p-1}\mathcal{I})) \text{ and } \Lambda^p\mathcal{E}^* = (\Lambda^p\mathcal{I}^*) \oplus (v^* \wedge (\Lambda^{p-1}\mathcal{I}^*)).$$

It is easily seen that the action of $\tau^{\mathcal{E}}_{\underline{w}}|_{\Lambda^p\mathcal{E}^*} : \Lambda^p\mathcal{E}^* \to \Lambda^{n-p}\mathcal{E}^{op}$ can be described with respect to the above decompositions in the following way

$$\tau^{\mathcal{E}}_{\underline{w}}|_{\Lambda^p\mathcal{I}^*} : \Lambda^p\mathcal{I}^* \to v \wedge (\Lambda^{n-p-1}\mathcal{I}^{op}),\ a \mapsto (-1)^{n-p-1}(v \wedge \tau^{\mathcal{I}}_{\underline{u}}(a)),$$
$$\tau^{\mathcal{E}}_{\underline{w}}|_{v^*\wedge(\Lambda^{p-1}\mathcal{I}^*)} : v^* \wedge (\Lambda^{p-1}\mathcal{I}^*) \to \Lambda^{n-p}\mathcal{I}^{op},\ v^* \wedge a \mapsto (-1)^{n-1}\tau^{\mathcal{I}}_{\underline{u}}(a)$$

(see the above explicit description of $\tau^{\mathcal{E}}_{\underline{w}}$).

For the following result recall that, if $\mathcal{X}$ is an object of the category $\mathcal{C}$ and $\rho : \mathcal{E} \to \mathcal{B}(\mathcal{X})$ is a representation, then its *dual representation* is a representation ρ^* of the opposite Lie algebra $\mathcal{E}^{op}$ defined in the following way

$$\rho^* : \mathcal{E}^{op} \to \mathcal{B}(\mathcal{X}^*),\ e \mapsto \rho(e)^*.$$

Also we recall the linear map $\theta_{\rho,v} : \mathcal{X} \otimes \Lambda\mathcal{I} \to \mathcal{X} \otimes \Lambda\mathcal{I}$ associated to a representation $\rho : \mathcal{E} \to \mathcal{B}(\mathcal{X})$ (see the notation introduced before Proposition 4 in § 10).

Proposition 1. *Let $\mathcal{E}$, $\mathcal{I}$, v, v^* and $\underline{w}$ be as in Remark 1. If $\mathcal{X}$ is an object of the category $\mathcal{C}$ and $\rho, \pi : \mathcal{E} \to \mathcal{B}(\mathcal{X})$ are representations such that*

$$\pi(v) = \rho(v) + Tr(ad\,v) \cdot I_{\mathcal{X}},$$

then the diagram

$$\begin{array}{ccc} \mathcal{X}^* \otimes \Lambda^p \mathcal{I}^* & \stackrel{(\theta^p_{\rho,v})^*}{\longrightarrow} & \mathcal{X}^* \otimes \Lambda^p \mathcal{I}^* \\ \downarrow I_{\mathcal{X}^*} \otimes \tau^{\mathcal{E}}_{\underline{w}} & & \downarrow I_{\mathcal{X}^*} \otimes \tau^{\mathcal{E}}_{\underline{w}} \\ \mathcal{X}^* \otimes (v \wedge (\Lambda^{n-p-1} \mathcal{I}^{op})) & \stackrel{\theta^{n-p}_{\pi^*,v}}{\longrightarrow} & \mathcal{X}^* \otimes (v \wedge (\Lambda^{n-p-1} \mathcal{I}^{op})) \end{array}$$

is commutative for $p = 0, \ldots, n$.

Proof. First observe that the horizontal arrows of the above diagram are well defined (see Remark 1). Next we have obviously

$$\theta_{\pi^*,v} = \pi^*(v) \otimes I_{\Lambda\mathcal{E}^{op}} + I_{\mathcal{X}^*} \otimes \Delta^{op}(v) \in \mathcal{B}(\mathcal{X}^* \otimes \Lambda\mathcal{E}^{op}).$$

Also $\theta_{\rho,v} = \rho(v) \otimes I_{\Lambda\mathcal{E}} + I_{\mathcal{X}} \otimes \Delta(v) \in \mathcal{B}(\mathcal{X} \otimes \Lambda\mathcal{E})$, hence $(\theta_{\rho,v})^* = \rho(v)^* \otimes I_{\Lambda\mathcal{E}^*} + I_{\mathcal{X}^*} \otimes (\Delta(v))^*$. So

$$(\theta_{\rho,v})^* = \rho(v)^* \otimes I_{\Lambda\mathcal{E}^*} - I_{\mathcal{X}^*} \otimes \Delta^*(v) \in \mathcal{B}(\mathcal{X}^* \otimes \Lambda\mathcal{E}^*).$$

This implies

$$(I_{\mathcal{X}^*} \otimes \tau^{\mathcal{E}}_{\underline{w}}) \circ (\theta_{\rho,v})^* = \rho(v)^* \otimes \tau^{\mathcal{E}}_{\underline{w}} - I_{\mathcal{X}^*} \otimes (\tau^{\mathcal{E}}_{\underline{w}} \circ \Delta^*(v)).$$

Consequently, if we denote $\lambda := Tr(ad\,v)$, by the above Corollary 1 we get

$$(I_{\mathcal{X}^*} \otimes \tau^{\mathcal{E}}_{\underline{w}}) \circ (\theta_{\rho,v})^* = \rho(v)^* \otimes \tau^{\mathcal{E}}_{\underline{w}} - I_{\mathcal{X}^*} \otimes ((\Delta(v) - \lambda) \circ \tau^{\mathcal{E}}_{\underline{w}}).$$

On the other hand, by the first formula of the present proof we get

$$\theta_{\pi^*,v} \circ (I_{\mathcal{X}^*} \otimes \tau^{\mathcal{E}}_{\underline{w}}) = \pi^*(v) \otimes \tau^{\mathcal{E}}_{\underline{w}} + I_{\mathcal{X}^*} \otimes (\Delta^{op}(v) \circ \tau^{\mathcal{E}}_{\underline{w}})$$

$$= (\rho(v)^* + \lambda \cdot I_{\mathcal{X}^*}) \otimes \tau^{\mathcal{E}}_{\underline{w}} - I_{\mathcal{X}^*} \otimes (\Delta(v) \circ \tau^{\mathcal{E}}_{\underline{w}})$$

(because $\Delta^{op}(v) = -\Delta(v)$ and $\pi(v) = \rho(v) + \lambda \cdot I_{\mathcal{X}}$).

Now, comparing the last two formulas we deduce that

$$\theta_{\pi^*,v} \circ (I_{\mathcal{X}^*} \otimes \tau^{\mathcal{E}}_{\underline{w}}) = (I_{\mathcal{X}^*} \otimes \tau^{\mathcal{E}}_{\underline{w}}) \circ (\theta_{\rho,v})^* \ (\in \mathcal{B}(\mathcal{X}^* \otimes \Lambda\mathcal{E}^*)),$$

hence the considered diagram is indeed commutative. □

Now, we are ready to prove the first of the main results of the present paragraphs. To simplify its statement let's fix some notation. For a solvable Lie algebra $\mathcal{E}$ with $\dim \mathcal{E} = n$ we consider a sequence of ideals

$$\{0\} = \mathcal{I}_0 \subset \mathcal{I}_1 \subset \cdots \subset \mathcal{I}_n = \mathcal{E}$$

such that $\dim \mathcal{I}_i = i$ $(0 \leq i \leq n)$ and $[\mathcal{E}, \mathcal{E}] = \mathcal{I}_{i_0}$ for a certain i_0. Moreover we chose $w_i \in \mathcal{I}_i \backslash \mathcal{I}_{i-1}$ $(1 \leq i \leq n)$ and denote $\underline{w} = w_1 \wedge \cdots \wedge w_n \in \Lambda^n \mathcal{E}$. Finally we define a character of $\mathcal{E}$ by

$$\chi_{\mathcal{E}} : \mathcal{E} \to \mathbb{C}, e \mapsto Tr(ad\, e).$$

Remark that for $i = 1, \ldots, n$ we have

$$\chi_{\mathcal{E}}(w_i) = Tr((ad\, w_i)|_{\mathcal{I}_i})$$

because $(ad\, w_i)\mathcal{E} \subseteq \mathcal{I}_i$.

Theorem 1. *Let $\rho : \mathcal{E} \to \mathcal{B}(\mathcal{X})$ be a representation of the solvable Lie algebra $\mathcal{E}$ with $\dim \mathcal{E} = n$ (where $\mathcal{X}$ is as usual an object of the category $\mathcal{C}$). Let $\underline{w} \in \Lambda^n \mathcal{E} \backslash \{0\}$ and $\chi_{\mathcal{E}} : \mathcal{E} \to \mathbb{C}$ be as above. Then we have a semi-isomorphism of complexes (in the category $\mathcal{C}$)*

$$\tau^{\mathcal{E}}_{\cdot} : (Kos\,(\rho))^* \to Kos\,(\rho^* + \chi_{\mathcal{E}})_{\cdot - n}$$

defined by $\tau^{\mathcal{E}}_{\cdot} = (I_{\mathcal{X}^*} \otimes \tau^{\mathcal{E}}_i)_{i \in \mathbb{Z}}$, where $\tau^{\mathcal{E}}_i := \tau^{\mathcal{E}}_{\underline{w}}|_{\Lambda^{n-i}\mathcal{E}^*} : \Lambda^{n-i}\mathcal{E}^* \to \Lambda^i \mathcal{E}$ for $i = 0, \ldots, n$ and $\tau^{\mathcal{E}}_i = 0$ for other i, and $\rho^* : \mathcal{E}^{op} \to \mathcal{B}(\mathcal{X}^*)$ is the dual representation to ρ.

Proof. Let's denote

$$Kos\,(\rho) : 0 \leftarrow \mathcal{X} \xleftarrow{\beta_1} \mathcal{X} \otimes \mathcal{E} \xleftarrow{\beta_2} \cdots \xleftarrow{\beta_p} \mathcal{X} \otimes \Lambda^p \mathcal{E} \xleftarrow{\beta_{p+1}} \cdots$$

and

$$Kos\,(\rho^* + \chi_{\mathcal{E}}) : 0 \leftarrow \mathcal{X}^* \xleftarrow{\beta'_1} \mathcal{X}^* \otimes \mathcal{E}^{op} \xleftarrow{\beta'_2} \cdots \xleftarrow{\beta'_p} \mathcal{X}^* \otimes \Lambda^p \mathcal{E}^{op} \xleftarrow{\beta'_{p+1}} \cdots.$$

We are going to prove by induction on $n(= \dim \mathcal{E})$ that for $p = 0, \ldots, n$ the diagram

$$\begin{array}{ccc} \mathcal{X}^* \otimes \Lambda^{n-p}\mathcal{E}^{op} & \xleftarrow{I_{\mathcal{X}^*} \otimes \tau^{\mathcal{E}}_{n-p}} & \mathcal{X}^* \otimes \Lambda^p \mathcal{E}^* \\ \downarrow (-1)^p \beta'_{n-p} & & \downarrow \beta^*_{p+1} \\ \mathcal{X}^* \otimes \Lambda^{n-p-1}\mathcal{E}^{op} & \xleftarrow{I_{\mathcal{X}^*} \otimes \tau^{\mathcal{E}}_{n-p-1}} & \mathcal{X}^* \otimes \Lambda^{p+1}\mathcal{E}^* \end{array} \qquad (*)$$

is commutative, and this fact will show that $\tau^{\mathcal{E}}_{\cdot}$ is a semi-morphism of complexes. Then the fact that it is semi-isomorphism follows since $\tau^{\mathcal{E}}_i : \Lambda^{n-i}\mathcal{E}^* \to \Lambda^i\mathcal{E}^{op}$ is a vector space isomorphism for $i = 0, \ldots, n$.

For $n = 1$ the desired fact is obvious (see Example 1 in § 10). Now assume that it holds for solvable Lie algebras of dimension strictly less than n. We shall use the notation preceding the statement of the theorem. Moreover we denote $\mathcal{I} := \mathcal{I}_{n-1}$ and $v := w_n$ and take $v^* \in \mathcal{E}^*$ such that $Ker v^* = \mathcal{I}$ and $v^*(v) = 1$. Then we have $\mathcal{E} = \mathcal{I} \oplus \mathbb{C}v$ and $\mathcal{E}^* = \mathcal{I}^* \oplus \mathbb{C}v^*$. More generally, for $q = 1, \ldots, n$ we have

$$\Lambda^q\mathcal{E} = (\Lambda^q\mathcal{I}) \oplus (v \wedge (\Lambda^{q-1}\mathcal{I})) \text{ and } \Lambda^q\mathcal{E}^* = (\Lambda^q\mathcal{I}^*) \oplus (v^* \wedge (\Lambda^{q-1}\mathcal{I}^*)).$$

Hence we can make the identifications $\Lambda^q\mathcal{E} \cong (\Lambda^q\mathcal{I}) \oplus (\Lambda^{q-1}\mathcal{I})$. Then, in view of Proposition 4 from § 10 (see the end of the proof of part b) of that Proposition) we can describe the map $\beta_{p+1} : \mathcal{X} \otimes \Lambda^{p+1}\mathcal{E} \to \mathcal{X} \otimes \Lambda^p\mathcal{E}$ from $Kos\,(\rho)$ in the following way

$$\beta_{p+1} : \begin{array}{c} \mathcal{X} \otimes \Lambda^{p+1}\mathcal{I} \\ \oplus \\ \mathcal{X} \otimes \Lambda^p\mathcal{I} \end{array} \to \begin{array}{c} \mathcal{X} \otimes \Lambda^{p}\mathcal{I} \\ \oplus \\ \mathcal{X} \otimes \Lambda^{p-1}\mathcal{I} \end{array}, \quad \beta_{p+1} = \begin{pmatrix} \alpha_{p+1} & \theta^p_{\rho,v} \\ 0 & -\alpha_p \end{pmatrix}$$

(where $\alpha_q : \mathcal{X} \otimes \Lambda^q\mathcal{I} \to \mathcal{X} \otimes \Lambda^{q-1}\mathcal{I}$ are the maps of the complex $Kos\,(\rho|_{\mathcal{I}})$). This implies that, in view of the identifications $\Lambda^q\mathcal{E}^* \cong (\Lambda^q\mathcal{I}^*) \oplus (\Lambda^{q-1}\mathcal{I}^*)$, we can describe the map $\beta^*_{p+1} : \mathcal{X} \otimes \Lambda^p\mathcal{E}^* \to \mathcal{X} \otimes \Lambda^{p+1}\mathcal{E}^*$ in the following way

$$\beta^*_{p+1} : \begin{array}{c} \mathcal{X}^* \otimes \Lambda^{p}\mathcal{I}^* \\ \oplus \\ \mathcal{X}^* \otimes \Lambda^{p-1}\mathcal{I}^* \end{array} \to \begin{array}{c} \mathcal{X}^* \otimes \Lambda^{p+1}\mathcal{I}^* \\ \oplus \\ \mathcal{X}^* \otimes \Lambda^{p}\mathcal{I}^* \end{array}, \quad \beta^*_{p+1} = \begin{pmatrix} \alpha^*_{p+1} & 0 \\ (\theta^p_{\rho,v})^* & -\alpha^*_p \end{pmatrix}.$$

Consequently, for $x^* \in \mathcal{X}^*$ and $\underline{u}^* \in \Lambda^p\mathcal{I}^*$, respectively $\underline{u}^* \in \Lambda^{p-1}\mathcal{I}^*$ we have

$$\beta^*_{p+1}(x^* \otimes \underline{u}^*) = \alpha^*_{p+1}(x^* \otimes \underline{u}^*) + v^* \wedge ((\theta^p_{\rho,v})^*(x^* \otimes \underline{u}^*)), \tag{4}$$

respectively

$$\beta^*_{p+1}(x^* \otimes v^* \wedge \underline{u}^*) = -v^* \wedge (\alpha^*_p(x^* \otimes \underline{u}^*)). \tag{5}$$

In the following, for the sake of simplicity we shall denote by $\tau^{\mathcal{I}}$ the vector space isomorphism $\tau^{\mathcal{I}}_{w_1\wedge\cdots\wedge w_{n-1}} : \Lambda\mathcal{I}^* \to \Lambda\mathcal{I}$ and moreover $\tau^{\mathcal{I}}_i := \tau^{\mathcal{I}}|_{\Lambda^{n-1-i}\mathcal{I}^*} : \Lambda^{n-1-i}\mathcal{I}^* \to \Lambda^i\mathcal{I}$ $(1 \le i \le n-1)$.

Then by (4) and Remark 1, for $x^* \in \mathcal{X}^*$ and $\underline{u}^* \in \Lambda^p\mathcal{I}^*$ we obtain

$$\begin{aligned} &((I_{\mathcal{X}^*} \otimes \tau^{\mathcal{E}}_{n-p-1}) \circ \beta^*_{p+1})(x^* \otimes \underline{u}^*) \\ &= ((I_{\mathcal{X}^*} \otimes \tau^{\mathcal{E}}_{n-p-1}) \otimes \alpha^*_{p+1})(x^* \otimes \underline{u}^*) \\ &+ (I_{\mathcal{X}^*} \otimes \tau^{\mathcal{E}}_{n-p-1})(v^* \wedge ((\theta^p_{\rho,v})^*(x^* \otimes \underline{u}^*))) \\ &= (-1)^{n-p-2}(v \wedge ((I_{\mathcal{X}^*} \otimes \tau^{\mathcal{I}}_{n-p-2}) \circ \alpha^*_{p+1})(x^* \otimes \underline{u}^*)) \\ &+ (-1)^{n-1}(((I_{\mathcal{X}^*} \otimes \tau^{\mathcal{I}}_{n-p-1}) \circ (\theta^p_{\rho,v})^*)(x^* \otimes \underline{u}^*)). \end{aligned} \tag{6}$$

But the induction hypothesis implies that a diagram like (*), written for $\mathcal{I}$ instead of $\mathcal{E}$, is commutative. That is,

$$(I_{\mathcal{X}^*} \otimes \tau^{\mathcal{I}}_{n-p-2}) \circ \alpha^*_{p+1} = (-1)^p \cdot \alpha'_{n-1-p} \circ (I_{\mathcal{X}^*} \otimes \tau^{\mathcal{I}}_{n-1-p}). \tag{7}$$

In view of this formula and of Proposition 1, from (6) we obtain

$$\begin{aligned}
&((I_{\mathcal{X}^*} \otimes \tau^{\mathcal{E}}_{n-p-1}) \circ \beta^*_{p+1})(x^* \otimes \underline{u}^*) \\
&= (-1)^{n-2}(v \wedge ((\alpha'_{n-1-p} \circ (I_{\mathcal{X}^*} \otimes \tau^{\mathcal{I}}_{n-1-p}))(x^* \otimes \underline{u}^*))) \\
&\quad +(-1)^{n-1}((\theta^{n-1-p}_{\rho'+\chi,v} \circ (I_{\mathcal{X}^*} \otimes \tau^{\mathcal{I}}_{n-1-p}))(x^* \otimes \underline{u}^*)),
\end{aligned} \tag{8}$$

where we denoted $\chi = \chi_{\mathcal{E}}$ for the sake of simplicity.

On the other hand, also by Remark 1 we get

$$\begin{aligned}
&(\beta'_{n-p} \circ (I_{\mathcal{X}^*} \otimes \tau^{\mathcal{E}}_{n-p}))(x^* \otimes \underline{u}^*) \\
&= \beta'_{n-p}((-1)^{n-p-1} v \wedge ((I_{\mathcal{X}^*} \otimes \tau^{\mathcal{I}}_{n-1-p})(x^* \otimes \underline{u}^*))).
\end{aligned}$$

Now, expressing $\beta'_{\cdot}$ in terms of $\alpha'_{\cdot}$ and $\theta^{\cdot}_{\rho'+\chi,v}$ (see above the similar expression of $\beta_{\cdot}$ in terms of $\alpha_{\cdot}$ and $\theta^{\cdot}_{\rho,v}$) we get

$$\begin{aligned}
&(\beta'_{n-p} \circ (I_{\mathcal{X}^*} \otimes \tau^{\mathcal{E}}_{n-p}))(x^* \otimes \underline{u}^*) \\
&= (-1)^{n-p-2}(v \wedge ((\alpha'_{n-p-1} \circ (I_{\mathcal{X}^*} \otimes \tau^{\mathcal{I}}_{n-p}))(x^* \otimes \underline{u}^*))) \\
&\quad +(-1)^{n-p-1}((\theta^{n-p-1}_{\rho'+\chi,v} \circ (I_{\mathcal{X}^*} \otimes \tau^{\mathcal{I}}_{n-1-p}))(x^* \otimes \underline{u}^*)).
\end{aligned}$$

Hence by (8) we get

$$(-1)^p(\beta'_{n-p} \circ (I_{\mathcal{X}^*} \otimes \tau^{\mathcal{E}}_{n-p})) = (I_{\mathcal{X}^*} \otimes \tau^{\mathcal{E}}_{n-p-1}) \circ \beta^*_{p+1} \text{ on } \mathcal{X}^* \otimes \Lambda^p \mathcal{I}^*. \tag{9}$$

Next, for $x^* \in \mathcal{X}^*$ and $\underline{u}^* \in \Lambda^{p-1}\mathcal{I}^*$, by (5) and Remark 1 we obtain

$$\begin{aligned}
&((I_{\mathcal{X}^*} \otimes \tau^{\mathcal{E}}_{n-p-1}) \circ \beta^*_{p+1})(x^* \otimes (v^* \wedge \underline{u}^*)) \\
&= -(I_{\mathcal{X}^*} \otimes \tau^{\mathcal{E}}_{n-p-1})(v^* \wedge (\alpha^*_p(x^* \otimes \underline{u}^*))) \\
&= (-1)^n((I_{\mathcal{X}^*} \otimes \tau^{\mathcal{I}}_{n-1-p}) \circ \alpha^*_p)(x^* \otimes \underline{u}^*).
\end{aligned} \tag{10}$$

On the other hand, also by Remark 1, we obtain

$$\begin{aligned}
(\beta'_{n-p} \circ (I_{\mathcal{X}^*} \otimes \tau^{\mathcal{E}}_{n-p}))(x^* \otimes (v^* \wedge \underline{u}^*)) &= (-1)^{n-1}(\beta'_{n-p} \circ (I_{\mathcal{X}^*} \otimes \tau^{\mathcal{I}}_{n-p}))(x^* \otimes \underline{u}^*) \\
&= (-1)^{n-1}(\alpha'_{n-p} \circ (I_{\mathcal{X}^*} \otimes \tau^{\mathcal{I}}_{n-p}))(x^* \otimes \underline{u}^*)
\end{aligned} \tag{11}$$

(where at the last step we used once again the above expression of $\beta_{\cdot}$ in terms of $\alpha_{\cdot}$ taken from the proof of Proposition 4 *b*) from § 10). Finally, the induction hypothesis (7) applied for $p-1$ instead of p gives

$$(I_{\mathcal{X}^*} \otimes \tau^{\mathcal{I}}_{n-p-1}) \circ \alpha^*_p = (-1)^{p-1} \cdot \alpha'_{n-p} \circ (I_{\mathcal{X}^*} \otimes \tau^{\mathcal{I}}_{n-p}),$$

so by (10) and (11) we get

$$(-1)^{n-p}(\beta'_{n-p} \circ (I_{\mathcal{X}^*} \otimes \tau^{\mathcal{E}}_{n-p})) = (I_{\mathcal{X}^*} \otimes \tau^{\mathcal{E}}_{n-p-1}) \circ \beta^*_{p+1}$$

on $\mathcal{X}^* \otimes (v^* \wedge (\Lambda^{p-1}\mathcal{I}^*))$. So in view of (9) and of the fact that $\mathcal{X}^* \otimes \Lambda^p\mathcal{E}^* = (\mathcal{X}^* \otimes \Lambda^p\mathcal{I}^*) \oplus (\mathcal{X}^* \otimes (v^* \wedge (\Lambda^{p-1}\mathcal{I}^*)))$ we obtain that $(*)$ is a commutative diagram and the proof is finished. □

Corollary 2. *Let $\rho : \mathcal{E} \to \mathcal{B}(\mathcal{X})$ be a representation of the nilpotent finite-dimensional Lie algebra $\mathcal{E}$ (where $\mathcal{X}$ is as usual an object of the category $\mathcal{C}$). Then the complexes $(Kos\,(\rho))^*$ and $Kos\,(\rho^*)$ are semi-isomorphic, where $\rho^* : \mathcal{E}^{op} \to \mathcal{B}(\mathcal{X}^*)$ is the representation dual to ρ.*

Proof. One applies Theorem 1 and the fact that $\chi_{\mathcal{E}} = 0$ because the Lie algebra $\mathcal{E}$ is nilpotent. □

As an application of Corollary 1 we prove the following useful result.

Corollary 3. *Let $\mathcal{X}$ be a finite-dimensional complex vector space, $\mathcal{E}$ be a finite-dimensional nilpotent Lie algebra and $\rho : \mathcal{E} \to \mathcal{B}(\mathcal{X})$ be a representation. If we denote $\dim \mathcal{E} = n$, then the following assertions are equivalent.*

(i) $\mathbf{H}_0(\rho) = \{0\}$.
(i′) $\rho(\mathcal{E})\mathcal{X} = \mathcal{X}$.
(ii) $\mathbf{H}_n(\rho) = \{0\}$.
(ii′) $\bigcap_{e \in \mathcal{E}} Ker\,(\rho(e)) = \{0\}$.
(iii) *The complex $Kos\,(\rho)$ is exact.*

Proof. Denote

$$Kos\,(\rho) : 0 \leftarrow \mathcal{X} \xleftarrow{\alpha_1} \mathcal{X} \otimes \mathcal{E} \xleftarrow{\alpha_2} \cdots \xleftarrow{\alpha_n} \mathcal{X} \otimes \Lambda^n\mathcal{E} \leftarrow 0.$$

We have $\mathbf{H}_0(Kos\,(\rho)) = \mathcal{X}/\rho(\mathcal{E})\mathcal{X}$ (see Example 2 in § 10), so obviously $(i) \Leftrightarrow (i')$. Next, since the Lie algebra $\mathcal{E}$ is nilpotent we can choose a basis $\{e_1, \ldots, e_n\}$ such that

$$[e_j, e_i] = \sum_{k<i} c^k_{ji} e_k \quad (1 \le i < j \le n).$$

Particularly $c^j_{ij} = 0$ for each i, j, hence

$$\mathbf{H}_n(\rho) = \bigcap_{i=1}^{n} Ker\,(\rho(e_i))$$

(see also Example 2 in § 10). Since $\{e_1, \ldots, e_n\}$ is a basis of $\mathcal{E}$ we get moreover

$$\mathbf{H}_n(\rho) = \bigcap_{e \in \mathcal{E}} Ker\,(\rho(e)),$$

so obviously $(ii) \Leftrightarrow (ii')$. Moreover, we have obviously $(iii) \Rightarrow (i)$ and $(iii) \Rightarrow (ii)$.

Next we prove *(ii)* $\Rightarrow$ *(iii)* by induction on $d := \dim \mathcal{X}$. For the case $d = 1$ see Example 4 from § 10. Now assume that the implication *(ii)* $\Rightarrow$ *(iii)* holds for representations in vector spaces of dimension strictly less than d. By Lie's Theorem we can find a linear subspace $\mathcal{X}_0$ of $\mathcal{X}$ which is invariant to each of the maps $\rho(e) \in \mathcal{B}(\mathcal{X})$ $(e \in \mathcal{E})$ and has the property that $\dim \mathcal{X}_0 = d - 1$. Then we consider the representation

$$\rho_0 : \mathcal{E} \to \mathcal{B}(\mathcal{X}_0),\ e \mapsto \rho(e)|_{\mathcal{X}_0},$$

and another representation

$$\rho_1 ; \mathcal{E} \to \mathcal{B}(\mathcal{X}/\mathcal{X}_0)$$

such that, for $e \in \mathcal{E}$, $\rho_1(e)$ is the map induced by $\rho(e)$ on $\mathcal{X}/\mathcal{X}_0$. But $\dim(\mathcal{X}/\mathcal{X}_0) = 1$, hence there exists a character $\widetilde{\lambda} : \mathcal{E} \to \mathbb{C}$ such that $\rho_1(e) = \widetilde{\lambda}(e) \cdot I_{\mathcal{X}/\mathcal{X}_0}$ $(e \in \mathcal{E})$.

On the other hand, by Proposition 1 from § 10 we have a long exact sequence

$$0 \leftarrow \mathbf{H}_0(\rho_1) \leftarrow \mathbf{H}_0(\rho) \leftarrow \mathbf{H}_0(\rho_0) \leftarrow \mathbf{H}_1(\rho_1) \leftarrow \cdots$$

$$\leftarrow \mathbf{H}_{n-1}(\rho_0) \leftarrow \mathbf{H}_n(\rho_1) \leftarrow \mathbf{H}_n(\rho) \leftarrow \mathbf{H}_n(\rho_0) \leftarrow 0.$$

Consequently, $\mathbf{H}_n(\rho) = \{0\}$ implies $\mathbf{H}_n(\rho_0) = \{0\}$. Hence we have $\mathbf{H}_j(\rho_0) = \{0\}$ for each j (by the induction hypothesis). In turn, this fact implies

$$\mathbf{H}_j(\rho_1) = \mathbf{H}_j(\rho) \text{ for each } j,$$

by the above long exact sequence. Particularly we have $\mathbf{H}_n(\rho_1) = \{0\}$. But this implies $\mathbf{H}_j(\rho_1) = \{0\}$ for each j, by the induction hypothesis. So $\mathbf{H}_j(\rho) = \{0\}$ for all j and the assertion *(iii)* is proved.

Finally we prove *(i)* $\Rightarrow$ *(iii)*. Since the dual of a linear surjective map between finite-dimensional vector spaces is injective, from $\mathbf{H}_0(Kos(\rho)) = \{0\}$ we deduce $\mathbf{H}_0((Kos(\rho))^*) = \{0\}$. So $\mathbf{H}_n(\rho^*) = \{0\}$ by Corollary 2 above (see also Theorem 1). Then the (already proved) implication *(ii)* $\Rightarrow$ *(iii)* shows that the complex $Kos(\rho^*)$ is exact. Then by Corollary 2 above and Theorem 1 from § 9 we deduce that $Kos(\rho)$ is also exact. □

To end this section we shall prove a basic result concerning the tensor product of two Koszul complexes (cf. Example 1 in § 8). In the following $\widetilde{\otimes}$ will be as in Example 1 from § 8. (In the applications $\widetilde{\otimes}$ will be either the algebraic tensor product $\otimes$ in the category of complex vector spaces or the completed Hilbert tensor product $\overline{\otimes}$ in the category of complex Hilbert spaces.) Before stating the following theorem let's recall two simple facts. First, if $\mathcal{E}_1, \mathcal{E}_2$ are Lie algebras then their *direct product* $\mathcal{E}_1 \times \mathcal{E}_2$ is the Lie algebra whose underlying vector space is the direct product

$$\mathcal{E}_1 \times \mathcal{E}_2 = \{(e_1, e_2) \mid e_1 \in \mathcal{E}_1, e_2 \in \mathcal{E}_2\}$$

and whose bracket is defined by

$$[(e_1, e_2), (e_1', e_2')] = ([e_1, e_1'], [e_2, e_2']) \quad (e_1, e_1' \in \mathcal{E}_1; e_2, e_2' \in \mathcal{E}_2).$$

Second, the *tensor product* (with respect to $\widetilde{\otimes}$) of two representations $\rho_j : \mathcal{E}_j \to \mathcal{B}(\mathcal{X}_j)$, $j = 1, 2$, is the representation

$$\rho_1 \widetilde{\otimes} \rho_2 : \mathcal{E}_1 \times \mathcal{E}_2 \to \mathcal{B}(\mathcal{X}_1 \widetilde{\otimes} \mathcal{X}_2),\ (e_1, e_2) \mapsto \rho_1(e_1) \widetilde{\otimes} I_{\mathcal{X}_2} + I_{\mathcal{X}_1} \widetilde{\otimes} \rho_2(e_2).$$

Now we can state:

Theorem 2. *Let $\rho_j : \mathcal{E}_j \to \mathcal{B}(\mathcal{X}_j)$ be representations of the finite-dimensional Lie algebras $\mathcal{E}_j$ $(j = 1, 2)$. Then we have*

$$Kos\,(\rho_1 \widetilde{\otimes} \rho_2) = Kos\,(\rho_1) \widetilde{\otimes} Kos\,(\rho_2).$$

Proof. Denote $\mathcal{E} = \mathcal{E}_1 \times \mathcal{E}_2$, $\mathcal{X} = \mathcal{X}_1 \widetilde{\otimes} \mathcal{X}_2$ and

$$Kos\,(\rho_1 \widetilde{\otimes} \rho_2) : 0 \leftarrow \mathcal{X} \xleftarrow{\alpha_1} \mathcal{X} \otimes \mathcal{E} \xleftarrow{\alpha_2} \cdots \xleftarrow{\alpha_p} \mathcal{X} \otimes \Lambda^p \mathcal{E} \xleftarrow{\alpha_{p+1}} \cdots,$$

$$Kos\,(\rho_1) : 0 \leftarrow \mathcal{X}_1 \xleftarrow{\beta_1} \mathcal{X}_1 \otimes \mathcal{E}_1 \xleftarrow{\beta_2} \cdots \xleftarrow{\beta_p} \mathcal{X}_1 \otimes \Lambda^p \mathcal{E}_1 \xleftarrow{\beta_{p+1}} \cdots,$$

$$Kos\,(\rho_2) : 0 \leftarrow \mathcal{X}_2 \xleftarrow{\delta_1} \mathcal{X}_2 \otimes \mathcal{E}_2 \xleftarrow{\delta_2} \cdots \xleftarrow{\delta_p} \mathcal{X}_2 \otimes \Lambda^p \mathcal{E}_2 \xleftarrow{\delta_{p+1}} \cdots.$$

By means of the canonical identifications $\mathcal{E}_1 \cong \mathcal{E}_1 \times \{0\}$, $\mathcal{E}_2 \cong \{0\} \times \mathcal{E}_2$, we can write $\mathcal{E}_1, \mathcal{E}_2 \subset \mathcal{E}$ and then for $m \geq 0$ we have

$$\Lambda^m \mathcal{E} = \bigoplus_{q=0}^{m} (\Lambda^q \mathcal{E}_1) \Lambda (\Lambda^{m-q} \mathcal{E}_2).$$

Now choose $p, q \geq 1$, $\underline{e} = e_1 \wedge \cdots \wedge e_q \in \Lambda^q \mathcal{E}_1$, $\underline{f} = f_1 \wedge \cdots \wedge f_q \in \Lambda^q \mathcal{E}_2$, $x_1 \in \mathcal{X}_1$ and $x_2 \in \mathcal{X}_2$. Then, since $[\mathcal{E}_1, \mathcal{E}_2] = \{0\}$, we obtain

$$\begin{aligned}
\alpha_{p+q}((x_1 \widetilde{\otimes} x_2) \otimes (\underline{e} \wedge \underline{f})) &= \sum_{i=1}^{q} (-1)^{i-1} (\rho_1(e_i) x_1 \widetilde{\otimes} x_2) \otimes (\overset{i}{\widehat{\underline{e}}} \wedge \underline{f}) \\
&+ \sum_{i=1}^{p} (-1)^{q+i-1} (x_1 \widetilde{\otimes} \rho_2(f_i) x_2) \otimes (\underline{e} \wedge \overset{i}{\widehat{\underline{f}}}) \\
&+ \sum_{1 \leq i < j \leq q} (-1)^{i+j-1} (x_1 \widetilde{\otimes} x_2) \otimes ([e_i, e_j] \wedge \overset{i,j}{\widehat{\underline{e}}} \wedge \underline{f}) \\
&+ \sum_{1 \leq i < j \leq p} (-1)^{i+j-1} (x_1 \widetilde{\otimes} x_2) \otimes ([f_i, f_j] \wedge \underline{e} \wedge \overset{i,j}{\widehat{\underline{f}}}) \\
&= \sum_{i=1}^{q} (-1)^{i-1} (\rho_1(e_i) x_1 \widetilde{\otimes} x_2) \otimes (\overset{i}{\widehat{\underline{e}}} \wedge \underline{f}) \\
&+ \sum_{1 \leq i < j \leq q} (-1)^{i+j-1} (x_1 \widetilde{\otimes} x_2) \otimes ([e_i, e_j] \wedge \overset{i,j}{\widehat{\underline{e}}} \wedge \underline{f})
\end{aligned}$$

$$+(-1)^q \left(\sum_{i=1}^{p} (-1)^{i-1} (x_1 \widetilde{\otimes} \rho_2(f_i) x_2) \otimes (\underline{e} \wedge \overset{i}{\widehat{\underline{f}}}) \right.$$

$$\left. + \sum_{1 \le i < j \le p} (-1)^{i+j-1} (x_1 \widetilde{\otimes} x_2) \otimes (\underline{e} \wedge [f_i, f_j] \wedge \overset{i,j}{\widehat{\underline{f}}}) \right)$$

$$= (\beta_q \widetilde{\otimes} I_{\mathcal{X}_2 \otimes \Lambda^p \mathcal{E}_2})((x_1 \widetilde{\otimes} x_2) \otimes (\underline{e} \wedge \underline{f}))$$

$$+(-1)^q (I_{\mathcal{X}_1 \otimes \Lambda^q \mathcal{E}_1} \widetilde{\otimes} \delta_p)((x_1 \widetilde{\otimes} x_2) \otimes (\underline{e} \wedge \underline{f})).$$

Consequently

$$\alpha_{p+q} = \beta_q \widetilde{\otimes} I_{\mathcal{X}_2 \otimes \Lambda^p \mathcal{E}_2} + (-1)^q \cdot I_{\mathcal{X}_1 \otimes \Lambda^q \mathcal{E}_1} \widetilde{\otimes} \delta_p,$$

and the desired conclusion follows by the definition of the tensor product of complexes (cf. Example 1 in § 8). □

C. Spectral Theory in Complex Banach Space

§ 12 General spectral theory and decomposable operators

The present book is intended as a synthesis of the results concerning Lie algebras of bounded operators. A part of these results concerns Lie algebras of general bounded operators. Another part of these results concerns Lie algebras of bounded operators which have "good spectral properties". In the following will be presented some classes of bounded operators with "good spectral properties".

The usual knowledge of general spectral theory is supposed. For instance, the Dunford analytic functional calculus in a unital complex Banach algebra and the spectral decomposition of the normal, unitary and self-adjoint operators on some complex Hilbert space are assumed.

Now let $\mathcal{B}$ be a Banach (associative) algebra with unit e. As we have already remarked, $\mathcal{B}$ has a Lie algebra structure given by $[T, U] = TU - UT$ for $U, T \in \mathcal{B}$ (cf. Example 1 in § 1). Hence we have for $T \in \mathcal{B}$,

$$adT : \mathcal{B} \to \mathcal{B}, \; (adT)A = TA - AT \text{ for } A \in \mathcal{B}.$$

We will denote by $\sigma(T)$ the *spectrum* of $T \in \mathcal{B}$, i.e.

$$\sigma(T) := \mathbb{C} \backslash \rho(T), \; \rho(T) := \{\lambda \in \mathbb{C} \mid \lambda e - T \text{ is invertible in } \mathcal{B}\}.$$

With this notation, $\rho(T)$ is called the *resolvent set* of T and it is an open subset of the complex plane $\mathbb{C}$. The spectrum $\sigma(T)$ is a non-empty compact subset of $\mathbb{C}$. One can prove that

$$\sup_{\lambda \in \sigma(T)} |\lambda| = \lim_{n \to \infty} \|T^n\|^{\frac{1}{n}} =: r(T),$$

and $r(T)$ is called the *spectral radius* of T.

We will denote as usual

$$R(\lambda, T) = (\lambda e - T)^{-1} \text{ for } \lambda \in \rho(T)$$

the *resolvent* of T. We recall that $\lambda \mapsto R(\lambda, T)$ is an analytic $\mathcal{B}$-valued function on the open set $\rho(T)$ and it satisfies the identity

$$(\mu - \lambda)R(\lambda, T)R(\mu, T) = R(\lambda, T) - R(\mu, T) \text{ for every } \lambda, \mu \in \rho(T).$$

This identity shows that

$$\frac{d}{d\lambda} R(\lambda, T) = -(R(\lambda, T))^2$$

and by induction

$$\frac{d^n}{d\lambda^n} R(\lambda, T) = (-1)^n (R(\lambda, T))^{n+1} \qquad (\lambda \in \rho(T), n \in \mathbb{N}).$$

If $\mathcal{O}(\sigma(T))$ denotes the set of analytic functions which are defined in suitable neighbourhoods of $\sigma(T)$, then for $f \in \mathcal{O}(\sigma(T))$ the Dunford analytic functional calculus is given by

$$f(T) = \frac{1}{2\pi i} \int_{\Gamma} f(\lambda) R(\lambda, T) d\lambda$$

where Γ is a finite union of Jordan curves (simple, closed, rectifiable and positively oriented) such that $\sigma(T) \subset int\,\Gamma$ and f is analytic on $\overline{int\,\Gamma}$. (Here $int\,\Gamma$ denotes the bounded open subset having Γ as boundary.)

If $\mathcal{B} = \mathcal{B}(\mathcal{X})$ is the associative algebra of all bounded linear operators on some complex Banach space $\mathcal{X}$, then the spectrum of an element $T \in \mathcal{B}(\mathcal{X})$ has a canonical decomposition in disjoint parts

$$\sigma(T) = \sigma_p(T) \cup \sigma_c(T) \cup \sigma_r(T)$$

which is defined in the following way. If we denote by I the identity operator on $\mathcal{X}$, then

$$\sigma_p(T) = \{\lambda \in \mathbb{C} \mid \lambda I - T \text{ is not injective}\},$$

$$\sigma_c(T) = \{\lambda \in \mathbb{C} \mid \lambda I - T \text{ is injective and } (\lambda I - T)\mathcal{X} \neq \overline{(\lambda I - T)\mathcal{X}} = \mathcal{X}\}$$

and

$$\sigma_r(T) = \{\lambda \in \mathbb{C} \mid \lambda I - T \text{ is injective and } \overline{(\lambda I - T)\mathcal{X}} \neq \mathcal{X}\}.$$

With this notation, $\sigma_p(T)$ is the *point spectrum* of T; it consists of the *eigenvalues* of T. Furthermore, $\sigma_c(T)$ is the *continuous spectrum* of T and $\sigma_r(T)$ is the *residual spectrum.*

Another interesting subset of $\sigma(T)$ is the following

$$\sigma_{pa}(T) = \{\lambda \in \mathbb{C} \mid \lim_{n \to \infty} \|(T - \lambda I)^n x_\lambda\|^{\frac{1}{n}} = 0 \text{ for some } 0 \neq x_\lambda \in \mathcal{X}\}.$$

The elements of $\sigma_{pa}(T)$ are called the *local approximate eigenvalues* of T. The fact that $\sigma_{pa}(T)$ is a subset of the spectrum of T follows from the next remark (which is actually an easy exercise).

Remark 1. If $\lim_{n\to\infty} \|T^n x\|^{\frac{1}{n}} = 0$ for some $0 \neq x \in \mathcal{X}$, then we have $0 \in \sigma(T)$. (Indeed, if $0 \notin \sigma(T)$, then there exists T^{-1} and we can write $\|x\| = \|(T^{-1})^n T^n x\| \leq \|T^{-1}\|^n \|T^n x\|$ and $\lim_{n\to\infty} \|x\|^{\frac{1}{n}} \leq \|T^{-1}\| \lim_{n\to\infty} \|T^n x\|^{\frac{1}{n}} = 0$. But this is impossible because $x \neq 0$.)

In the following we will write simply λ instead of $\lambda I_{\mathcal{X}}$ for $\lambda \in \mathbb{C}$. Another very simple exercise is the fact that, if $\lim_{n\to\infty} \|(T-\lambda)^n x'_\lambda\|^{\frac{1}{n}} = 0$ and $\lim_{n\to\infty} \|(T-\lambda)^n x''_\lambda\|^{\frac{1}{n}} = 0$, then $\lim_{n\to\infty} \|(T-\lambda)^n (x'_\lambda + x''_\lambda)\|^{\frac{1}{n}} = 0$. On the other hand, if U is a bounded operator which commutes with T, then the following implication holds,

$$\lim_{n\to\infty} \|(T-\lambda)^n x_\lambda\|^{\frac{1}{n}} = 0 \Rightarrow \lim_{n\to\infty} \|(T-\lambda)^n U x_\lambda\|^{\frac{1}{n}} = 0.$$

Therefore we can state the following fact.

Remark 2. The set

$$\mathcal{X}_T(\lambda) = \{x \in \mathcal{X} \mid \lim_{n\to\infty} \|(T-\lambda)^n x\|^{\frac{1}{n}} = 0\}$$

is a linear (not necessarily closed) subspace of $\mathcal{X}$ which is invariant for any operator belonging to the commutant $\{T\}'$ of T.

If the dimension of $\mathcal{X}$ is finite (i.e. $\mathcal{X}$ is isomorphic to $\mathbb{C}^m$ for a certain positive integer m) then the following classical theorem holds.

Theorem 1. *If T is a linear operator on the finite-dimensional complex vector space $\mathcal{X}$, then $\sigma(T) = \sigma_p(T) = \{\lambda_1, \ldots, \lambda_n\}$ and*

$$\mathcal{X} = \mathcal{X}(\lambda_1) \oplus \cdots \oplus \mathcal{X}(\lambda_n),$$

where the linear subspaces

$$\mathcal{X}(\lambda_i) = \mathcal{X}_T(\{\lambda_i\}) = \{x \in \mathcal{X} \mid (T-\lambda_i)^k x = 0 \textit{ for some } k \in \mathbb{N}\}$$

are invariant for T.

Now, in the conditions of the above theorem let $F = \{\lambda_{i_1}, \ldots, \lambda_{i_k}\} \subseteq \sigma(T)$. We denote

$$\mathcal{X}_T(F) = \mathcal{X}(\lambda_{i_1}) \oplus \cdots \oplus \mathcal{X}(\lambda_{i_k})$$

and as usual $(adT)A = TA - AT$ for $A \in \mathcal{B}(\mathcal{X})$. Recall that a subspace $\mathcal{X}_0$ of $\mathcal{X}$ is *invariant* to $U \in \mathcal{B}(\mathcal{X})$ if $U\mathcal{X}_0 \subseteq \mathcal{X}_0$. In this case we denote $U|_{\mathcal{X}_0}$ the restriction of U to $\mathcal{X}_0$. The following proposition describes the properties of the above introduced subspace $\mathcal{X}_T(F)$.

Proposition 1. *In the conditions of Theorem 1, for every $F \subseteq \sigma(T)$ the following assertions hold.*

1. *The subspace $\mathcal{X}_T(F)$ is invariant to T.*
2. *If a subspace $\mathcal{X}_0$ of $\mathcal{X}$ satisfies*

$$T\mathcal{X}_0 \subseteq \mathcal{X}_0 \,(\textit{i.e. } \mathcal{X}_0 \textit{ is invariant to } T),\ \sigma(T|_{\mathcal{X}_0}) \subseteq \sigma(T|_{\mathcal{X}_T(F)}),$$

then

$$\mathcal{X}_0 \subseteq \mathcal{X}_T(F).$$

3. *The subspace $\mathcal{X}_T(F)$ is invariant to any $U \in \mathcal{B}(\mathcal{X})$ which satisfies*

$$(adT)^m U = 0 \textit{ for some } m \in \mathbb{N}.$$

Such results have been described in an arbitrary complex Banach space $\mathcal{X}$ (not necessarily of finite dimension) for the class of decomposable operators which was introduced by C. Foiaş. In what follows we present the properties of this class. The basic concept of the theory of decomposable operators is that of spectral maximal subspace.

Definition 1. Let $T \in \mathcal{B}(\mathcal{X})$. A closed subspace $\mathcal{Y}$ of $\mathcal{X}$ is a *spectral maximal subspace* for T if the following conditions are fulfilled.

1. The subspace $\mathcal{Y}$ is invariant to T.
2. If $\mathcal{X}_0$ is another closed invariant subspace of T and $\sigma(T|_{\mathcal{X}_0}) \subseteq \sigma(T|_{\mathcal{Y}})$ then $\mathcal{X}_0 \subseteq \mathcal{Y}$.

Remark 3. If $\mathcal{Y}_1, \mathcal{Y}_2$ are two spectral maximal subspaces for T, then $\mathcal{Y}_1 \subseteq \mathcal{Y}_2$ iff $\sigma(T|_{\mathcal{Y}_1}) \subseteq \sigma(T|_{\mathcal{Y}_2})$.

Remark 4. If $\mathcal{Y}$ is a spectral maximal subspace for T, then $\mathcal{Y}$ is invariant for the *commutant* $\{T\}'$ of T. (Recall that $\{T\}' = \{U \in \mathcal{B}(\mathcal{X}) \mid UT = TU$, i.e. $[T,U] = 0\}$.)

In general there is no characterisation of these spectral maximal subspaces, but one can describe a large class of operators for which such a characterisation is possible. This is the class of operators having the single-valued extension property.

Definition 2. We say that the operator $T \in \mathcal{B}(\mathcal{X})$ has the *single-valued extension property* if the following implication holds:

> If f is an analytic $\mathcal{X}$-valued function defined on an open subset $G \subseteq \mathbb{C}$ and $(T - \lambda)f(\lambda) = 0$ for every $\lambda \in G$, then $f(\lambda) = 0$ for every $\lambda \in G$.

If $T \in \mathcal{B}(\mathcal{X})$ has the single-valued extension property, then for every $x \in \mathcal{X}$ we can define the following set, which is denoted $\rho_T(x)$ and is called the *local resolvent set of x with respect to T*:

$$\rho_T(x) := \{\xi \in \mathbb{C} \mid \exists G = \overset{\circ}{G} \subseteq \mathbb{C}, \xi \in G, \exists x_T(\cdot) : G \to \mathcal{X} \text{ analytic},$$

$$(\lambda - T)x_T(\lambda) = x,\ \forall \lambda \in G\}.$$

We introduce also the *local spectrum of x with respect to T*:

$$\sigma_T(x) := \mathbb{C} \backslash \rho_T(x).$$

Remark 5. The set $\rho_T(x)$ is open and contains the resolvent of T, i.e. $\rho(T) \subseteq \rho_T(x)$ for every $x \in \mathcal{X}$. We have also

$$\sigma_T(x) \subseteq \sigma(T)$$

for every $x \in \mathcal{X}$.

Remark 6. The analytic $\mathcal{X}$-valued function $x_T(\cdot)$ is uniquely determined and $x_T(\lambda) = R(\lambda, T)x$ for every $\lambda \in \rho(T)$.

Notation. If $T \in \mathcal{B}(\mathcal{X})$ is an operator with the single-valued extension property, then for every subset F of the complex plane $\mathbb{C}$ we shall denote

$$\mathcal{X}_T(F) = \{x \in \mathcal{X} \mid \sigma_T(x) \subseteq F\}.$$

For operators having the single-valued extension property there is the following characterization of the spectral maximal subspaces.

Theorem 2. *If the operator $T \in \mathcal{B}(\mathcal{X})$ has the single-valued extension property and F is a closed subset of $\mathbb{C}$ such that $\mathcal{X}_T(F)$ is a closed subspace of $\mathcal{X}$, then $\mathcal{X}_T(F)$ is a spectral maximal subspace for T and $\sigma(T|_{\mathcal{X}_T(F)}) \subseteq F$.*

Conversely, if for every closed subset F of $\mathbb{C}$ the subspace $\mathcal{X}_T(F)$ is closed, then every spectral maximal subspace $\mathcal{Y}$ of T has the form

$$\mathcal{Y} = \mathcal{X}_T(\sigma(T|_{\mathcal{Y}})).$$

The class of decomposable operators is a class of operators with "enough" spectral maximal subspaces.

Definition 3. The operator $T \in \mathcal{B}(\mathcal{X})$ is called *decomposable* if for any finite open covering $(G_j)_{1 \le j \le n}$ of the spectrum $\sigma(T)$ of T there exists a family $(\mathcal{Y}_j)_{1 \le j \le n}$ of spectral maximal subspaces for T such that:

1) $\sigma(T|_{\mathcal{Y}_j}) \subseteq G_j$ for every $j = 1, \dots, n$.

2) $\mathcal{X} = \mathcal{Y}_1 + \dots + \mathcal{Y}_n$ (i.e. every $x \in \mathcal{X}$ has a decomposition $x = x_1 + \dots + x_n$ with $x_j \in \mathcal{Y}_j$ for $j = 1, \dots, n$).

Remark 7. If $\mathcal{X}$ is finite-dimensional, then any linear operator $T \in \mathcal{B}(\mathcal{X})$ is decomposable. In this case, with the notation of Definition 3 we can take $\mathcal{Y}_j = \mathcal{X}_T(F_j)$, where $F_j := \sigma(T) \cap G_j$ (and $\mathcal{Y}_j = \{0\}$ if $G_j \cap \sigma(T) = \emptyset$), see the notation introduced before Proposition 1.

In general it is useful to know when the contribution of G_j to the decomposition of $\mathcal{X}$ (cf. Definition 3) is zero. We have:

Proposition 2. *If G is an open subset of $\mathbb{C}$ and $T \in \mathcal{B}(\mathcal{X})$ is a decomposable operator, then the following assertions are equivalent:*

(i) *There is no non-zero spectral maximal subspace $\mathcal{Y}$ for T with the property $\sigma(T|_{\mathcal{Y}}) \subset G$.*

(ii) $G \cap \sigma(T) = \emptyset$.

Corollary 1. *For every decomposable operator $T \in \mathcal{B}(\mathcal{X})$ the following assertions are equivalent:*

(i) *There exists $\lambda \in \mathbb{C}$ such that $\sigma(T) = \{\lambda\}$ (i.e. $T = \lambda I + Q$ with $Q \in \mathcal{B}(\mathcal{X})$ quasinilpotent).*

(ii) *The operator T has only two (trivial) spectral maximal subspaces: $\{0\}$ and $\mathcal{X}$.*

For decomposable operators the above characterization of the spectral maximal subspaces holds:

Theorem 3. *Every decomposable operator $T \in \mathcal{B}(\mathcal{X})$ has the single-valued extension property and for any closed subset F of $\mathbb{C}$ the subspace $\mathcal{X}_T(F)$ is closed. Hence every spectral maximal subspace $\mathcal{Y}$ equals $\mathcal{X}_T(\sigma(T|_{\mathcal{Y}}))$.*

§ 13 The transformation of the spectral maximal subspaces by bounded operators

It is possible to describe the case when a bounded operator transforms spectral maximal subspaces into spectral maximal subspaces.

Let $\mathcal{X}, \mathcal{Y}$ be two complex Banach spaces and $\mathcal{Z} := \mathcal{B}(\mathcal{X}, \mathcal{Y})$ the Banach space of all bounded operators from $\mathcal{X}$ into $\mathcal{Y}$. If $S \in \mathcal{B}(\mathcal{X})$ and $T \in \mathcal{B}(\mathcal{Y})$ one denotes by $C(T, S)$ that element of $\mathcal{B}(\mathcal{Z})$ which is defined by

$$C(T,S)X = TX - XS \qquad \text{for every } X \in \mathcal{Z} = \mathcal{B}(\mathcal{X}, \mathcal{Y}).$$

A fundamental result concerning the spectral theory of the *commutator* $C(T, S)$ is the following evaluation of its spectrum:

Theorem 1 (Rosenblum). $\sigma(C(T, S)) = \sigma(T) - \sigma(S)$.

Proposition 1. *If $S \in \mathcal{B}(\mathcal{X}), T \in \mathcal{B}(\mathcal{Y})$ are decomposable operators, then the operator $C := C(T, S)$ has the single-valued extension property.*

Proof. Let us consider a $\mathcal{B}(\mathcal{X}, \mathcal{Y})$-valued analytic function $X(\cdot)$ defined on an open subset $G \subset \mathbb{C}$ such that $(\lambda - C)X(\lambda) = 0$ for every $\lambda \in G$. We can consider an open covering $\{G_j\}_{1 \le j \le n}$ of $\sigma(T)$ such that

$$\sup_{\lambda,\mu \in G_j} |\lambda - \mu| < \sup_{\lambda,\mu \in G} |\lambda - \mu| \qquad (1 \le j \le n).$$

By the decomposability of S we have for every $x \in \mathcal{X}$ a decomposition

$$x = \sum_{j=1}^{n} x_j, \quad \sigma_S(x_j) \subset G_j \; (1 \le j \le n).$$

Hence we have

$$\bigcap_{\lambda \in G} (\sigma_S(x_j) + \lambda) = \emptyset \quad (1 \leq j \leq n).$$

The function $X(\cdot)$ verifies the equality:

$$(T - \lambda)X(\lambda)x = X(\lambda)Sx \text{ for every } x \in \mathcal{X} \text{ and } \lambda \in G. \tag{1}$$

It suffices to prove that $X(\cdot) \equiv 0$ on any connected component of G. Therefore we can suppose that G is connected. If we consider an $\mathcal{X}$-valued analytic function $x_S(\cdot)$ on a neighbourhood V of $\xi_0 \in \rho_S(x)$ such that $(\xi - S)x_S(\xi) = x$ for $\xi \in V$, then we have by (1)

$$(T - \lambda)X(\lambda)x_S(\xi) = X(\lambda)Sx_S(\xi) = X(\lambda)(\xi x_S(\xi) - x)$$

or

$$(\xi - (T - \lambda))X(\lambda)x_S(\xi) = X(\lambda)x \text{ for } \xi \in V, \lambda \in G.$$

But the $\mathcal{X}$-valued function $\xi \mapsto X(\lambda)x_S(\xi)$ is analytic on the neighbourhood V of $\xi_0 \in \rho_S(x)$, hence

$$\sigma_{T-\lambda}(X(\lambda)x) \subseteq \sigma_S(x).$$

But by a reasoning as in the spectral mapping theorem we deduce

$$\sigma_{T-\lambda}(X(\lambda)x) = \sigma_T(X(\lambda)x) - \lambda.$$

Hence we have

$$\sigma_T(X(\lambda)x) \subseteq \sigma_S(x) + \lambda \text{ for every } x \in \mathcal{X} \text{ and } \lambda \in G.$$

Now let's fix $\lambda \in G$ for the moment and let $r > 0$ be such that $D(\lambda, r) := \{\mu \in \mathbb{C} \mid |\mu - \lambda| \leq r\} \subset G$. Then we can write

$$X(\mu)x \in \mathcal{Y}_T(\sigma_S(x) + D(\lambda, r)) \quad \text{for every } \mu \in D(\lambda, r).$$

Hence we have for all μ in G

$$X(\mu)x \in \mathcal{Y}_T(\sigma_S(x) + D(\lambda, r))$$

because $X(\cdot)x$ is an analytic function on the connected set G. Therefore for every μ, λ in G we have

$$X(\mu)x \in \bigcap_r \mathcal{Y}_T(\sigma_S(x) + D(\lambda, r)) = \mathcal{Y}_T\left(\bigcap_r (\sigma_S(x) + D(\lambda, r))\right) = \mathcal{Y}_T(\sigma_S(x) + \lambda).$$

Then

$$\sigma_T(X(\mu)x) \subseteq \bigcap_{\lambda \in G} (\sigma_S(x) + \lambda).$$

For $x = x_j$ we have

$$\sigma_T(X(\mu)x_j) = \emptyset \quad \text{for every } \mu \in G,$$

and by Liouville's Theorem we have

$$X(\mu)x_j = 0.$$

Hence

$$X(\mu)x = 0 \text{ for every } \mu \in G$$

because $x = \sum_{j=1}^{n} x_j$. □

Now we can describe a "good case" when a bounded operator transforms spectral maximal subspaces into spectral maximal subspaces.

Proposition 2. *Let us consider the decomposable operators $S \in \mathcal{B}(\mathcal{X}), T \in \mathcal{B}(\mathcal{Y})$. Then we have*

$$\sigma_T(Xx) \subseteq \sigma_S(x) + \sigma_{C(T,S)}(X)$$

for every $x \in \mathcal{X}$ and $X \in \mathcal{B}(\mathcal{X}, \mathcal{Y})$.

Proof. If $\lambda_0 \notin \sigma_{C(T,S)}(X) + \sigma_S(x)$ then we can find a compact neighbourhood V_0 of λ_0 disjoint of $\sigma_{C(T,S)}(X) + \sigma_S(x)$. Then there exist an open set $G_0 \supset \sigma_S(x)$ such that $V_0 \cap (\sigma_{C(T,S)} + \bar{G}_0) = \emptyset$ and a suitable contour Γ surrounding $\sigma_S(x)$ in G_0 such that for $\lambda \in V_0, \xi \in \Gamma \subset \rho_S(x)$ we have $\lambda - \xi \notin \sigma_{C(T,S)}(X)$.

We can consider

$$\lambda \longmapsto \frac{1}{2\pi i} \int_\Gamma X_{C(T,S)}(\lambda - \xi) x_S(\xi) d\xi$$

which defines an analytic function in V_0. We recall that $X_{C(T,S)}(\cdot)$ is defined as a $\mathcal{B}(\mathcal{X}, \mathcal{Y})$-valued function which is analytic in $\rho_{C(T,S)}(X)$ and satisfies

$$(\mu - T)X_{C(T,S)}(\mu) = X - X_{C(T,S)}(\mu)S \text{ for every } \mu \in \rho_{C(T,S)}(X),$$

and $x_S(\cdot)$ is an $\mathcal{X}$-valued function which is analytic in $\rho_S(x)$ and

$$(\theta - S)x_S(\theta) = x \text{ for every } \theta \in \rho_S(x).$$

Then we obtain as in the proof of Proposition 1,

$$(\lambda - T)\frac{1}{2\pi i}\int_\Gamma X_{C(T,S)}(\lambda - \xi)x_S(\xi)d\xi = \frac{1}{2\pi i}\int_\Gamma (\lambda - T)X_{C(T,S)}(\lambda - \xi)x_S(\xi)d\xi = Xx$$

for any $\lambda \in V_0$. Hence $\lambda_0 \in \rho_T(Xx)$. □

Corollary 1. *In the conditions of Proposition 2 we have*

$$X\mathcal{X}_S(F) \subseteq \mathcal{Y}_T(\sigma_{C(T,S)} + F)$$

for every $F = \bar{F} \subseteq \mathbb{C}$.

One can obtain a more complete result.

Theorem 2. *Let $S \in \mathcal{B}(\mathcal{X})$ and $T \in \mathcal{B}(\mathcal{Y})$ be two decomposable operators, $X \in \mathcal{B}(\mathcal{X},\mathcal{Y})$ and $K \subset \mathbb{C}$ a compact set. The following assertions are equivalent:*

(i) $\sigma_{C(T,S)} \subseteq K$.

(ii) $\sigma_T(Xx) \subseteq K + \sigma_S(x)\ \forall x \in \mathcal{X}$.

(iii) $X\mathcal{X}_S(F) \subseteq \mathcal{Y}_T(F+K)\ \forall F = \bar{F} \subseteq \mathbb{C}$.

If $\mathcal{X} = \mathcal{Y}$ this theorem gives a characterization of the operators which have as invariant subspaces the spectral maximal subspaces of a certain decomposable operator.

Theorem 3. *If $T \in \mathcal{B}(\mathcal{X})$ is decomposable and $X \in \mathcal{B}(\mathcal{X})$, then the following assertions are equivalent:*

(j) $\sigma_{adT}(X) = \{0\}$ *(i.e. $\lim_{n\to\infty} \|(adT)^n X\|^{\frac{1}{n}} = 0$).*

(jj) *Every spectral maximal subspace of T is invariant for X.*

§ 14 Special classes of decomposable operators

1. The class of the generalized scalar operators

The operators of this class have a $\mathcal{C}^\infty$-functional calculus and have some finiteness properties similar to those of the linear operators acting on finite-dimensional spaces.

We will denote

$$\mathcal{C}^\infty(\mathbb{R}^2) := \{f : \mathbb{R}^2 \to \mathbb{C} \mid f \text{ is indefinitely } \mathbb{R}\text{-differentiable}\}.$$

For the sake of simplicity we will denote sometimes $\mathcal{C}^\infty$ instead of $\mathcal{C}^\infty(\mathbb{R}^2)$.

Definition 1. A $\mathcal{B}(\mathcal{X})$-valued distribution

$$\mathbf{U} : \mathcal{C}^\infty(\mathbb{R}^2) \to \mathcal{B}(\mathcal{X}), f \mapsto \mathbf{U}_f$$

is called a *spectral distribution* if

$$\mathbf{U}_{fg} = \mathbf{U}_f\mathbf{U}_g \text{ for every } f,g \in \mathcal{C}^\infty(\mathbb{R}^2)$$

and

$$\mathbf{U}_1 = I_{\mathcal{X}} \text{ where 1 denotes the constant function 1 on } \mathbb{R}^2.$$

A *generalized scalar operator* on $\mathcal{X}$ is a bounded operator $T \in \mathcal{B}(\mathcal{X})$ which has a spectral distribution $\mathbf{U}$ with the property

$$\mathbf{U}_\lambda = T$$

(where λ is an abuse of notation for the identity function on $\mathbb{R}^2$).

Remark 1. We can identify $\mathbb{R}^2$ with $\mathbb{C}$ (as real vector spaces). Then $\mathcal{C}^\infty(\mathbb{R}^2)$ is the complex vector space of all indefinitely $\mathbb{R}$-differentiable complex functions defined on $\mathbb{C}$ and can be denoted also $\mathcal{C}^\infty(\mathbb{C})$. This explains the notation λ from the above definition.

There exists an interesting result concerning the commutativity of a bounded operator with the values of spectral distributions.

Theorem 1. *Let* $\mathbf{U}$ *and* $\mathbf{V}$ *be two* $\mathcal{B}(\mathcal{X})$*-valued spectral distributions of orders* m*, respectively* n*. Then for* $A \in \mathcal{B}(\mathcal{X})$ *the following implication holds*

$$C(\mathbf{V}_\lambda, \mathbf{U}_\lambda)A = 0 \Rightarrow C(\mathbf{V}_{f_1}, \mathbf{U}_{f_1}) \dots C(\mathbf{V}_{f_{m+n+1}}, \mathbf{U}_{f_{m+n+1}})A = 0$$

for every system of functions $f_1, \dots, f_{m+n+1} \in \mathcal{C}^\infty(\mathbb{R}^2)$.

Now let us consider a spectral distribution $\mathbf{U}$. Then we denote

$$\mathcal{X}(\mathbf{U}; D) := \overline{sp}\Big(\bigcup_{\substack{f \in \mathcal{C}^\infty \\ supp\, f \subset D}} \mathbf{U}_f \mathcal{X}\Big) \quad \text{for every } D = \overset{o}{D} \subseteq \mathbb{C}$$

and

$$\mathcal{X}(\mathbf{U}; F) := \bigcap_{D = \overset{o}{D} \supset F} \mathcal{X}(\mathbf{U}; D) \quad \text{for every } F = \bar{F} \subseteq \mathbb{C}.$$

We recall that $\overline{sp}(\mathcal{E})$ is the closed vector subspace spaned by a subset $\mathcal{E}$ of $\mathcal{X}$.

Theorem 2. *If* T *is a generalized scalar operator, then it has the single-valued extension property and we have*

$$\mathcal{X}_T(F) = \mathcal{X}(\mathbf{U}; F)$$

for every $F = \bar{F} \subseteq \mathbb{C}$ *and every spectral distribution* $\mathbf{U}$ *of* T.

Remark 2. By means of the $\mathcal{C}^\infty$-partition of unity in $\mathbb{R}^2$ and by the above theorem it follows that any generalized scalar operator is decomposable.

A very useful nilpotence property is the following.

Proposition 1. *If* $T \in \mathcal{B}(\mathcal{X})$ *is a generalized scalar operator and* $\sigma(T) = \{0\}$ *(i.e.* T *is quasinilpotent), then* T *is a nilpotent operator.*

An analogue of Corollary 1 from § 12 is the following:

Corollary 1. *If* $T \in \mathcal{B}(\mathcal{X})$ *is a generalized scalar operator, then the following assertions are equivalent:*

(j) $T = \lambda I + Q$ *for certain* $\lambda \in \mathbb{C}$ *and bounded nilpotent operator* Q.

(jj) T *has the only trivial spectral maximal subspaces* $\{0\}$ *and* $\mathcal{X}$.

We can recognize the structure of the subspaces $\mathcal{X}_T(\{\lambda\})$ as in the finite-dimensional case. Proposition 2 is a consequence of Proposition 1 because the restriction of a generalized scalar operator to a closed subspace which is invariant to the values of its spectral distribution, is also a generalized scalar operator.

Proposition 2. *If $T \in \mathcal{B}(\mathcal{X})$ is a generalized scalar operator then*

$$\mathcal{X}_T(\{\lambda\}) = \{x \in \mathcal{X} \mid \exists n \in \mathbb{N},\ (\lambda I - T)^n x = 0\}$$

for every $\lambda \in \sigma(T)$.

An interesting class of generalized scalar operators is given by the following theorem.

Theorem 3. *For an operator $T \in \mathcal{B}(\mathcal{X})$ the following assertions are equivalent.*

(1) *T is a generalized scalar operator whose spectrum lies on the real line.*

(2) *T satisfies for a certain positive integer n and certain positive constants a, M the inequalities*

$$\|p(T)\| \leq \sup_{|\rho| \leq a, 0 \leq k \leq n} \left| \left(\frac{d}{d\rho}\right)^k p(\rho) \right|$$

for every polynomial p.

(3) *The spectrum of T lies on $\mathbb{R}$ and*

$$\|R(\lambda, T)\| = O(|Im\, \lambda|^{-\beta}) \textit{ for } Im\, \lambda \neq 0, Im\, \lambda \to 0$$

for a certain constant $\beta \geq 1$.

(4) *We have*

$$\|e^{itT}\| = O(|t|^{\gamma}) \textit{ for } |t| \to \infty$$

for a certain constant $\gamma \geq 0$.

Corollary 2. *If S is a quasi-skew-adjoint operator (i.e. $\sup_{t \in \mathbb{R}} \|e^{tS}\| < \infty$) then iS is a generalized scalar operator whose spectrum lies on the real line.*

2. The class of the Dunford scalar operators

Definition 2. Let $E : \mathcal{B} \to \mathcal{B}(\mathcal{X})$ be a function defined on the σ-algebra $\mathcal{B}$ of all Borel subsets of $\mathbb{C}$. This function is called a *spectral measure* if it verifies the following conditions:

1) The function

$$E(\cdot)x : \mathcal{B} \to \mathcal{X},\ a \mapsto E(a)x,$$

is σ-additive for every $x \in \mathcal{X}$.

2) $E(a \cap b) = E(a)E(b)$ for every $a, b \in \mathcal{B}$.

3) $E(\mathbb{C}) = I$.

Definition 3. A bounded operator $T \in \mathcal{B}(\mathcal{X})$ is called *(D)-scalar (Dunford scalar operator)* if there exists a spectral measure

$$E : \mathcal{B} \to \mathcal{B}(\mathcal{X})$$

with compact support such that the equality

$$T = \int_{\mathbb{C}} \lambda dE(\lambda)$$

holds.

Remark 3. Any normal (particularly any unitary or self-adjoint) bounded operator on a complex Hilbert space is (D)-scalar.

Proposition 3. *If $T \in \mathcal{B}(\mathcal{X})$ and $\mathcal{X}$ is a Hilbert space (or even a weakly complete Banach space) then the following assertions are equivalent:*
a) *T is (D)-scalar.*
b) *T is a generalized scalar operator having a spectral distribution of order 0.*

3. The class of the generalized spectral operators

Definition 4. A bounded operator $T \in \mathcal{B}(\mathcal{X})$ is called a *generalized spectral operator* if there exists a spectral distribution $\mathbf{U} : \mathcal{C}^\infty \to \mathcal{B}(\mathcal{X})$ with the following properties:
1) $T\mathbf{U}_f = \mathbf{U}_f T$ for every $f \in \mathcal{C}^\infty$.
2) $\sigma(T|\mathcal{X}_{\mathbf{U}_\lambda(F)}) \subseteq F$ for every $F = \bar{F} \subseteq \mathbb{C}$.

If the distribution $\mathbf{U}$ is given by a spectral measure then T is called a *(D)-spectral operator (Dunford spectral operator)* and this property is equivalent to the existence of a spectral measure $E : \mathcal{B} \to \mathcal{B}(\mathcal{X})$ with the properties
1′) $TE(a) = E(a)T$ for every $a \in \mathcal{B}$
and
2′) $\sigma(T|_{E(a)\mathcal{X}}) \subseteq \bar{a}$ for every $a \in \mathcal{B}$.

Then we have the following diagram where the arrows are inclusions:

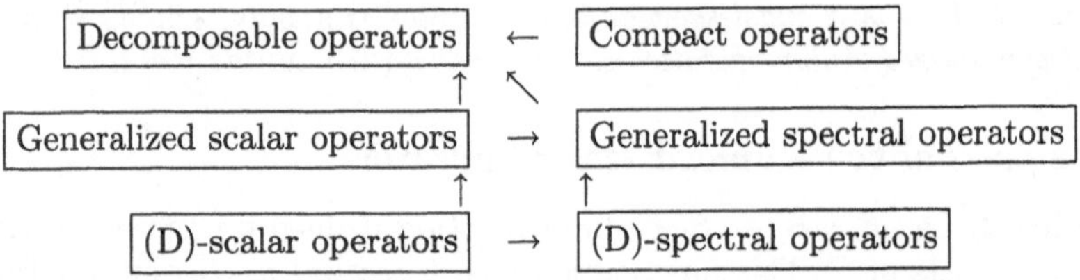

The theorem concerning the transformation of spectral maximal subspaces has a particular form for generalized scalar operators because, if T, S are generalized scalar operators, then $C(T, S)$ is also a generalized scalar operator.

Theorem 4. *Let $T \in \mathcal{B}(\mathcal{X})$ be a generalized scalar operator. Let us consider also $X \in \mathcal{B}(\mathcal{X})$ and $\lambda \in \mathbb{C}$. The following assertions are equivalent:*
(j) $(\lambda I - adT)^n X = 0$ *for a certain positive integer n.*
(jj) $X\mathcal{X}_T(F) \subseteq \mathcal{X}_T(F + \lambda)$ *for every $F = \bar{F} \subseteq \mathbb{C}$.*

In the following we present some special classes of generalized spectral operators.

Definition 5. A *hermitian operator* is an operator $A \in \mathcal{B}(\mathcal{X})$ such that $\|e^{itA}\| \leq 1$ for every $t \in \mathbb{R}$.

Remark 4. If $A \in \mathcal{B}(\mathcal{X})$ is hermitian, then we have actually $\|e^{itA}\| = 1$ for every $t \in \mathbb{R}$. (Indeed, $1 = \|e^{itA} \cdot e^{-itA}\| \leq \|e^{itA}\| \cdot \|e^{-itA}\| \leq 1$.)

Remark 5. If $\mathcal{A}$ is the set of all hermitian operators on $\mathcal{X}$, then $i\mathcal{A}$ is a closed real Lie subalgebra of $\mathcal{B}(\mathcal{X})$. (This follows from the well-known formulas

$$e^{A+B} = \lim_{n\to\infty} \left(e^{\frac{1}{n}A}e^{\frac{1}{n}B}\right)^n, \; e^{[A,B]} = \lim_{n\to\infty} \left(e^{\frac{1}{n}A}e^{\frac{1}{n}B}e^{-\frac{1}{n}A}e^{-\frac{1}{n}B}\right)^{n^2},$$

which hold for every $A, B \in \mathcal{B}(\mathcal{X})$.) This Lie algebra can reduce to $i\mathbb{R}I_{\mathcal{X}}$ on certain Banach spaces. Furthermore we have $\mathcal{B}(\mathcal{X}) = \mathcal{A} + i\mathcal{A}$ iff $\mathcal{X}$ is a Hilbert space.

Theorem 5. *If $A \in \mathcal{B}(\mathcal{X})$ is a hermitian operator, then the spectral radius of A equals $\|A\|$.*

Corollary 3. *If $A \in \mathcal{B}(\mathcal{X})$ is a hermitian operator and $\sigma(A) = \{0\}$, then $A = 0$.*

Now we introduce two more general classes of operators.

Definition 6. An operator $H \in \mathcal{B}(\mathcal{X})$ is *hermitian-equivalent* if

$$\sup_{t\in\mathbb{R}} \|e^{itH}\| < \infty$$

(i.e. iH is quasi-skew-adjoint in the sense of the Corollary 2 above). Moreover an operator $S \in \mathcal{B}(\mathcal{X})$ is called *normal-equivalent* if there exist two hermitian-equivalent operators $A, B \in \mathcal{B}(\mathcal{X})$ such that $S = A + iB$ and $[A, B] = 0$.

Remark 6. (a) Every hermitian-equivalent operator is generalized scalar and has real spectrum (cf. Corollary 2 above). Consequently every normal-equivalent operator is also generalized scalar.

(b) Every $\mathcal{C}(\mathbb{R})$-scalar operator (i.e. possessing a spectral distribution of order 0 and with the support in $\mathbb{R}$) is hermitian-equivalent and every $\mathcal{C}(\mathbb{C})$-scalar operator (i.e. possessing a spectral distribution of order 0) is normal-equivalent.

Obviously every hermitian operator is hermitian-equivalent. Conversely we have the following result (which also justifies the term "hermitian-equivalent").

Lemma 1. *If $A \in \mathcal{B}(\mathcal{X})$ is a hermitian-equivalent operator, then there exists on $\mathcal{X}$ a norm $\|\cdot\|_1$ equivalent to the initial one $\|\cdot\|$ such that A is a hermitian operator on the Banach space $(\mathcal{X}, \|\cdot\|_1)$.*

Proof. For $x \in \mathcal{X}$ define

$$\|x\|_1 := \sup_{t\in\mathbb{R}} \|e^{itA}x\|.$$

Then $\|x\| \leq \|x\|_1 \leq M\|x\|$ for every $x \in \mathcal{X}$, where $M := \sup_{t\in\mathbb{R}} \|e^{itA}\|$. Moreover it is easy to check that $\|e^{itA}x\|_1 = \|x\|_1$ for every $x \in \mathcal{X}$ and $t \in \mathbb{R}$. □

We have the following extension of Corollary 3.

Theorem 6. *If $S \in \mathcal{B}(\mathcal{X})$ is a normal-equivalent operator and $\sigma(S) = \{0\}$ then $S = 0$.*

Proof. Let $A, B \in \mathcal{B}(\mathcal{X})$ be hermitian-equivalent operators such that $S = A + iB$ and $[A, B] = 0$. There exists a spectral distribution $\mathbf{U} : \mathcal{C}^\infty \to \mathcal{B}(\mathcal{X})$ such that

$\mathbf{U}_\lambda = S$, $\mathbf{U}_{Re\,\lambda} = A$ and $\mathbf{U}_{Im\,\lambda} = B$ (see Remark 6 (a)). Then by the spectral mapping theorem for non-analytic functional calculus (see Theorem 3.2.1 from the book of I. Colojoară and C. Foiaş [1]) it follows that

$$\sigma(A) = \{Re\, z \mid z \in \sigma(S)\},\ \sigma(B) = \{Im\, z \mid z \in \sigma(S)\}.$$

Now $A = B = 0$ by Lemma 1 and Corollary 3, hence $S = 0$. □

Corollary 4. *If $S \in \mathcal{B}(\mathcal{X})$ is a normal-equivalent operator, then for every $\lambda \in \mathbb{C}$ we have*

$$\mathcal{X}_S(\{\lambda\}) = Ker\,(S - \lambda) = Ker(A - Re\,\lambda) \cap Ker(B - Im\,\lambda)$$

(where $S = A + iB$ as in Definition 6).

Proof. The operator S is decomposable by Remark 6 (a). Hence $\mathcal{X}_S(\{\lambda\})$ is a closed subspace of $\mathcal{X}$ (cf. Theorem 3 in § 12). Moreover the spectral maximal subspace $\mathcal{X}_S(\{\lambda\})$ is invariant to A and B because $[A, S] = [B, S] = 0$. The restriction of a hermitian-equivalent operator to an invariant subspace is obviously hermitian-equivalent, hence the restrictions of A and B to $\mathcal{X}_S(\{\lambda\})$ are hermitian-equivalent operators. Consequently the restriction of T to $\mathcal{X}_S(\{\lambda\})$ is a normal-equivalent operator. Then $S - \lambda$ restricted to $\mathcal{X}_S(\{\lambda\})$ is obviously a normal-equivalent operator with the spectrum $\{0\}$. Then by Theorem 6 and its proof we deduce that $S - \lambda = 0$ on $\mathcal{X}_S(\{\lambda\})$, so the desired equalities follow. □

Another useful fact is the following.

Proposition 4. *If $S \in \mathcal{B}(\mathcal{X})$ is a normal-equivalent operator, then the operator $ad\,S : \mathcal{B}(\mathcal{X}) \to \mathcal{B}(\mathcal{X})$ is also normal-equivalent.*

Proof. Let $A, B \in \mathcal{B}(\mathcal{X})$ be hermitian-equivalent operators such that $S = A + iB$ and $[A, B] = 0$. Then $ad\,S = ad\,A + i\,ad\,B$ and $[ad\,A, ad\,B] = 0$ hence it suffices to prove that $ad\,A$ and $ad\,B$ are hermitian-equivalent operators. To this end remark that for every $D \in \mathcal{B}(\mathcal{X})$ and $t \in \mathbb{R}$ we have

$$(\exp(it \cdot ad\,A))D = (\exp(ad\,(itA)))D = e^{itA} D e^{-itA}$$

(see e.g. Remark 1 in § 15) hence $\|\exp(it \cdot ad\,A)\| \le M^2$ for every $t \in \mathbb{R}$, where $M := \sup_{t \in \mathbb{R}} \|e^{itA}\|$. One proves similarly that $ad\,B$ is hermitian-equivalent. □

Now we introduce another class of operators.

Definition 7. A *Jordan operator* is an operator $T \in \mathcal{B}(\mathcal{X})$ such that there exist a normal-equivalent operator $S \in \mathcal{B}(\mathcal{X})$ and a quasinilpotent operator $Q \in \mathcal{B}(\mathcal{X})$ such that $[S, Q] = 0$ and $T = S + Q$. This last formula is called the *Jordan decomposition* of T.

Theorem 7. *Let $T \in \mathcal{B}(\mathcal{X})$ be a Jordan operator with the Jordan decomposition $T = S + Q$. Moreover let $A, B \in \mathcal{B}(\mathcal{X})$ be hermitian-equivalent operators such that $S = A + iB$ and $[A, B] = 0$. Then the operators A, B and Q belong to the bicommutant of T.*

Proof. Since $[adT, adS] = 0$ and $adT - adS = adQ$ is a quasinilpotent operator by Rosenblum's Theorem (Theorem 1 in § 13), it follows that the operators adT and adS are quasinilpotent equivalent. (Recall that two operators A and B are quasinilpotent equivalent whenever $\lim_{n\to\infty} \|C(A,B)^n I\|^{1/n} = 0 = \lim_{n\to\infty} \|C(B,A)^n I\|^{1/n}$.) But adS is a decomposable operator (cf. Proposition 4 and Remark 6 (b)). Hence adT will be also decomposable and moreover we have $\mathcal{B}(\mathcal{X})_{adT}(F) = \mathcal{B}(\mathcal{X})_{adS}(F)$ for every closed subset F of $\mathbb{C}$ (see Theorem 2.2.1 from the book of I. Colojoară and C. Foiaş [1]). Particularly by Proposition 4 and Corollary 4 we deduce

$$\mathcal{B}(\mathcal{X})_{adT}(\{0\}) = \mathcal{B}(\mathcal{X})_{adS}(\{0\}) = Ker(adS) = Ker(adA) \cap Ker(adB). \quad (1)$$

Now let's come back to the proof and choose an arbitrary $D \in \mathcal{B}(\mathcal{X})$ with $[D,T] = 0$. We must prove that $[D,A] = [D,B] = [D,Q] = 0$. We have $D \in Ker(adT) \subseteq \mathcal{B}(\mathcal{X})_{adT}(\{0\})$, hence by (1) we deduce $(adA)D = (adB)D = 0$. Since $T = A + iB + Q$ and $[D,T] = 0$ it then follows also $[Q,T] = 0$. □

Corollary 5. *Every Jordan operator is a completely regular generalized spectral operator.*

Proof. Let $T \in \mathcal{B}(\mathcal{X})$ be a Jordan operator and $T = S + Q$ be its Jordan decomposition. Since S is particularly a Jordan operator and $[S,Q] = 0$, by Theorem 7 it then follows that Q commutes with every value of the spectral distribution of S. Consequently T is a generalized spectral operator. The same Theorem 7 shows that it is completely regular, since T itself commutes with the values of the spectral distribution of S. □

Corollary 6. *If $T_1, T_2 \in \mathcal{B}(\mathcal{X})$ are commuting normal-equivalent operators then $T_1 + T_2$ is also normal-equivalent.*

Proof. Let $A_j, B_j \in \mathcal{B}(\mathcal{X})$ be hermitian-equivalent operators such that $T_j = A_j + iB_j$ and $[A_j, B_j] = 0$ for $j = 1, 2$. Since $[T_1, T_2] = 0$, by Theorem 7 one easily gets $[A_1, A_2] = [B_1, B_2] = [A_1, B_2] = [A_2, B_1] = 0$. Then $A_1 + A_2$ is hermitian-equivalent since

$$\sup_{t\in\mathbb{R}} \|e^{it(A_1+A_2)}\| = \sup_{t\in\mathbb{R}} \|e^{itA_1} e^{itA_2}\| \le (\sup_{t\in\mathbb{R}} \|e^{itA_1}\|) \cdot (\sup_{t\in\mathbb{R}} \|e^{itA_2}\|) < \infty.$$

Similarly, $B_1 + B_2$ is hermitian-equivalent so $T_1 + T_2 = (A_1 + A_2) + i(B_1 + B_2)$ is a normal-equivalent operator. □

Notes

The facts concerning finite-dimensional Lie algebras are classical and can be found for instance in Séminaire Sophus Lie année 1954–1955 [1], N. Jacobson [2] or J.-P. Serre [1]. The countably solvable (∞-solvable) Lie algebras were introduced in M. Şabac [1] in order to prove an infinite-dimensional variant of Lie's Theorem. As a generalization of this concept, F.-H. Vasilescu [2] introduced the quasisolvable Lie algebras. Locally finite, locally solvable, ideally finite Lie algebras and the radical splitting theorem for ideally finite Lie algebras were given in I. Stewart [1].

The S^p-classes were introduced by M. Şabac [11]. We note that the Lie algebras of class S^3 were used long time ago by M. Şabac [4] under the name of "hypersolvable Lie algebras", whereas the ones of class S^2 are called "strongly hypersolvable" by D. Beltiţă [4].

The LM-decomposability was introduced by M. Şabac [7].

Lemmas 1–2 and Propositions 3–4 from § 5 were proved by D. Beltiţă [3]. For more complete results related to Proposition 5 see D. Winter [1]. Theorems 2–3 summarize well-known facts; see e.g. J. Dixmier [1].

Theorem 1 of § 7 is an extension to infinite-dimensional case of the result of T. Sherman [1]. This extension, the characterization of Lie *-algebras and the possibility to have an implicitly contained Lie *-algebra structure are due to M. Şabac [8].

Most of the facts contained in § 8 are well known. A classical reference is S. MacLane [1]. For Theorem 1 see also the Appendix C in J. Eschmeier [1]. Propositions 2, 3 were stated also in D. Beltiţă [4]. Proposition 4 was proved in A.S. Fainshtein [1].

For the general theory of Banach space complexes see C.-G. Ambrozie and F.-H. Vasilescu [1]. In the present book, we need only four results from that theory. It seems that Theorem 1 from § 9 was essentially proved for the first time in Z. Słodkowski [1]. See also A.Ya. Helemskiĭ [1] for a detailed proof and discussion. Theorem 2 was proved by J.L. Taylor [2]; see also F.-H. Vasilescu [5]. Theorem 3 (Słodkowski's Lemma) was proved by Z. Słodkowski [1]. We note that C. Ott [3] calls "Słodkowski's Lemma" another result, namely our Lemma 2. Finally, Theorem 4 is a special case of the results of C. Grosu and F.-H. Vasilescu [1].

The Koszul complexes were introduced by J.L. Koszul [1]. They are described also in D.W. Barnes [1], A.S. Fainshtein [1], E. Boasso and A. Larotonda [1], C. Ott [1], [3] etc. For the cohomological version of the Koszul complexes see e.g. A. Verona [1]. Propositions 1–3 from § 10 were proved by A.S. Fainshtein [1]. Proposition 5 was proved by E. Boasso [3]; see also the proof of Proposition 10 from C. Ott [2]. All the facts concerning the bicomplex $\mathcal{B}_\lambda(\mathcal{I}, \mathcal{F}, \mathcal{X})$ are taken from A.S. Fainshtein [1].

Lemma 1 and Corollary 1 from § 11 are related to the classical Poincaré duality in algebraic framework, see J. L. Koszul [1]. Theorem 1 was proved by E. Boasso [1]; see also C. Ott [3]. Corollary 1 has been proved for the first time by A.S. Fainshtein [1]. The facts contained in Corollary 2 were first obtained by D.W. Barnes [1]; see also E. Boasso [4]. Theorem 2 can be found also in C. Ott [3].

The results concerning general spectral theory and Dunford scalar operators can be found in N. Dunford and J.T. Schwartz [1], [2], [3]. All the facts concerning decomposable operators, generalized scalar operators or generalized spectral operators can be found in I. Colojoară and C. Foiaş [1].

Theorem 1 from § 13 was proved by M. Rosenblum [1] and G. Lumer and M. Rosenblum [1]. The theorem of transformation of the spectral maximal subspaces is due to C. Foiaş and F.-H. Vasilescu [1].

For the properties of hermitian operators on Banach spaces see F.F. Bonsall and J. Duncan [1], [2]. Theorems 6, 7, Proposition 4 and Corollaries 4–6 from § 14 are taken from E. Albrecht [1]; see also Şt. Frunză [3] and J.-Ph. Labrousse [1].

Chapter II

The Commutators and Nilpotence Criteria

As we have seen (Theorem 1 from § 13) the spectral properties related to the commutators as operators on $\mathcal{B}(\mathcal{X},\mathcal{Y})$ (particularly on $\mathcal{B}(\mathcal{X})$) can give interesting properties of the elements of $\mathcal{B}(\mathcal{X},\mathcal{Y})$ (particularly on $\mathcal{B}(\mathcal{X})$ for $\mathcal{X}=\mathcal{Y}$). In what follows we prove that it is possible to describe some nilpotence criteria for elements in an associative algebra (particularly in $\mathcal{B}(\mathcal{X})$) as consequences of some properties of commutators. We begin with an asymptotic formula for the commutators.

§ 15 An asymptotic formula for the commutators

Let us consider $\mathcal{X}$, $\mathcal{Y}$ two Banach spaces. We will denote $Q=(Q_1,\dots,Q_n)$ a commuting n-tuple from $\mathcal{B}(\mathcal{X})$ and $Q'=(Q'_1,\dots,Q'_n)$ a commuting n-tuple from $\mathcal{B}(\mathcal{Y})$, i.e. $Q_i\in\mathcal{B}(\mathcal{X})$ (resp. $Q'_i\in\mathcal{B}(\mathcal{Y})$) and $Q_iQ_j=Q_jQ_i$ (resp. $Q'_iQ'_j=Q'_jQ'_i$) for any $i,j=1,\dots,n$. T will be a bounded linear operator from $\mathcal{X}$ to $\mathcal{Y}$ i.e. $T\in\mathcal{B}(\mathcal{X},\mathcal{Y})$. We shall use the well-known notations related to the *commutator system* $C(Q,Q')=(C(Q_1,Q'_1),\dots,C(Q_n,Q'_n))$, where $C(Q_i,Q'_i)$ are the operators defined on $\mathcal{B}(\mathcal{X},\mathcal{Y})$ by the formula

$$C(Q_i,Q'_i)T:=Q_iT-TQ'_i \qquad (1\le i\le n).$$

If $k=(k_1,\dots,k_n)\in\mathbb{N}^n$ we denote also

$$C^k(Q,Q')=C^{k_1}(Q_1,Q'_1)\cdots C^{k_n}(Q_n,Q'_n),$$

where $C^{k_i}(Q_i,Q'_i)$ denotes the k_i-th power of the operator $C(Q_i,Q'_i)$. (We shall use similar notation for other commuting n-tuples of operators.)

For an open set $U\subseteq\mathbb{C}^n$, $\mathcal{O}(U)$ will be the set of all complex analytic functions on U. For a compact set $K\subseteq\mathbb{C}^n$, $\mathcal{O}(K)$ will be the set of functions which are analytic in suitable neighbourhoods of K. For $f\in\mathcal{O}(U)$ ($U\subset\mathbb{C}^n$, U open) and $\alpha=(\alpha_1,\dots,\alpha_n)\in\mathbb{N}^n$ we use the well-known notation

$$\partial^\alpha f=\frac{\partial^{|\alpha|}f}{\partial z_1^{\alpha_1}\cdots\partial z_n^{\alpha_n}},\ |\alpha|=\alpha_1+\cdots+\alpha_n \ \text{ and }\ \alpha!=\alpha_1!\cdots\alpha_n!.$$

A standard result of analytic functional calculus is the following:

Lemma 1. *Let us consider the bounded open subsets $D_j \subset \mathbb{C}$, $j = 1, \dots, n$, such that their boundaries γ_j consist of a finite number of piecewise smooth Jordan curves and $\sigma(Q_j) \subset D_j$. If we denote $D := D_1 \times \cdots \times D_n (\subset \mathbb{C}^n)$, then for any $f \in \mathcal{O}(\overline{D})$ and $\alpha = (\alpha_1, \dots, \alpha_n) \in \mathbb{N}^n$ we have*

$$(\partial^\alpha f)(Q) = \frac{\alpha!}{(2\pi i)^n} \int_{\gamma_1} \cdots \int_{\gamma_n} f(\lambda_1, \dots, \lambda_n)$$
$$\times R(\lambda_1, Q_1)^{\alpha_1+1} \cdots R(\lambda_n, Q_n)^{\alpha_n+1} d\lambda_n \dots d\lambda_1,$$

where $R(\lambda_j, Q_j) = (\lambda_j - Q_j)^{-1}$ $(1 \leq j \leq n)$.

We will prove the following asymptotic formulas for the commutator system given by values of the analytic functional calculus. For $A \subseteq \mathbb{C}$ and $r \in [0, \infty)$ we will denote $\bar{B}(A, r) := \{z \in \mathbb{C} \mid dist(z, A) \leq r\}$.

Theorem 1. *Let $\rho \in [0, \infty)$ be such that*

$$\limsup_{\substack{|\alpha| \to \infty \\ \alpha \in \mathbb{N}^n}} \|C^\alpha(Q, Q')T\|^{\frac{1}{|\alpha|}} \leq \rho.$$

We denote also,

$$K' := \bar{B}(\sigma(Q'_1), \rho) \times \cdots \times \bar{B}(\sigma(Q'_n), \rho)$$

and

$$K := \bar{B}(\sigma(Q_1), \rho) \times \cdots \times \bar{B}(\sigma(Q_n), \rho).$$

1′. *If $\sigma(Q_j) \subset \bar{B}(\sigma(Q'_j), \rho)$, $j = 1, \dots, n$, then for any f in $\mathcal{O}(K')$ we have*

$$C(f(Q), f(Q'))T = \sum_{\alpha \in \mathbb{N}^n \setminus \{0\}} \frac{1}{\alpha!} \{C^\alpha(Q, Q')T\} (\partial^\alpha f)(Q'). \tag{1'}$$

1. *If $\sigma(Q'_j) \subset \bar{B}(\sigma(Q_j), \rho)$, $j = 1, \dots, n$, then for any f in $\mathcal{O}(K)$ we have*

$$C(f(Q), f(Q'))T = \sum_{\alpha \in \mathbb{N}^n \setminus \{0\}} \frac{(-1)^{|\alpha|+1}}{\alpha!} (\partial^\alpha f)(Q) \{C^\alpha(Q, Q')T\}. \tag{1}$$

Before proceeding to the proof of this theorem we wish to note some of its consequences.

Corollary 1. *With the notation of Theorem 1, let us assume that $\mathcal{X} = \mathcal{Y}$ and $Q = Q'$. Moreover for any $\alpha \in \mathbb{N}^n$ we denote $(ad\, Q)^\alpha := C^\alpha(Q, Q')$. Then for any f in $\mathcal{O}(K)$ the following Taylor type formulas hold.*

$$(ad\, f(Q))T = \sum_{\alpha \in \mathbb{N}^n \setminus \{0\}} \frac{1}{\alpha!} \{(ad\, Q)^\alpha T\} (\partial^\alpha f)(Q), \tag{2'}$$

$$(ad\, f(Q))T = \sum_{\alpha \in \mathbb{N}^n \setminus \{0\}} \frac{(-1)^{|\alpha|+1}}{\alpha!} (\partial^\alpha f)(Q) \{(ad\, Q)^\alpha T\}. \tag{2}$$

If f is an integer function, obviously $f \in \mathcal{O}(K)$ and the formulas (2) and (2′) hold. Particularly we deduce for $n = 1$ the following corollary.

Corollary 2. *If $Q, T \in \mathcal{B}(\mathcal{X})$ and f is an integer function of one complex variable then*

$$(ad\, f(Q))T = \sum_{n=1}^{\infty} \frac{(-1)^{n+1}}{n!} (ad\, Q)^n (T) f^{(n)}(Q).$$

Remark 1. If $f(z) = \exp(tz)$ for some fixed $t \in \mathbb{C}$, then the formula from Corollary 2 becomes

$$\exp(tQ)T = (exp(t \cdot ad\, Q))(T) \cdot \exp(tQ),$$

or

$$(\exp(t \cdot ad\, Q))(T) = \exp(tQ) T \exp(-tQ),$$

which is the well-known *Rosenblum's formula.*

For $n = 1$ we can easily deduce also the following result.

Corollary 3. *If $Q, T \in \mathcal{B}(\mathcal{X})$ and*

$$\lim_{p \to \infty} \|(ad\, Q)^p T\|^{\frac{1}{p}} = 0,$$

then the formulas (2) and (2′) hold for any $f \in \mathcal{O}(\sigma(Q))$. Particularly, if

$$(ad\, Q)^n T = 0$$

for some $n \in \mathbb{N}$, then

$$(ad\, f(Q))T = \sum_{0<\alpha<n} \frac{1}{\alpha!} \{(ad\, Q)^\alpha T\} f^{(\alpha)}(Q)$$

$$= \sum_{0<\alpha<n} \frac{(-1)^{\alpha+1}}{\alpha!} f^{(\alpha)}(Q) \{(ad\, Q)^\alpha T\}$$

for every $f \in \mathcal{O}(\sigma(Q))$.

The proof of Theorem 1 requires the following lemmas.

Lemma 4. *Let's suppose that $S_0 \in \mathcal{B}(\mathcal{X}), \varepsilon \in (0, \infty)$ and K is a compact subset of $\mathbb{C} \backslash \bar{B}(\sigma(S_0), \varepsilon)$. Then one can find positive numbers δ and M which depend only on S_0, ε and K such that for any $S \in \mathcal{B}(\mathcal{X})$ with $\|S - S_0\| < \delta$ we have $\sigma(S) \subset \mathbb{C} \backslash K$ and $\|R(\xi, S)^m\| \leq M \varepsilon^{-m}$ for any $m \in \mathbb{N}$ and $\xi \in K$.*

Proof. It is well known that for $\varepsilon_1 < \varepsilon$ there exists $\delta > 0$ such that $\|S - S_0\| < \delta$ implies $\sigma(S) \subset B(\sigma(S_0), \varepsilon_1)$ and

$$\|R(\lambda, S) - R(\lambda, S_0)\| < \varepsilon_1 \text{ for } \lambda \notin B(\sigma(S_0), \varepsilon_1).$$

There exists an open set U containing $\bar{B}(\sigma(S_0), \varepsilon_1)$ having as boundary B a union of a finite number of piecewise smooth Jordan curves such that $|\lambda - \mu| > \varepsilon$ for every $\lambda \in U \cup B$ and $\mu \in K$. Then we can write for $\xi \in K$,

$$\|R(\xi, S)^m\| = \|\frac{1}{2\pi i}\int_B (\xi - \lambda)^{-m} R(\lambda, S) d\lambda\| \leq \varepsilon^{-m} \cdot \frac{1}{2\pi}\int_B \|R(\lambda, S)\| d\lambda$$

$$\leq \varepsilon^{-m} \cdot \frac{1}{2\pi}\int_B (\|R(\lambda, S_0)\| + \varepsilon_1) d\lambda = M\varepsilon^{-m} \ ,$$

and the proof is finished. □

The following auxiliary result is basic for the proof of Theorem 1.

Lemma 5. *Assume that* $\lambda_j \in \mathbb{C}\backslash(\sigma(Q_j) \cup \sigma(Q'_j))$ *and denote* $R_j := R(\lambda_j, Q_j)$, $R'_j := R(\lambda_j, Q'_j)$ *for* $j = 1, \dots, n$. *Moreover we denote* $\lambda := (\lambda_1, \dots, \lambda_n)$, $R := (R_1, \dots, R_n)$, $R' := (R'_1, \dots, R'_n)$. *Then for any positive integer* p *we have*

$$C(\prod_{j=1}^{n} R_j, \prod_{j=1}^{n} R'_j)T = \sum_{1 \leq |\alpha| \leq p} \{C^\alpha(Q, Q')T\} R'^\alpha(\prod_{j=1}^{n} R'_j) + \rho'_p(\lambda) \qquad (3')$$

$$C(\prod_{j=1}^{n} R_j, \prod_{j=1}^{n} R'_j)T = \sum_{1 \leq |\alpha| \leq p} (-1)^{|\alpha|+1} R^\alpha(\prod_{j=1}^{n} R_j)\{C^\alpha(Q, Q')T\} + (-1)^p \rho_p(\lambda), \qquad (3)$$

where

$$\rho'_p(\lambda) := \sum_{1 \leq i_0 \leq \cdots \leq i_p \leq n} (\prod_{k=1}^{i_0} R_k)\{C(Q_{i_0}, Q'_{i_0}) \cdots C(Q_{i_p}, Q'_{i_p})T\}(\prod_{l=1}^{p} R'_{i_l})(\prod_{j=i_0}^{n} R'_j) \qquad (4')$$

and

$$\rho_p(\lambda) := \sum_{1 \leq i_1 \leq \cdots \leq i_p \leq i^0 \leq n} (\prod_{l=1}^{p} R_{i_l})(\prod_{j=1}^{i^0} R_j)$$

$$\times \{C(Q_{i_1}, Q'_{i_1}) \cdots C(Q_{i_p}, Q'_{i_p}) C(Q_{i^0}, Q'_{i^0})T\}(\prod_{k=i^0}^{n} R'_k). \qquad (4)$$

Proof. We shall prove only (3′) because (3) can be proved similarly. We shall use induction on p and the following formulas

$$C(R_j, R'_j)T = R_j\{C(Q_j, Q'_j)T\}R'_j, \qquad (5)$$

$$C(AB, A'B')D = A\{C(B, B')D\} + \{C(A, A')D\}B', \qquad (6)$$

where $A, B \in \mathcal{B}(\mathcal{X})$, $A', B' \in \mathcal{B}(\mathcal{Y})$ and $D \in \mathcal{B}(\mathcal{X}, \mathcal{Y})$. The formula (3′) is true for $p = 1$. Indeed, using (5) and (6) we can write:

$$\begin{aligned}
C(\prod_{j=1}^{n} R_j, \prod_{j=1}^{n} R'_j)T &\overset{(6)}{=} \sum_{j=1}^{n} (\prod_{i<j} R_i)\{C(R_j, R'_j)T\}(\prod_{i>j} R'_i) \\
&\overset{(5)}{=} \sum_{j=1}^{n} (\prod_{i\le j} R_i)\{C(Q_j, Q'_j)T\}(\prod_{i\ge j} R'_i) \\
&= \sum_{j=1}^{n} \{C(Q_j, Q'_j)T\}(\prod_{i\le j} R'_i)(\prod_{i\ge j} R'_i) \\
&\quad + \sum_{j=1}^{n} \{C(\prod_{i\le j} R_i, \prod_{i\le j} R'_i)C(Q_j, Q'_j)T\}(\prod_{q\ge j} R'_q) \\
&\overset{(6)}{=} \sum_{j=1}^{n} \{C(Q_j, Q'_j)T\}R'_j(\prod_{i=1}^{n} R'_i) \\
&\quad + \sum_{1\le i\le j\le n} (\prod_{k<i} R_k)\{C(R_i, R'_i)C(Q_j, Q'_j)T\} \\
&\quad \times (\prod_{i<k\le j} R'_k)(\prod_{q\ge j} R'_q) \\
&\overset{(5)}{=} \sum_{j=1}^{n} \{C(Q_j, Q'_j)T\}R'_j(\prod_{i=1}^{n} R'_i) \\
&\quad + \sum_{1\le i\le j\le n} (\prod_{k\le i} R_k)\{C(Q_i, Q'_i)C(Q_j, Q'_j)T\} \\
&\quad \times R'_j(\prod_{q=i}^{n} R'_q).
\end{aligned}$$

Hence (3′) holds for $p = 1$. For an arbitrary p we have

$$\begin{aligned}
\rho'_p(\lambda) &\overset{(4')}{=} \sum_{1\le i_1\le\cdots\le i_{p+1}\le n} (\prod_{k=1}^{i_1} R_k)\{C(Q_{i_1}, Q'_{i_1})\cdots C(Q_{i_{p+1}}, Q'_{i_{p+1}})T\} \\
&\quad \times (\prod_{l=2}^{p+1} R'_{i_l})(\prod_{j=i_1}^{n} R'_j)
\end{aligned}$$

$$\begin{aligned}
&= \sum_{1\le i_1\le\cdots\le i_{p+1}\le n} \{C(Q_{i_1}, Q'_{i_1})\cdots C(Q_{i_{p+1}}, Q'_{i_{p+1}})T\}(\prod_{k=1}^{i_1} R'_k)(\prod_{l=2}^{p+1} R'_{i_l})(\prod_{j=i_1}^{n} R'_j) \\
&\quad + \sum_{1\le i_1\le\cdots\le i_{p+1}\le n} \{C(\prod_{k=1}^{i_1} R_k, \prod_{k=1}^{i_1} R'_k)C(Q_{i_1}, Q'_{i_1})\cdots C(Q_{i_{p+1}}, Q'_{i_{p+1}})T\} \\
&\qquad \times (\prod_{l=2}^{p+1} R'_{i_l})(\prod_{j=i_1}^{n} R'_j).
\end{aligned}$$

Now an alternative use of (6) and (5) (as in the case $p = 1$) shows that the last sum equals $\rho'_{p+1}(\lambda)$. Hence the above formula can be written

$$\rho'_p(\lambda) = \sum_{1\le i_1\le\cdots\le i_{p+1}\le n} \{C(Q_{i_1}, Q'_{i_1})\cdots C(Q_{i_{p+1}}, Q'_{i_{p+1}})T\} \times(\prod_{l=1}^{p+1} R'_{i_l})(\prod_{j=1}^{n} R'_j) + \rho'_{p+1}(\lambda).$$

This is equivalent to

$$\rho'_p(\lambda) = \sum_{|\alpha|=p+1} \{C^\alpha(Q, Q')T\}R'^\alpha(\prod_{j=1}^{n} R'_j) + \rho'_{p+1}(\lambda).$$

The last formula obviously justifies the induction step. Hence (3′) is proved. □

Now we can prove Theorem 1.

Proof of Theorem 1. Only the part 1′ will be proved since part 1 can be proved similarly. Let us consider two positive numbers η and δ such that $\rho < \eta < \delta$. We choose also, for $1 \le j \le n$, a bounded open subset D_j of $\mathbb{C}$ such that its boundary γ_j consists of a finite number of piecewise smooth Jordan curves and the following conditions are fulfilled

$(i)\bar{B}(\sigma(Q'_j), \delta) \subset D_j$;
$(ii) f \in \mathcal{O}(\overline{D_1} \times \cdots \times \overline{D_n})$.
By the hypothesis we have

$$\sigma(Q_j) \subset \bar{B}(\sigma(Q'_j), \rho) \subset D_j.$$

In view of the condition (i) above, Lemma 4 implies the existence of a constant M such that

$$\|R(\lambda_1, Q'_1)^{\alpha_1}\cdots R(\lambda_n, Q'_n)^{\alpha_n}\| \le M\delta^{-|\alpha|} \tag{7}$$

for any $\alpha = (\alpha_1, \ldots, \alpha_n) \in \mathbb{N}^n$ and $\lambda_j \in \gamma_j (1 \le j \le n)$. Since $\rho < \eta$, by hypothesis follows the existence of a positive integer p_0 such that

$$\|C^\alpha(Q, Q')T\| < \eta^{|\alpha|} \tag{8}$$

for any $\alpha \in \mathbb{N}^n$ with $|\alpha| \ge p_0$.

We have also, $\sigma(Q'_1)\times\cdots\times\sigma(Q'_n) \subset D_1\times\cdots\times D_n$ and $f \in \mathcal{O}(\bar{D}_1\times\cdots\times\bar{D}_n)$. Hence, by integration, from Lemmas 1 and 5 we easily obtain (for any positive integer p)

$$C(f(Q), f(Q'))T = \sum_{1\le|\alpha|\le p} \{C^\alpha(Q, Q')T\}\frac{1}{\alpha!}(\partial^\alpha f)(Q) + \mathcal{R}_p \tag{9}$$

where

$$\mathcal{R}_p := \frac{1}{(2\pi i)^n}\int_{\gamma_1}\cdots\int_{\gamma_n} f(\lambda_1, \ldots, \lambda_n)\left\{\sum_{1\le i_0\le\cdots\le i_p\le n}\left(\prod_{k=1}^{i_0} R(\lambda_k, Q_k)\right)\right.$$

$$\times\{C(Q_{i_0},Q'_{i_0})\cdots C(Q_{i_p},Q'_{i_p})T\}\left(\prod_{l=1}^{p} R(\lambda_l,Q'_{i_l})\right)$$

$$\times\left(\prod_{j=i_0}^{n} R(\lambda_j,Q'_j)\right)\Bigg\}\, d\lambda_n\cdots d\lambda_1. \tag{10}$$

Then by (7), (8) and (10) we deduce that for any $p\in\mathbb{N}$,

$$\|\mathcal{R}_p\|\le C\cdot U_n(p)\cdot\left(\frac{\eta}{\delta}\right)^p$$

where $U_n(p)$ denotes the number of terms from the sum which appears in (10) and C is a positive constant independent of p.

It is well known that

$$U_n(p)=\frac{n(n+1)\cdots(n+p)}{(p+1)!}=\frac{(n+p)!}{(p+1)!(n-1)!}=\frac{(p+n)\cdots(p+2)}{(n-1)!}.$$

Hence $U_n(\cdot)$ is a polynomial with the degree $n-1$. Since $\eta<\delta$, the above estimate of $\|\mathcal{R}_p\|$ shows that $\lim_{p\to\infty}\mathcal{R}_p=0$. So (1′) follows from (9) for $p\to\infty$. □

Remark 2. A similar proof is possible if we replace $\mathcal{B}(\mathcal{X})$, $\mathcal{B}(\mathcal{Y})$, $\mathcal{B}(\mathcal{X},\mathcal{Y})$ respectively by $\mathcal{B}_1$, $\mathcal{B}_2$, $\mathcal{B}$ a triplet of Banach spaces such that $\mathcal{B}_1,\mathcal{B}_2$ are Banach algebras and $\mathcal{B}$ is a $\mathcal{B}_1-\mathcal{B}_2$ bimodule such that

$$\|AT\|_{\mathcal{B}}\le\|T\|_{\mathcal{B}}\|A\|_{\mathcal{B}_1}\text{ for any }T\in\mathcal{B}\text{ and }A\in\mathcal{B}_1,$$

$$\|TB\|_{\mathcal{B}}\le\|T\|_{\mathcal{B}}\|B\|_{\mathcal{B}_2}\text{ for any }T\in\mathcal{B}\text{ and }B\in\mathcal{B}_2.$$

Obviously we can define all the elements of Theorem 1 ($C(A,B)T=AT-TB$ for $T\in\mathcal{B},A\in\mathcal{B}_1,B\in\mathcal{B}_2$ etc.) and the statement of Theorem 1 can be rewritten in an obvious manner.

Particularly we obtain an asymptotic formula for the commutators in a complex Banach algebra when $\mathcal{B}_1=\mathcal{B}_2=\mathcal{B}$ is acomplex Banach algebra. This result contains as a special case the situation $\mathcal{B}=\mathcal{B}(\mathcal{X})$:

Theorem 2. *Let us consider a complex Banach algebra $\mathcal{B}$ and $Q,T\in\mathcal{B}$. If $\rho\in[0,\infty)$ is such that*

$$\limsup_{n\to\infty}\|(ad\,Q)^nT\|^{\frac{1}{n}}\le\rho$$

and $K=\bar{B}(\sigma(Q),\rho),f\in\mathcal{O}(K)$, then the following formula holds:

$$(ad\,f(Q))T=\sum_{n=1}^{\infty}\frac{1}{n!}\{(ad\,Q)^nT\}f^{(n)}(Q)$$

$$=\sum_{n=1}^{\infty}\frac{(-1)^{n+1}}{n!}f^{(n)}(Q)\{(ad\,Q)^nT\}.$$

Corollary 4. *With the notation of the above theorem, if*

$$\lim_{n\to\infty} \|(ad\,Q)^n T\|^{\frac{1}{n}} = 0,$$

then the formula stated in that theorem hold for any $f \in \mathcal{O}(\sigma(Q))$. *Particularly, if* $(ad\,Q)^m T = 0$ *for a certain positive integer* m*, then we have*

$$(ad\,f(Q))T = \sum_{n=1}^{m-1} \frac{1}{n!}\{(ad\,Q)^n T\} f^{(n)}(Q)$$

$$= \sum_{n=1}^{m-1} \frac{(-1)^{n+1}}{n!} f^{(n)}(Q)\{(ad\,Q)^n T\}.$$

We are going to show that the hypothesis $\lim_{n\to\infty} \|(ad\,Q)^n T\|^{\frac{1}{n}} = 0$ is essential in the above corollary.

Example 1. We consider $\mathcal{X} = \mathbb{C}^2, \mathcal{B} = \mathcal{B}(\mathcal{X}), Q = \begin{pmatrix} b+\gamma & 0 \\ 0 & b \end{pmatrix}, T = \begin{pmatrix} 0 & 1 \\ 0 & 0 \end{pmatrix}$, $b, \gamma \in \mathbb{C}$ and $\gamma \neq 0$. In this case we have $(ad\,Q)T = \gamma T$ and the general term of the above series is

$$\frac{1}{n!}\{(ad\,Q)^n T\} f^{(n)}(Q) = \frac{f^{(n)}(b)}{n!}\gamma^n T \qquad n \in \mathbb{N}$$

if $f \in \mathcal{O}(\sigma(Q)) = \mathcal{O}(\{b, b+\gamma\})$. Hence we have successively: The above series is convergent ($\Longleftrightarrow \sum_{m=1}^{\infty} \frac{f^{(m)}(b)}{m!}\gamma^m$ is convergent) $\Longrightarrow f$ has an analytic extension to the open disk $B(b, |\gamma|)$. So the first series from Theorem 2 is divergent if f is a holomorphic function defined in a neighbourhood of the set $\{b, b+\gamma\}$ such that f cannot be analytically extended to $B(b, |\gamma|)$. On the other hand Q is even a decomposable operator (if b and γ are real numbers then it is even self-adjoint on a bi-dimensional Hilbert space) but

$$\limsup_{n\to\infty} \|(ad\,Q)^n T\|^{\frac{1}{n}} = |\gamma| > 0.$$

Remark 3. One can obtain a multi-dimensional variant of the above example. We can put $b = (b_1, \ldots, b_n) \in \mathbb{C}^n$, $\gamma = (\gamma_1, \ldots, \gamma_n) \in \mathbb{C}^n \setminus \{0\}$, $\mathcal{X} = \mathbb{C}^2$, $T = \begin{pmatrix} 0 & 1 \\ 0 & 0 \end{pmatrix}$, $Q_j = \begin{pmatrix} b_j+\gamma_j & 0 \\ 0 & b_j \end{pmatrix}$ and $Q = (Q_1, \ldots, Q_n)$. Then $\sigma(Q) = \{b, b+\gamma\}$ and for $f \in \mathcal{O}(\{b, b+\gamma\})$ we have

$$\frac{1}{\alpha!}\{(ad\,Q)^\alpha T\}(\partial^\alpha f)(Q) = \frac{1}{\alpha!}(\partial^\alpha f)(b)\gamma^\alpha T \quad (\alpha \in \mathbb{N}^n)$$

and

$$\limsup_{|\alpha|\to\infty} \|(ad\,Q)^\alpha T\|^{\frac{1}{|\alpha|}} = \max_{1\le j\le n} |\gamma_j|.$$

§ 16 Nilpotence criteria in an associative algebra

Let us consider an associative algebra $\mathcal{A}$ over the complex field $\mathbb{C}$ (or a field of characteristic 0). We denote as usual

$$(ad\,a)b = [a,b] = ab - ba \text{ for } a,b \in \mathcal{A}.$$

Proposition 1. *Let $D : \mathcal{A} \to \mathcal{A}$ be a derivation of $\mathcal{A}$, $\lambda \in \mathbb{C}$, $\lambda \neq 0$ and $a \in \mathcal{A}$ so that $Da = \lambda a$. If the set of all eigenvalues of $D|_{span\{a^n | n \in \mathbb{N}\}}$ is bounded, then a is nilpotent.*

Proof. We observe that $\lambda \neq 0$ and by induction

$$Da^n = n\lambda a^n, \quad n = 1,2,\ldots$$

because D is a derivation. □

Remark 1. Proposition 1 holds if $D = ad\,x$ for some $x \in \mathcal{A}$.

Definition 1. We say that $\mathcal{L} \subseteq \mathcal{A}$ has the *property* (m) if for every $x, a \in \mathcal{L}$ with $ad(span\{a^n \mid n \in \mathbb{N}\}) \subseteq span\{a^n \mid n \in \mathbb{N}\}$ the set of all eigenvalues of $ad\,x|_{span\{a^n | n \in \mathbb{N}\}}$ is bounded.

Remark 2. $\mathcal{A}$ has the property (m) in the following cases:

1. $\dim \mathcal{A} < \infty$;
2. $\mathcal{A} = \mathcal{B}(\mathcal{X})$, the associative algebra of all bounded linear operators on some complex Banach space $\mathcal{X}$.

Corollary 1. *Let $\mathcal{L} \subseteq \mathcal{A}$ be a Lie subalgebra of $\mathcal{A}$ such that $\mathcal{L}$ has the property (m). Then the following implication*

$$x, a \in \mathcal{L},\ [x,a] = \lambda a,\ \lambda \neq 0 \Rightarrow a \textit{ nilpotent}$$

is true.

Corollary 2. *If $X, A \in \mathcal{B}(\mathcal{X})$ are bounded operators on some complex Banach space $\mathcal{X}$ and $[X,A] = \lambda A$ for some $\lambda \neq 0$, then A is a nilpotent operator.*

Definition 2. For $a, b \in \mathcal{A}$ we say that b *polynomially commutes* with a if there exists a non-constant polynomial $p \in \mathbb{C}[X]$ so that $p(b)$ commutes with a, i.e.

$$[p(b), a] = 0.$$

The element b is called *polynomially central* in $\mathcal{M} \subseteq \mathcal{A}$ if b polynomially commutes with a for every $a \in \mathcal{M}$.

Proposition 2. *Let $x, q \in \mathcal{A}$. If $(ad\,q)^2 x = 0$ and q polynomially commutes with x, then $(ad\,q)x$ is nilpotent.*

Proof. We denote $a = [x, q]$. We have by induction $[x, q^n] = nq^{n-1}a$ for every positive integer n, because $[q, a] = 0$ and $ad\,x$ is a derivation of $\mathcal{A}$. Hence we can write

$$[x, p(q)] = p'(q)a \quad \text{for every } p \in \mathbb{C}[X]$$

where p' denotes the derivative of p.

But q polynomially commutes with x. Then there exists $p \in \mathbb{C}[X]$, $[x, p(q)] = 0$ and by the above equality we have

$$p'(q)a = 0. \tag{1}$$

If $p_1 \in \mathbb{C}[X]$, k is a positive integer and $p_1(q)a^k = 0$, then we can write

$$[x, p_1(q)a^k] = [x, p_1(q)]a^k + p_1(q)[x, a^k],$$

consequently

$$0 = p_1'(q)a^{k+1} + p_1(q)ma^k$$

We multiply the last equation at left by a^k and we have

$$p_1(q)a^k = 0 \Rightarrow p_1'(q)a^{2k+1} = 0$$

because $[a, q] = 0$. Hence we deduce

$$p'(q)a = 0 \Rightarrow p''(q)a^{2+1} = 0 \Rightarrow \cdots \Rightarrow n!a^{\sum_{i=0}^{n-1} 2^i} = 0$$

where n is the degree of the polynomial p. □

Corollary 3. *If $x, q \in \mathcal{A}$, $(ad\,q)^2x = 0$ and q is nilpotent, then $(ad\,q)x$ is nilpotent.*

Proof. Since q is nilpotent, it polynomially commutes with every element of $\mathcal{A}$. □

Corollary 4. *If $\mathcal{L}$ is a Lie subalgebra of $\mathcal{A}$ and C is an ideal of $\mathcal{L}$ such that*

$$\mathcal{N} := \{q \in C \mid q \text{ is nilpotent}\} \subseteq Z_C,$$

then $\mathcal{N}$ is an ideal of $\mathcal{L}$. (In particular if C is commutative then $\mathcal{N}$ is an ideal of $\mathcal{L}$.)

Corollary 5. *If $\mathcal{A}$ has finite dimension, $x, q \in \mathcal{A}$ and $(ad\,q)^2x = 0$, then $(ad\,q)x$ is nilpotent.*

Proof. It suffices to observe that in this case q polynomially commutes with every element of $\mathcal{A}$, since there exists a polynomial $p \in \mathbb{C}[X]$ such that $p(q) = 0$. □

§ 17 Quasinilpotence and nilpotence criteria in complex Banach algebras

In what follows $(\mathcal{U}, \|\cdot\|)$ will be a complex Banach algebra, i.e. it is a complex associative algebra endowed with a Banach space norm which is submultiplicative (that is, $\|xy\| \le \|x\| \cdot \|y\|$ for every $x, y \in \mathcal{U}$). Particularly, $\mathcal{U}$ may be $\mathcal{B}(\mathcal{X})$, the Banach (associative) algebra of all bounded linear operators on the complex Banach space $\mathcal{X}$.

We recall that a *quasinilpotent element* of $\mathcal{U}$ is an element $x \in \mathcal{U}$ which satisfies one of the following equivalent properties:

(q1) $\lim_{n\to\infty} \|x^n\|^{\frac{1}{n}} = 0$;

(q2) $\sigma(x) = \{0\}$.

Theorem 1. *Let D be a bounded derivation of $\mathcal{U}$, i.e. D is a bounded linear operator on $\mathcal{U}$ such that $D(ab) = (Da)b + a(Db)$ for every $a, b \in \mathcal{U}$. If $D^2x = 0$ for some $x \in \mathcal{U}$, then Dx is quasinilpotent.*

Proof. By the hypothesis we deduce

$$D^n(x^n) = n!(Dx)^n.$$

Obviously we can write

$$\|(Dx)^n\| \le \frac{1}{n!}\|D\|^n \|x\|^n$$

because D is a bounded linear operator on $\mathcal{U}$. Therefore

$$\|(Dx)^n\|^{\frac{1}{n}} \le \frac{1}{\sqrt[n]{n!}}\|D\| \cdot \|x\|.$$

Hence Dx is quasinilpotent. □

Remark 1. For $\mathcal{U} = \mathcal{B}(\mathcal{X})$ and $D = ad\,A$ for some $A \in \mathcal{B}(\mathcal{X})$, Theorem 1 is the Kleinecke-Shirokov Theorem.

Lemma 1. *Let $D : \mathcal{U} \to \mathcal{U}$ be a derivation of $\mathcal{U}$, $\lambda, \mu \in \mathbb{C}$ and $x, y \in \mathcal{U}$ be such that*

$$\lim_{n\to\infty} \|(D-\lambda)^n x\|^{\frac{1}{n}} = 0 \text{ and } \lim_{n\to\infty} \|(D-\mu)^n y\|^{\frac{1}{n}} = 0.$$

Then we have

$$\lim_{n\to\infty} \|(D-(\lambda+\mu))^n (xy)\|^{\frac{1}{n}} = 0.$$

Proof. First we observe that

$$(D-(\lambda+\mu))(xy) = [(D-\lambda)x]y + x[(D-\mu)y]$$

and by induction we deduce that

$$(D-(\lambda+\mu))^n(xy) = \sum_{j=0}^{n} \binom{n}{j} [(D-\lambda)^j x] \cdot [(D-\mu)^{n-j} y].$$

Now, let us consider $\varepsilon > 0$. By the hypothesis we can find $k_\varepsilon \in \mathbb{N}$ such that

$$\|(D-\lambda)^k\| \leq \varepsilon^k,\ \|(D-\mu)^k y\| \leq \varepsilon^k \text{ for every } k \geq k_\varepsilon.$$

Therefore there exists $M_\varepsilon \geq 1$ such that

$$\|(D-\lambda)^n\| \leq M_\varepsilon \varepsilon^n,\ \|(D-\mu)^n y\| \leq M_\varepsilon \varepsilon^n \text{ for every } n \in \mathbb{N}.$$

Then we can write

$$\|(D-(\lambda+\mu))^p(xy)\| \leq \sum_{j=0}^{p} \binom{p}{j} M_\varepsilon \varepsilon^j M_\varepsilon \varepsilon^{p-j} = M_\varepsilon^2 \varepsilon^p 2^p$$

for every $p \in \mathbb{N}$. Hence

$$\limsup_{p\to\infty} \|(D-(\lambda+\mu))^p(xy)\|^{\frac{1}{p}} \leq 2\varepsilon \text{ for every } \varepsilon > 0$$

and the lemma has been proved. □

The following theorem is an easy consequence of Lemma 1.

Theorem 2. *Let $\lambda, \mu \in \mathbb{C}$ be two locally approximate eigenvalues of some derivation $D : \mathcal{U} \to \mathcal{U}$, corresponding to $0 \neq x_\lambda \in \mathcal{U}, 0 \neq x_\mu \in \mathcal{U}$. Then one of the following assertions is true:*

1) $x_\lambda x_\mu = 0$
or
2) $x_\mu x_\lambda = 0$
or
3) $x_\lambda x_\mu \neq 0$ *and* $\lambda + \mu$ *is a local approximate eigenvalue of* D *corresponding to* $x_\lambda x_\mu$.

Corollary 1 (Nilpotence criterion). *Let λ be a locally approximate eigenvalue of some derivation $D : \mathcal{U} \to \mathcal{U}$ and $x_\lambda \neq 0$ the corresponding approximate eigenvector, i.e.*

$$\lim_{n\to\infty} \|(D-\lambda)^n x_\lambda\|^{\frac{1}{n}} = 0.$$

If D is bounded (as a linear operator) and $\lambda \neq 0$ then x_λ is nilpotent.

Proof. By Theorem 2, $k\lambda$ is a local approximate eigenvalue of D corresponding to $(x_\lambda)^k$ if $(x_\lambda)^k \neq 0$. Particularly $k\lambda \in \sigma(D)$ if $(x_\lambda)^k \neq 0$. But $\sigma(D)$ is a compact subset because D is bounded. Then there exists $k \in \mathbb{N}$ such that $(x_\lambda)^k = 0$ because it is imposible that $k\lambda \in \sigma(D)$ for every $k \in \mathbb{N}$. □

Corollary 2. *Let D be a bounded derivation of $\mathcal{U}$ and $0 \neq x_\lambda \in \mathcal{U}$, $0 \neq x_\mu \in \mathcal{U}$ two approximate eigenvectors of D corresponding to the locally approximate eigenvalues λ and μ.*

If $\lambda + \mu \neq 0$ then $x_\lambda x_\mu$, $x_\mu x_\lambda$ and $[x_\lambda, x_\mu] = x_\lambda x_\mu - x_\mu x_\lambda$ are nilpotent elements.

If $\lambda \neq 0$ then

$$\mathcal{U}_\lambda := \{x \in \mathcal{U} \mid \lim_{n\to\infty} \|(D-\lambda)^n x\|^{\frac{1}{n}} = 0\}$$

is a linear subspace of $\mathcal{U}$ consisting of nilpotent elements. $\mathcal{U}_\lambda$ is invariant for every bounded linear operator on $\mathcal{U}$ commuting with D.

The above results can be rewritten for $D = ad\,t$ ($t \in \mathcal{U}$), which is obviously a bounded derivation of $\mathcal{U}$. For instance we have the following theorem:

Theorem 3. *Let us consider $t \in \mathcal{U}$,*

$$ad\,t : \mathcal{U} \to \mathcal{U},\ (ad\,t)x = tx - xt,$$

$\lambda \in \mathbb{C}$, $x_\lambda \in \mathcal{U}$ such that

$$\lim_{n\to\infty} \|(ad\,t - \lambda)^n x_\lambda\|^{\frac{1}{n}} = 0.$$

If $\lambda \neq 0$, then x_λ is nilpotent and

$$\{x \in \mathcal{U} \mid \lim_{n\to\infty} \|(ad\,t-\lambda)^n x\|^{\frac{1}{n}} = 0\}$$

is a linear subspace of $\mathcal{U}$.

If $\mu \in \mathbb{C}$ and

$$\lim_{n\to\infty} \|(ad\,t - \mu)^n x_\mu\|^{\frac{1}{n}} = 0,$$

then

$$\lim_{n\to\infty} \|(ad\,t - (\lambda+\mu))^n (x_\lambda x_\mu)\|^{\frac{1}{n}} = 0.$$

In the case when $\mathcal{U} = \mathcal{B}(\mathcal{X})$ we can also rewrite the above results for $ad\,T$, where T is a bounded operator on the complex Banach space $\mathcal{X}$.

Theorem 4. *If T, Q are bounded operators on $\mathcal{X}$ and λ is a complex number different from zero such that*

$$\lim_{n\to\infty} \|(ad\,T - \lambda)^n Q\|^{\frac{1}{n}} = 0,$$

then Q is a nilpotent operator on $\mathcal{X}$. Moreover

$$\mathcal{N}_\lambda := \{Q \in \mathcal{B}(\mathcal{X}) \mid \lim_{n\to\infty} \|(ad\,T - \lambda)^n Q\|^{\frac{1}{n}} = 0\}$$

is a linear subspace of $\mathcal{B}(\mathcal{X})$ which is invariant to any bounded linear operator on $\mathcal{B}(\mathcal{X})$ commuting with $ad\,T$.

We can also give a nilpotence criterion in the Banach algebra $\mathcal{B}(\mathcal{X})$ if we replace polynomial commutativity by a weaker condition, namely analytic commutativity.

Definition 1. For $A, B \in \mathcal{B}(\mathcal{X})$ we say that B *analytically commutes* with A if there exists $f \in \mathcal{O}(\sigma(B))$, f non-constant on every connected open set D with

$D \cap \sigma(B) \neq \emptyset$, so that $[A, f(B)] = Af(B) - f(B)A = 0$. The operator B is called *analytically central* in $\mathcal{M} \subseteq \mathcal{B}(\mathcal{X})$ if B polynomially commutes with A for every $a \in \mathcal{M}$.

Theorem 5. *Let $X, Q \in \mathcal{B}(\mathcal{X})$ be such that $(\operatorname{ad} Q)^2 X = 0$. If Q analytically commutes with X, then $(\operatorname{ad} Q)X$ is nilpotent.*

The proof of Theorem 5 is an easy consequence of Lemmas 2 and 3 below.

Lemma 2. *Let $X, Q \in \mathcal{B}(\mathcal{X})$ be such that $(\operatorname{ad} Q)^2 X = 0$. Then*

$$[X, f(Q)] = f'(Q)[X, Q]$$

for every $f \in \mathcal{O}(\sigma(Q))$.

Proof. It is an easy consequence of the asymptotic formula for the commutators (see Corollary 3 from § 15). □

Lemma 3. *Let $X, Q \in \mathcal{B}(\mathcal{X})$ be such that $(\operatorname{ad} Q)^2 X = 0$. If there exists an open set $V \supset \sigma(Q)$ and $f \in \mathcal{O}(V)$ such that $[X, f(Q)] = 0$, then the following statement holds:*

> *if W is a connected component of of V such that $f|_W$ is non constant, then for every spectral set σ of Q, $\sigma \subset W$ (particularly $\sigma = W \cap \sigma(Q)$), the operator $\{(\operatorname{ad} X)Q\}|_{E(\sigma)\mathcal{X}}$ is nilpotent.*

Proof. Let σ be a spectral set for Q. We recall that this means that σ is a subset of $\sigma(Q)$ such that there exists an analytic function e_σ defined in a neighbourhood of $\sigma(Q)$ with $e_\sigma|_\sigma \equiv 1$ and $e_\sigma|_{\sigma(Q)\setminus\sigma} \equiv 0$. We denote by

$$E(\sigma) = e_\sigma(Q)$$

the spectral projection of Q associated to σ by the analytic functional calculus of Q.

By Lemma 2 we have

$$[X, E(\sigma)] = e'_\sigma(Q) = 0.$$

Let us denote

$$A := [X, Q]$$

hence $[A, Q] = 0$ by the hypothesis. Then it is well known that $E(\sigma)$ commutes with A and with every value of the analytic functional calculus of Q. Therefore $E(\sigma)\mathcal{X} =: \mathcal{X}_\sigma$ is invariant to X, A and $\varphi(Q)$ for every $\varphi \in \mathcal{O}(\sigma(Q))$.

Now we consider $V \supset \sigma(Q)$, $f \in \mathcal{O}(V)$, $[X, f(Q)] = 0$. We consider also a connected component W of V such that $f|_W$ is not constant and $\sigma = \sigma(Q) \cap W$.

Let W_1 be a relatively compact open set such that

$$W \supset W_1 \supset \sigma = \sigma(Q) \cap W.$$

Obviously the derivative f' has in W_1 a finite number of zeros because f is not constant on W. Therefore $f'|_{W_1} = p \cdot g$, where $g \in \mathcal{O}(W_1)$, $g(z) \neq 0$ for every $z \in W_1$ and $p \in \mathbb{C}[X]$ is a polynomial. Using the above introduced notation $A = [X, Q]$, by Lemma 2 we have

$$0 = [X, f(Q)] = f'(Q)A.$$

But it is well known that

$$f'(Q)|_{\mathcal{X}_\sigma} = f'(Q|_{\mathcal{X}_\sigma}) = g(Q|_{\mathcal{X}_\sigma}) \cdot p(Q|_{\mathcal{X}_\sigma}).$$

Hence

$$0 = f'(Q)A|_{\mathcal{X}_\sigma} = g(Q|_{\mathcal{X}_\sigma})p(Q|_{\mathcal{X}_\sigma})A|_{\mathcal{X}_\sigma}.$$

On the other hand $g(Q|_{\mathcal{X}_\sigma})$ is invertible because $g(z) \neq 0$ for every $z \in \sigma = \sigma(Q|_{\mathcal{X}_\sigma})$. Therefore we have

$$p(Q|_{\mathcal{X}_\sigma})A|_{\mathcal{X}_\sigma} = 0,\ [X|_{\mathcal{X}_\sigma}, Q|_{\mathcal{X}_\sigma}] = A|_{\mathcal{X}_\sigma},\ [A|_{\mathcal{X}_\sigma}, Q|_{\mathcal{X}_\sigma}] = 0.$$

As in the last part of the proof of the nilpotence criterion in an associative algebra (see the proof of Proposition 2 from § 16) we deduce that $A|_{\mathcal{X}_\sigma}$ is nilpotent and $(ad\,X)Q|_{\mathcal{X}_\sigma} = -A|_{\mathcal{X}_\sigma}$ is also nilpotent. □

Now the proof of Theorem 5 is obvious.

§ 18 Nilpotent elements in LM-decomposable Lie subalgebras of an associative algebra

Let us consider an associative algebra $\mathcal{A}$. The property of polynomial commutativity is related to existence of the nilpotent elements as follows.

Theorem 1. *Let $\mathcal{L}$ be a Lie subalgebra of $\mathcal{A}$ and $Z_{\mathcal{L}}$ be the center of $\mathcal{L}$ (i.e. $Z_{\mathcal{L}} = \{z \in \mathcal{L} | [z, x] = 0$ for every $x \in \mathcal{L}\}$). If $\mathcal{I}$ is a finite-dimensional ideal of $\mathcal{L}$ such that $ad\,x|_{\mathcal{I}}$ is nilpotent for every $x \in \mathcal{L}$, then one of the following assertions is true:*

(*i*) *$[q, \mathcal{I}] = \{0\}$ for every polynomially central element q in $\mathcal{L}$;*
(*ii*) *there exists $q \neq 0$, q nilpotent, $q \in \mathcal{I} \cap Z_{\mathcal{L}}$.*

Proof. Using Engel's Theorem we can find a chain

$$\{0\} = \mathcal{I}_0 \subset \mathcal{I}_1 \subset \cdots \subset \mathcal{I}_k \subset \cdots \subset \mathcal{I}_n = \mathcal{I}$$

$\mathcal{I}_k$ ideal in $\mathcal{L}$, dim $\mathcal{I}_k = k$, $(ad\,x)\mathcal{I}_k \subseteq \mathcal{I}_{k-1}$ for every $k = 1, \ldots, n$ and $x \in \mathcal{L}$. Obviously $(ad\,x)\mathcal{I}_1 = \{0\}$ for every $x \in \mathcal{L}$ and we have either

(*i*) $[q, \mathcal{I}_1] = \{0\}$ for every polynomially central element q in $\mathcal{L}$,
or

(ii) there exists $q \neq 0$, q nilpotent, $q \in \mathcal{I}_1 \cap Z_{\mathcal{L}}$.

Now we suppose that for $j \leq k$ either (i) or (ii) holds for $\mathcal{I}_j$, i.e. either

(i) $[q, \mathcal{I}_j] = \{0\}$ for every polynomially central q in $\mathcal{L}$

or

(ii) there exists $q \neq 0$, q nilpotent, $q \in \mathcal{I}_j \cap Z_{\mathcal{L}}$.

Obviously if (ii) holds for $\mathcal{I}_k$, then (ii) holds for $\mathcal{I}$. It remains to study the case when (i) holds for $\mathcal{I}_k$. In this case let a_{k+1} be so that $\mathcal{I}_{k+1} = \mathcal{I}_k + \mathbb{C}a_{k+1}$.

We can write

$$[q, a_{k+1}] = q_k \in \mathcal{I}_k \text{ for every } q \in \mathcal{L},$$

because $[q, \mathcal{I}_{k+1}] \subset \mathcal{I}_k$. If q is polynomially central in $\mathcal{L}$ then $[q, q_k] = 0$ because $\mathcal{I}_k$ verifies (i).

Obviously we have either

(1) $[q, a_{k+1}] = 0$ for every polynomially central q in $\mathcal{L}$

or

(2) there exists q polynomially central in $\mathcal{L}$ so that $[q, a_{k+1}] = q_k \neq 0$.

If we have (1), then (i) holds for $\mathcal{I}_{k+1}$ because $\mathcal{I}_{k+1} = \mathcal{I}_k + \mathbb{C}a_{k+1}$ and $\mathcal{I}_k$ verifies (i).

If we have (2), then there exists q polynomially central in $\mathcal{L}$ with

$$0 \neq [q, a_{k+1}] = q_k \in \mathcal{I}_k, \ [q, q_k] = 0$$

and by the nilpotence criterion given by Proposition 2 from § 16 we deduce that q_k is nilpotent. Hence in the case (2) there exists $q_k \in \mathcal{I}_k$, $q_k \neq 0$, q_k nilpotent. For any $x \in \mathcal{L}$ we have $[x, q_k] = q_{k-1} \in \mathcal{I}_{k-1}$ and one of the following assertions is true:

α) $[x, q_k] = 0$ for every $x \in \mathcal{L}$, i.e. (ii) is true for $\mathcal{I}_k$, hence also for $\mathcal{I}$,

or

β) there exists $x \in \mathcal{L}$, $[x, q_k] = q_{k-1} \neq 0$, q_k nilpotent.

But $\mathcal{I}_k$ verifies (i), hence also $\mathcal{I}_{k-1}$ verifies (i) because $\mathcal{I}_{k-1} \subset \mathcal{I}_k$. We have $[q_k, q_{k-1}] = 0$ because q_k is nilpotent (hence polynomially central in $\mathcal{L}$). By β) and by the nilpotence criterion given by Corollary 3 from § 16, we deduce that q_{k-1} is nilpotent, $0 \neq q_{k-1} \in \mathcal{I}_{k-1}$ and we can repeat the proof with α) and β) for q_{k-1} etc. Clearly we obtain that (ii) holds for some $j \in \{1, \ldots, k\}$. Hence, in the case (2), (ii) holds for $\mathcal{I}$. □

Corollary 1. *Let $\mathcal{L}$ be a quasinilpotent Lie subalgebra of $\mathcal{A}$ (i.e. $\mathcal{L} = \sum_\alpha \mathcal{I}_\alpha$, where each $\mathcal{I}_\alpha$ is a finite-dimensional nilpotent ideal of $\mathcal{L}$). If $Z_{\mathcal{L}}$ (the center of $\mathcal{L}$) contains no nilpotent elements, then every polynomially central element in $\mathcal{L}$ is central in $\mathcal{L}$.*

Proof. We apply Theorem 1 for $\mathcal{I} = \mathcal{I}_\alpha$. □

We can prove also an analogue of Theorem 1 if the adjoint representation of $\mathcal{L}$ restricted to an ideal $\mathcal{I}$ is quasisolvable and the algebra $\mathcal{L}$ has the property (m). We have the following theorem.

Theorem 2. *Let $\mathcal{L} \subseteq \mathcal{A}$ be a Lie subalgebra with the property (m). If $N \in \mathbb{N} \cup \{\aleph_0\}$ and*

$$\mathcal{I}_1 \subset \mathcal{I}_2 \subset \cdots \subset \mathcal{I}_n \subset \cdots \subset \mathcal{L}$$

is an increasing chain of ideals in $\mathcal{L}$ such that $\dim \mathcal{I}_n = n$ *for any $1 \leq n < N$, then for every n one of the following assertions holds:*

(i) $[q, \mathcal{I}_n] = \{0\}$ *for every polynomially central element q in $\mathcal{L}$;*
(ii) there exists $q \neq 0$, q nilpotent, $q \in \mathcal{I}_n \cap Z_{\mathcal{L}}$;
(iii) there exists a commutative ideal $\mathcal{N}$ of $\mathcal{L}$, $\{0\} \neq \mathcal{N} \subseteq [\mathcal{L}, \mathcal{I}_n]$, such that every $a \in \mathcal{N}$ is nilpotent.

Proof. We will denote by (γ) the following assertion:

(γ) $ad x|_{\mathcal{I}_n}$ is nilpotent for every $x \in \mathcal{L}$.

We observe that one of $(\gamma), (ii), (iii)$ true for $\mathcal{I}_n$ implies by Theorem 1 for $\mathcal{I}_n$ one of $(i), (ii), (iii)$. We shall prove by induction that for every n one of $(\gamma), (ii), (iii)$ holds for $\mathcal{I}_n$.

We have $\mathcal{I}_1 = \mathbb{C}a_1$, $a_1 \in \mathcal{L}$, $[x, a_1] = \lambda(x)a_1$, $\lambda(x) \in \mathbb{C}$ for every $x \in \mathcal{L}$. If $\lambda(x) = 0$ for every $x \in \mathcal{L}$, then (γ) holds for $\mathcal{I}_1$. If $\lambda(x) \neq 0$ for some $x \in \mathcal{L}$ then by the nilpotence criterion for associative algebra with the property (m) (see Corollary 1 from § 16), we deduce that $\mathcal{N} = \mathcal{I}_1$ verifies (iii).

Now let suppose that one of $(\gamma), (ii), (iii)$ holds for $\mathcal{I}_n$. The same fact will be proved for $\mathcal{I}_{n+1}$. Obviously (ii) or (iii) for $\mathcal{I}_n$ implies (ii) or (iii) for $\mathcal{I}_{n+1}$. Hence it remains to be studied the case when (γ) holds for $\mathcal{I}_n$, (ii) and (iii) are not true for any $k \leq n$.

Let $a_{n+1} \in \mathcal{L}$ be such that $\mathcal{I}_{n+1} = \mathcal{I}_n + \mathbb{C}a_{n+1}$. For any $x \in \mathcal{L}$ we can write

$$[x, a_{n+1}] = b_n + \lambda_{n+1}(x)a_{n+1}, \; b_n \in \mathcal{I}_n, \; \lambda_{n+1}(x) \in \mathbb{C}.$$

From (γ) we deduce $(ad x)^m|_{\mathcal{I}_n} = 0$ and

$$(ad x)^{m+1} a_{n+1} = \lambda_{n+1}(x)(ad x)^m a_{n+1}$$

for a certain positive integer m. If for every $x \in \mathcal{L}$ we have either $\lambda_{n+1}(x) = 0$ or $(ad x)^m a_{n+1} \in \mathcal{I}_n$, then (γ) holds for $\mathcal{I}_{n+1}$. The following case remains to be studied:

$$\exists x_0 \in \mathcal{L}, \; 0 \neq (ad x_0)^{m+1} a_{n+1} = \lambda_{n+1}(x_0)(ad x_0)^m a_{n+1} \notin \mathcal{I}_n.$$

Hence $a'_{n+1} := (ad x_0)^m a_{n+1} \neq 0$ is nilpotent (see Corollary 1 from § 16) and $a'_{n+1} \notin \mathcal{I}_n$. But (iii) is not true for $\mathcal{I}_n$ and $ad x|_{\mathcal{I}_n}$ is nilpotent for every $x \in \mathcal{L}$; by Theorem 1 we have

$$q \text{ polynomially central in } \mathcal{L} \Rightarrow q \text{ commutes with } \mathcal{I}_n.$$

Particularly,

(*) $q \in \mathcal{L}$, q nilpotent $\Rightarrow q$ commutes with $\mathcal{I}_n$.

Hence a'_{n+1} commutes with $\mathcal{I}_n$ and a'_{n+1} commutes with $\mathcal{I}_{n+1}$, because $a'_{n+1} \in \mathcal{I}_{n+1} \backslash \mathcal{I}_n$ implies $\mathcal{I}_{n+1} = \mathcal{I}_n + \mathbb{C}a'_{n+1}$. We can write

$$q \in \mathcal{I}_n,\ q \text{ nilpotent } \Rightarrow q \text{ commutes with } \mathcal{I}_{n+1}.$$

By (*) we deduce:

$$q \in \mathcal{I}_{n+1},\ q \text{ nilpotent } \Rightarrow q \text{ commutes with } \mathcal{I}_{n+1}.$$

Hence

$$0 \neq a'_{n+1} \in \mathcal{N} := \{q \mid q \in [\mathcal{L}, \mathcal{I}_{n+1}],\ q \text{ nilpotent}\} \subseteq Z_{\mathcal{I}_{n+1}}.$$

If $u \in \mathcal{N}$, $x \in \mathcal{L}$ we have $[x,u] = u' \in \mathcal{I}_{n+1}$ and $[u,u'] = 0$. By Corollary 3 from § 16, u' will be nilpotent. We deduce that $\mathcal{N}$ is a commutative ideal of the Lie algebra $\mathcal{L}$, because $[\mathcal{L}, \mathcal{I}_{n+1}]$ is an ideal of $\mathcal{L}$. Hence $\mathcal{N}$ is a commutative ideal of $\mathcal{L}$ consisting of nilpotent elements, $\{0\} \neq \mathcal{N} \subseteq [\mathcal{L}, \mathcal{I}_{n+1}]$, and (iii) holds for $\mathcal{I}_{n+1}$. □

We obtain also the following result.

Theorem 3. *For the same conditions as in Theorem 2, one of the following statements is true:*

(γ) *$ad\,x|_{\mathcal{I}_n}$ is nilpotent for every $x \in \mathcal{L}$;*

(ii) *there exists $q \neq 0$, q nilpotent, $q \in \mathcal{I}_n \cap Z_{\mathcal{L}}$;*

(iii) *there exists a commutative ideal $\mathcal{N}$ of $\mathcal{L}$, $\{0\} \neq \mathcal{N} \subseteq [\mathcal{L}, \mathcal{I}_n]$, such that every $a \in \mathcal{N}$ is nilpotent.*

Definition 1. Let $\mathcal{L}$ be a Lie subalgebra of the associative algebra $\mathcal{A}$. A *nil-ideal* of $\mathcal{L}$ is an ideal consisting of nilpotent elements.

Corollary 2. *Let $\mathcal{R}$ be a quasisolvable Lie subalgebra of $\mathcal{A}$ having the property (m). If $\mathcal{R}$ contains no nonzero nil-ideal, then every polynomially central element in $\mathcal{R}$ is central.*

Corollary 3. *Let $\mathcal{R}$ be a quasisolvable Lie subalgebra of $\mathcal{A}$ having the property (m). If $\mathcal{R}$ contains no nonzero nil-ideal, then $\mathcal{R}$ is quasinilpotent (i.e. $\mathcal{R} = \sum_\alpha \mathcal{I}_\alpha$, where each $\mathcal{I}_\alpha$ is a finite-dimensional nilpotent ideal in $\mathcal{R}$).*

An analogue of this result can be proved for an LM-decomposable Lie subalgebra $\mathcal{U}$ of $\mathcal{A}$ having the property (m) (see Definition 3 from § 4).

Let us consider $\mathcal{U} = \mathcal{R} + \mathcal{G}$, $\mathcal{R} = \sum_{\alpha \in \Lambda} \mathcal{I}_\alpha$, where $\mathcal{I}_\alpha$ are finite-dimensional solvable ideals of $\mathcal{U}$ and $\mathcal{G}$ is a Lie algebra so that every ideal of $\mathcal{G}$ is primitive, the decomposition of $\mathcal{U}$ given by the property of LM-decomposability.

We will denote

$$\mathcal{U}_\alpha = ad\,\mathcal{U}|_{\mathcal{I}_\alpha} = \{ad\,x|_{\mathcal{I}_\alpha} \mid x \in \mathcal{U}\}$$

and

$$\mathcal{R}_\alpha = ad\,\mathcal{R}|_{\mathcal{I}_\alpha} = \{ad\,x|_{\mathcal{I}_\alpha} \mid x \in \mathcal{R}\}.$$

Because dim $\mathcal{U}_\alpha < \infty$, dim $\mathcal{R}_\alpha < \infty$ and $\mathcal{R}_\alpha$ is a solvable ideal of $\mathcal{U}_\alpha$, it is well known (see Corollary 2 Theorem 8 sect. 5 from Chapter II of the book of N. Jacobson [2]) that the nil-radical of the associative algebra generated by $\mathcal{U}_\alpha$ contains $[\mathcal{U}_\alpha, \mathcal{R}_\alpha]$. Then $[\mathcal{I}_\alpha, \mathcal{U}]$ is a finite-dimensional nilpotent ideal of $\mathcal{U}$. We have

$$\mathcal{B}_\alpha = [\mathcal{I}_\alpha, \mathcal{U}] \subseteq \mathcal{I}_\alpha$$

and the representation of $\mathcal{G}$ defined by

$$g \mapsto ad\, g|_{\mathcal{B}_\alpha}$$

is semisimple because every ideal of $\mathcal{G}$ is primitive. By Weyl's Theorem, $\mathcal{B}_\alpha$ splits into the direct sum

$$\mathcal{B}_\alpha = \mathcal{B}_\alpha^0 \oplus \mathcal{B}_\alpha^1$$

with $\mathcal{B}_\alpha^0, \mathcal{B}_\alpha^1$ vector subspaces which are invariant to $ad\, g$ for any $g \in \mathcal{G}$, $ad\, g|_{\mathcal{B}_\alpha^0} = 0$ for any $g \in \mathcal{G}$ and $\bigcap_{g \in \mathcal{G}} Ker(ad\, g|_{\mathcal{B}_\alpha^1}) = \{0\}$. If Z_α denotes the center of $\mathcal{B}_\alpha$, then by Theorem 1 we deduce that one of the following assertions holds:

1. $q \in Z_\alpha$ for every element q polynomially central in $\mathcal{B}_\alpha$;
2. There exists q nilpotent with $0 \neq q \in Z_\alpha$.

We discuss the following two cases : $\mathcal{B}_\alpha^1 \neq \{0\}$, $\mathcal{B}_\alpha^1 = \{0\}$.

If $\mathcal{B}_\alpha^1 \neq \{0\}$, by the nilpotence criterion in an associative algebra with the property (m) (see Corollary 1 from § 16) we deduce that $\mathcal{B}_\alpha^1$ contains non-zero nilpotent elements because $\mathcal{U}$ has the property (m). In this case each of the above assertions 1 and 2 proves that there exists non-zero nilpotent elements in the center Z_α of $\mathcal{B}_\alpha$. Then

$$\mathcal{N}_\alpha := \{q \in Z_\alpha \mid q \text{ is nilpotent}\} \neq \{0\}$$

is an ideal in $\mathcal{U}$. This is an easy consequence of Corollary 4 from § 16 because Z_α is an ideal in $\mathcal{U}$. Therefore $\mathcal{B}_\alpha^1 \neq \{0\}$ implies the following assertion:

(N) There exists a non-zero commutative finite-dimensional ideal of $\mathcal{U}$ consisting of nilpotent elements.

If $\mathcal{B}_\alpha^1 = \{0\}$, we have $\mathcal{B}_\alpha = \mathcal{B}_\alpha^0$ and

$$[\mathcal{G}, [\mathcal{U}, \mathcal{I}_\alpha]] = \{0\}.$$

Since $\mathcal{R}$ is quasisolvable and $\mathcal{U}$ has the property (m), Theorem 2 implies that for every $\alpha \in \Lambda$ one of the following assertions is true:

a) $ad\, x|_{\mathcal{I}_\alpha}$ is nilpotent for every $x \in \mathcal{R}$;
b) there exists a commutative nil-ideal $\mathcal{N}$ of $\mathcal{R}$, $\{0\} \neq \mathcal{N} \subseteq [\mathcal{R}, \mathcal{I}_\alpha]$;
c) there exists $q \neq 0$, q nilpotent, $q \in Z_{\mathcal{R}} \cap \mathcal{I}_\alpha$.

By Theorem 1 we deduce in the case *a*) one of the following assertions:

a_1) $[q, \mathcal{I}_\alpha] = \{0\}$ for every polynomially central q in $\mathcal{R}$;
a_2) there exists $q \neq 0$, q nilpotent, $q \in Z_{\mathcal{R}} \cap \mathcal{I}_\alpha$.

Therefore, if $\mathcal{B}_\alpha^1 = \{0\}$ then one of a_1), *b*), *c*) is true. Obviously *b*) $\Rightarrow$ (N) because $\mathcal{B}_\alpha^1 = \{0\}$ shows that $\mathcal{N}$ is an ideal of $\mathcal{U}$.

Also we have $c) \Rightarrow$ (N) because

$$\{0\} \neq \{q \mid q \text{ nilpotent}, q \in Z_{\mathcal{I}_\alpha}\}$$

is a finite-dimensional ideal in $\mathcal{U}$ by Corollary 4 from § 16 as above.

Hence there exist two possibilities:

A) the assertion (N) is true;

B) $[\mathcal{G}, [\mathcal{U}, \mathcal{I}_\alpha]] = \{0\}$, $ad\, x|_{\mathcal{I}_\alpha}$ is nilpotent for every $x \in \mathcal{R}, \alpha \in \Lambda$ and $\mathcal{R}$ verifies a_1).

In case *B*) we can eliminate the situation when $\mathcal{R}$ contains non-zero nilpotent elements. Indeed, by a_1) for any $\alpha \in \Lambda$ we can find $\mathcal{I} = \mathcal{I}_{\alpha_1} + \cdots + \mathcal{I}_{\alpha_p}$, a finite-dimensional ideal of $\mathcal{U}$ so that $Z_{\mathcal{R}} \cap \mathcal{I}$ contains nonzero nilpotent elements and by Corollary 4 from § 16,

$$\{0\} \neq \{q \mid q \text{ nilpotent}, q \in Z_{\mathcal{I}}\} =: \mathcal{N}$$

verifies (N).

If $\mathcal{R}$ contains no non-zero nilpotent elements, then we obtain $[\mathcal{G}, \mathcal{I}_\alpha] = \{0\}$ for any $\alpha \in \Lambda$. Indeed, by the semisimplicity of the Lie algebra $ad\mathcal{G}|_{\mathcal{I}_\alpha}$ we can write $\mathcal{I}_\alpha = \mathcal{I}_\alpha^0 \oplus \mathcal{I}_\alpha^1$ with $ad\mathcal{G}|_{\mathcal{I}_\alpha^0} = \{0\}$ and $\bigcap_{g \in \mathcal{G}} Ker(ad\, g|_{\mathcal{I}_\alpha^1}) = \{0\}$. By Corollary 1 from § 16 we deduce $\mathcal{I}_\alpha^1 = \{0\}$ so $[\mathcal{G}, \mathcal{I}_\alpha] = \{0\}$.

We have proved the following result:

Theorem 4. *Let $\mathcal{U}$ be an LM-decomposable Lie subalgebra of $\mathcal{A}$ having the property (m). If $\mathcal{U} = \mathcal{R} + \mathcal{G}$ is the decomposition of $\mathcal{U}$ (cf. Definition 3 from § 4) then one of the following statements is true:*

(I) *there exists a finite-dimensional commutative non-zero ideal $\mathcal{N}$ of $\mathcal{U}$ consisting of nilpotent elements;*

(II) *$[\mathcal{G}, \mathcal{R}] = \{0\}$, $\mathcal{R}$ is a quasinilpotent Lie algebra, $\mathcal{R}$ contains no non-zero nilpotent element and every polynomially central element in $\mathcal{R}$ is central in $\mathcal{R}$.*

Corollary 4. *Let $\mathcal{U}$ be a finite-dimensional or ideally finite Lie subalgebra of $\mathcal{A}$ and $\mathcal{U} = \mathcal{R} + \mathcal{G}$ its Levi-Malçev decomposition. If $\mathcal{U}$ contains no non-zero commutative nil-ideals, then $[\mathcal{G}, \mathcal{R}] = \{0\}$, $\mathcal{R}$ is a quasinilpotent Lie algebra which contains no non-zero nilpotent elements and every polynomially central element in $\mathcal{R}$ is central in $\mathcal{R}$.*

Proof. It suffices to recall that every ideally finite or finite-dimensional Lie algebra is LM-decomposable. □

§ 19 Nilpotent elements in LM-decomposable Lie algebras of bounded linear operators

In that which follows we extend the results of the preceding paragraph to Lie subalgebras of $\mathcal{B}(\mathcal{X})$, the associative algebra of the bounded linear operators on the complex Banach space $\mathcal{X}$, when we replace the polynomial commutativity by analytic commutativity.

The results of § 18 can be extended to a Lie subalgebra $\mathcal{L}$ of $\mathcal{B}(\mathcal{X})$ when we replace "polynomial commutativity" by "analytic commutativity" and "polynomially central" by "analytically central", by using the same proofs where we replace Proposition 2 from § 16 with Theorem 5 from § 17 and Corollary 1 from § 16 with Corollary 2 from § 16. One uses also the fact that $\mathcal{B}(\mathcal{X})$, hence every of its Lie subalgebras, has the property (m). Hence we will give the following results without proof.

Theorem 1. *Let $\mathcal{L}$ be a Lie subalgebra of $\mathcal{B}(\mathcal{X})$ with the center $Z_{\mathcal{L}}$. If $\mathcal{I}$ is an ideal of $\mathcal{L}$, $\dim \mathcal{I} < \infty$ and $ad x|_{\mathcal{I}}$ is nilpotent for every $x \in \mathcal{L}$, then one of the following statements is true:*

(*i*) *$[Q, \mathcal{I}] = \{0\}$ for every Q analytically central in $\mathcal{L}$.*

(*ii*) *There exists $Q \neq 0$, Q nilpotent, $Q \in \mathcal{I} \cap Z_{\mathcal{L}}$.*

Theorem 2. *Let $\mathcal{L} \subset \mathcal{B}(\mathcal{X})$ be a Lie algebra of bounded operators. If $N \in \mathbb{N} \cup \{\aleph_0\}$ and*

$$\mathcal{I}_1 \subset \mathcal{I}_2 \subset \cdots \subset \mathcal{I}_n \subset \cdots$$

is an increasing chain of ideals in $\mathcal{L}$ with $\dim \mathcal{I}_n = n$ for $1 \leq n < N$, then for every n one of the following statements holds:

(*i*) *$ad X|_{\mathcal{I}_n}$ is nilpotent for every $X \in \mathcal{L}$ and $[Q, \mathcal{I}_n] = \{0\}$ for every Q analytically central in $\mathcal{L}$.*

(*ii*) *There exists $Q \neq 0$, Q nilpotent, $Q \in \mathcal{I}_n \cap Z_{\mathcal{L}}$.*

(*iii*) *There exists a commutative ideal $\mathcal{N}$ of $\mathcal{L}$, $\{0\} \neq \mathcal{N} \subseteq [\mathcal{L}, \mathcal{I}_n]$, such that every $Q \in \mathcal{N}$ is nilpotent.*

Theorem 3. *Let $\mathcal{U}$ be an LM-decomposable Lie subalgebra of $\mathcal{B}(\mathcal{X})$, $\mathcal{U} = \mathcal{R} + \mathcal{G}$ the decomposition of $\mathcal{U}$. One of the following statements holds:*

(I) *There exists a finite-dimensional commutative non-zero ideal $\mathcal{N}$ of $\mathcal{U}$, consisting of nilpotent operators.*

(II) *$[\mathcal{G}, \mathcal{R}] = \{0\}$, $\mathcal{R}$ is a quasinilpotent Lie algebra, $\mathcal{R}$ contains no non-zero nilpotent operators and every analytically central element in $\mathcal{R}$ is central in $\mathcal{R}$.*

Notes

The asymptotic formula of § 15 was given by D. Beltita and M. Şabac [1]; it unifies similar formulas of C. Apostol [1], [2] and M. Şabac [12]. For related formulas see also S. G. Krein and A.M. Shihvatov [1] (formula (2.10) from that paper), R.Ya. Grabovskaya and S. G. Krein [1], R. Ya Grabovskaya [1], Sh. Kantorovitz [2], J.-Ph. Labrousse [1] (formula (5.3.5) from that paper) and lemma 2.2.22 from J. Dixmier [1]; see also Yu.B. Farforovskaya [1].

The nilpotence criteria of § 16 were given in M. Şabac [10] and are extensions of nilpotence criteria of N. Jacobson [1].

Lemma 1, Theorems 2–4 and Corollaries 1–2 of § 17 appear here for the first time. Theorem 5 from § 17, as well as the results of § 18 and § 19 have been published in M. Şabac [10].

Chapter III

Infinite-dimensional Variants of the Lie and Engel Theorems

§ 20 Weights for operator algebras

The classical Lie's Theorem (see Theorem 2 from § 2) has the following equivalent form.

Theorem 1 (Theorem of weight). *If $\mathcal{V} \neq \{0\}$ is a finite-dimensional complex vector space and $\mathcal{L}$ is a solvable Lie subalgebra of End$(\mathcal{V})$, then there exist $v \in \mathcal{V}\backslash\{0\}$ and a functional $\varphi : \mathcal{L} \to \mathbb{C}$ such that $(T - \varphi(T))v = 0$ for every $T \in \mathcal{L}$.*

The functional φ is called a (classical) *weight* for $\mathcal{L}$.

This result has an analogue for solvable Lie algebras of operators on a complex Banach space. In order to obtain an infinite-dimensional variant of this result it was necessary (see the papers of D. Gurarie and Yu.I. Lyubich [1] and Şt. Frunză [2]) to redefine the concept of weight. In the following we present the variant from Şt. Frunză [2], based on Żelazko's Theorem concerning joint topological divisors of zero.

1. Shilov boundary and joint topological divisors of zero in commutative Banach algebras

Let $\mathcal{B}$ be a commutative complex Banach algebra with unity e. We denote by $\mathcal{M}$ the compact space of maximal ideals of $\mathcal{B}$ endowed with the Gelfand topology. We denote also by $\hat{x}(M) = x(M)$ the Gelfand transform of $x \in \mathcal{B}$ for $M \in \mathcal{M}$, i.e. $\hat{x}(M) = \varphi_M(\tilde{x}_M)$ where φ_M is the isomorphism of $\mathcal{B}/M$ onto $\mathbb{C}$ and $\tilde{x}_M$ is the equivalence class of x in $\mathcal{B}/M$ for $M \in \mathcal{M}$.

Definition 1. The *Shilov boundary* of $\mathcal{B}$ will be denoted by $\Gamma(\mathcal{B})$ or Γ and is the smallest closed subset F of $\mathcal{M}$ having the property

$$\sup_{\omega\in F} |x(\omega)| = \sup_{\omega\in\mathcal{M}} |x(\omega)|$$

for every $x \in \mathcal{B}$.

Lemma 1. *There exist minimal (with respect to inclusion) closed subsets F of $\mathcal{M}$ having the property*

$$\sup_{\omega\in F} |x(\omega)| = \sup_{\omega\in\mathcal{M}} |x(\omega)|$$

for every $x \in \mathcal{B}$.

Proof. We observe that the set of all closed subsets of $\mathcal{M}$ having the desired property is inductively ordered by inclusion. The conclusion results by Zorn's Lemma. □

Lemma 2. *If Γ_1 and Γ_2 are two minimal closed subsets of $\mathcal{M}$ as in Lemma 1, then $\Gamma_1 = \Gamma_2$. Hence the Shilov boundary is well defined.*

Proof. It suffices to prove that $\Gamma_1 \subseteq \overline{\Gamma_2}$. Therefore it suffices to prove that $V(M_1)\cap \Gamma_2 \neq \emptyset$ for every $M_1 \in \Gamma_1$ and $V(M_1) := V(M_1; x_1, \ldots, x_n, \varepsilon)$ a neighbourhood of M_1 (in the Gelfand topology) given by $x_1, \ldots, x_n \in \mathcal{B}$ and $\varepsilon > 0$,

$$\begin{aligned} V &:= V(M_1; x_1, \ldots, x_n, \varepsilon) \\ &:= \{M \in \mathcal{M} \mid |x_k(M) - x_k(M_1)| < \varepsilon, 1 \le k \le n\} \\ &= \{M \in \mathcal{M} \mid |z_k(M)| < \varepsilon, z_k := x_k - x_k(M_1)e, 1 \le k \le n\}. \end{aligned}$$

The set Γ_1 is a minimal closed subset of $\mathcal{M}$ with the property

$$m := \sup_{M\in\mathcal{M}} |x(M)| = \sup_{M\in\Gamma_1} |x(M)| \text{ for every } x \in \mathcal{B}. \tag{1}$$

Then there exists $y \in \mathcal{B}$ such that

$$\sup_{M\in\mathcal{M}} |y(M)| = \sup_{M\in\Gamma_1\cap V} |y(M)| \text{ and } |y(M)| < m \;\forall M \in \Gamma_1\backslash(\Gamma_1 \cap V). \tag{2}$$

Indeed, if this is not true, then for every $y \in \mathcal{B}$ one of the following assertions holds: either

$$\sup_{M\in\Gamma_1\cap V} |y(M)| < m,$$

or

$$\exists M \in \Gamma_1\backslash(\Gamma_1 \cap V) : |y(M)| \ge m, \text{ i.e. } |y(M)| = m.$$

In any case it results that

$$\sup_{M\in\Gamma_1\backslash(\Gamma_1\cap V)} |y(M)| = m,$$

for every $y \in \mathcal{B}$, which is impossible because Γ_1 is minimal with the same property (1).

Replacing y of (2) by $\frac{1}{m}y$ we can suppose that $m = 1$. Hence $y \in \mathcal{B}$ and

$$\sup_{M \in \Gamma_1 \cap V} |y(M)| = 1 \text{ and } |y(M)| < 1 \, \forall M \in \Gamma_1 \backslash (\Gamma_1 \cap V).$$

Now, replacing y by y^n (with sufficiently large n) we can suppose also that

$$|y(M)| < \frac{\varepsilon}{\max_{1 \leq k \leq n} \|z_k\|} \qquad \forall M \in \Gamma_1 \backslash (\Gamma_1 \cap V).$$

Then we have for $M \in \Gamma_1$

$$|(z_k y)(M)| = |z_k(M) y(M)| < \varepsilon.$$

But Γ_1 verifies (1) and we deduce

$$|z_k(M) y(M)| < \varepsilon \quad \forall M \in \mathcal{M}.$$

On the other hand Γ_2 verifies (1). Then there exists $M_2 \in \Gamma_2$ with the property $\sup_{M \in \mathcal{M}} |y(M)| = |y(M_2)|$. Hence

$$|z_k(M_2)| < \varepsilon$$

and we have $M_2 \in \Gamma_2 \cap V(M_1; x_1, \ldots, x_n, \varepsilon)$. □

We obtain also the following local characterization of the Shilov boundary.

Proposition 1. *Let $\mathcal{B}$ be a commutative complex Banach algebra with unity e. Let Γ be the Shilov boundary of $\mathcal{B}$. The following assertions are equivalent:*

1) $M_0 \in \Gamma$;
2) *for every neighbourhood $U = U(M_0)$ of M_0 there exists $y \in \mathcal{B}$ such that* $\sup_{M \in U} |y(M)| > \sup_{M \in \Gamma \backslash U} |y(M)|$.

Proof. The implication 1) $\Rightarrow$ 2) is contained in the proof of Lemma 1.

The implication 2) $\Rightarrow$ 1) holds because if we have 2) then we can find (as in the proof of Lemma 1) an element of Γ in every neighbourhood of M_0, which implies $M_0 \in \overline{\Gamma} = \Gamma$. □

In order to prove the existence of the joint topological divisors of zero, first we consider as in W. Żelazko [1] the following special case. Let Ω be a compact Hausdorff topological space and $\mathcal{A} = \mathcal{C}(\Omega)$ the commutative Banach algebra of all complex-valued continuous functions on Ω. Then every algebra morphism $\mathcal{A} \to \mathbb{C}$ is of the form

$$\mathcal{A} \ni x \mapsto f_\omega(x) = x(\omega) \text{ for some fixed } \omega \in \Omega$$

and every maximal ideal M of $\mathcal{A}$ corresponds to some $\omega \in \Omega$, i.e. it equals

$$M_\omega = \{x \in \mathcal{A} \mid x(\omega) = 0\}.$$

As a special case of Definition 1 we have:

Definition 2. The *Shilov boundary* of $\mathcal{A}$ will be denoted $\Gamma(\mathcal{A})$ or Γ and is the smallest closed subset F of Ω with the property

$$\sup_{\omega\in F}|x(\omega)| = \sup_{\omega\in\Omega}|x(\omega)|$$

for every $x \in \mathcal{A}$.

We observe also as a special case of Proposition 1:

Remark 1. We have $\omega_0 \in \Gamma(\mathcal{A})$ iff for each neighbourhood U of ω_0 we can find a function $x \in \mathcal{A}$ such that

$$\sup_{\omega\in U}|x(\omega)| > \sup_{\omega\in\Omega\setminus U}|x(\omega)|. \tag{3}$$

The following result was given by W. Żelazko in order to characterize the points of the Shilov boundary as sets of joint topological divisors of zero.

Proposition 2. *If* $\omega_0 \in \Gamma(\mathcal{A})$ *then there exists a net* $(x_\alpha)_\alpha$ *from* $\mathcal{A}$ *such that* $\|x_\alpha\| = 1$ *for every* α *and* $\lim_\alpha ax_\alpha = 0$ *for every* $a \in M_{\omega_0}$.

Proof. By Remark 1 let's consider $x \in \mathcal{A}$ and a neighbourhood U of ω_0 satisfying (3). We can suppose $\|x\| = 1$ and for each fixed $\varepsilon > 0$ we can find a power x^N of x, such that

$$\sup_{\omega\in\Omega\setminus U}|x^N(\omega)| < \varepsilon.$$

For every neighbourhood U of ω_0 and every natural number n we denote $\alpha = (U, n)$. We can find $x_\alpha \in \mathcal{A}$, $\|x_\alpha\| = 1$, such that

$$\sup_{\omega\in\Omega\setminus U}|x_\alpha(\omega)| < \frac{1}{n}.$$

For $a \in M_{\omega_0}$ we have $a(\omega_0) = 0$ and for a given $\varepsilon > 0$ there exists a neighbourhood U_ε of ω_0, such that

$$\sup_{\omega\in U_\varepsilon}|a(\omega)| < \varepsilon.$$

Let us consider also a natural number n_ε with the property

$$\frac{\|a\|}{n_\varepsilon} < \varepsilon.$$

Then for $\alpha_\varepsilon = (U_\varepsilon, n_\varepsilon)$ we have

$$\begin{array}{lclcl} \sup_{\omega\in U_\varepsilon}|a(\omega)x_{\alpha_\varepsilon}(\omega)| & < & \varepsilon\cdot\|x_{\alpha_\varepsilon}\| & = & \varepsilon, \\ \sup_{\omega\in\Omega\setminus U_\varepsilon}|a(\omega)x_{\alpha_\varepsilon}(\omega)| & < & \|a\|\cdot\frac{1}{n_\varepsilon} & < & \varepsilon. \end{array}$$

If $\alpha = (U, n)$ is such that $U \subseteq U_\varepsilon$ and $n > n_\varepsilon$ then we have also

$$\begin{array}{rcccl} \sup_{\omega\in U} |a(\omega)x_\alpha(\omega)| & < & \varepsilon \cdot \|x_\alpha\| & = & \varepsilon, \\ \sup_{\omega\in\Omega\setminus U} |a(\omega)x_\alpha(\omega)| & < & \|a\| \cdot \frac{1}{n} & < & \varepsilon, \end{array}$$

i.e. $\|ax_\alpha\| = \sup_{\omega\in\Omega} |a(\omega)x_\alpha(\omega)| < \varepsilon$.

Then $(x_\alpha)_\alpha$ is a net of elements of $\mathcal{A}$ if we define $\alpha_1 \leq \alpha_2$ by $U_2 \subseteq U_1$ and $n_1 \leq n_2$ (where $\alpha_i = (U_i, n_i)$ for $i = 1, 2$). For every $a \in M_{\omega_0}$ we have:

$$\forall \varepsilon > 0\, \exists \alpha_\varepsilon : \|ax_\alpha\| = \sup_{\omega\in\Omega} |a(\omega)x_\alpha(\omega)| < \varepsilon\ \forall \alpha \geq \alpha_\varepsilon.$$

This means that there exists a net $(x_\alpha)_\alpha$ from $\mathcal{A}$ such that $\|x_\alpha\| = 1$ for every α and $\lim_\alpha ax_\alpha = 0$ for every $a \in M_{\omega_0}$. □

Using the above result, W. Żelazko extends it to any commutative complex Banach algebra.

Theorem 2. *Let $\mathcal{A}$ be an arbitrary (which is not necessarily a function algebra) commutative complex Banach algebra with unity. Then any maximal ideal belonging to the Shilov boundary $\Gamma(\mathcal{A})$ of $\mathcal{A}$ consists of joint topological divisors of zero, i.e. if $M \in \Gamma(\mathcal{A})$, then there exists a net $(x_\alpha)_\alpha$ from $\mathcal{A}$ such that $\|x_\alpha\| = 1$ for every α and $\lim_\alpha ax_\alpha = 0$ for every $a \in M$.*

Proof. We shall use the following lemma which is an easy exercise.

Lemma 3. *A set $S \subseteq \mathcal{A}$ consists of joint topological divisors of zero (i.e. there exists a net $(x_\alpha)_\alpha$ from $\mathcal{A}$ such that $\|x_\alpha\| = 1$ for every α and $\lim_\alpha ax_\alpha = 0$ for every $a \in S$) iff we have*

$$\delta(a_1, \ldots, a_n) := \inf_{x\in\mathcal{A}, \|x\|=1} \sum_{i=1}^{k} \|a_i x\| = 0$$

for every finite system $(a_1, \ldots, a_n) \in S^n$.

Then for proving the theorem it suffices to prove that the following implication holds for every $a_1, \ldots, a_n \in \mathcal{A}$:

$$\delta(a_1, \ldots, a_n) > 0 \Rightarrow \begin{array}{c} a_1, \ldots, a_n \text{ cannot belong to the same} \\ \text{maximal ideal } M \in \Gamma(\mathcal{A}). \end{array}$$

Let $a_1, \ldots, a_n \in \mathcal{A}$ be such that $\delta(a_1, \ldots, a_n) > 0$. Without loss of generality we can suppose $\delta(a_1, \ldots, a_n) \geq 1$ or, equivalently,

$$\sum_{i=1}^{k} \|a_i x\| \geq \|x\| \text{ for every } x \in \mathcal{A}. \tag{4}$$

It is easy to verify that the algebra $\mathcal{B}$ of all formal power series

$$s = \sum_{i_1,\ldots,i_n\geq 0} a_{i_1\ldots i_n} (t_1)^{i_1} \cdots (t_n)^{i_n}$$

with coefficients in $\mathcal{A}$, verifying

$$\|s\| := \sum_{i_1,\ldots,i_n \geq 0} \|a_{i_1\ldots i_n}\| < \infty,$$

is an extension of $\mathcal{A}$ (a *superalgebra* of $\mathcal{A}$); an isometric embedding φ of $\mathcal{A}$ in $\mathcal{B}$ is given by

$$\varphi(x) = x + \sum_{i_1+\cdots+i_n>0} 0 \cdot (t_1)^{i_1} \cdots (t_n)^{i_n}.$$

Then $w := a_1t_1 + \cdots + a_nt_n \in \mathcal{B}$ and by (4) we have obviously

$$\|wx\| \geq \|x\| \text{ for every } x \in \mathcal{A}.$$

One can obtain by induction for $k = 1, 2, \ldots$

$$\|w^kx\| \geq \|x\| \text{ for every } x \in \mathcal{A}.$$

Indeed,

$$w^{k-1} = \sum_{i_1+\cdots+i_n=k-1} c^{k-1}_{i_1\ldots i_n} (a_1)^{i_1} \cdots (a_n)^{i_n} (t_1)^{i_1} \cdots (t_n)^{i_n}$$

with $c^{k-1}_{i_1\ldots i_n} > 0$ and

$$\|w^{k-1}\| = \sum_{i_1+\cdots+i_n=k-1} c^{k-1}_{i_1\ldots i_n} \|(a_1)^{i_1} \cdots (a_n)^{i_n}\|.$$

On the other hand, using (4) we can write

$$\|w^kx\| = \|w \sum_{i_1+\cdots+i_n=k-1} c^{k-1}_{i_1\ldots i_n} (a_1)^{i_1} \cdots (a_n)^{i_n} x (t_1)^{i_1} \cdots (t_n)^{i_n}\|$$

$$= \|\sum_{s=1}^{n} \sum_{i_1+\cdots+i_n=k-1} c^{k-1}_{i_1\ldots i_n} (a_1)^{i_1} \cdots (a_s)^{i_s+1} \cdots (a_n)^{i_n} x (t_1)^{i_1} \cdots (t_n)^{i_n}\|$$

$$= \sum_{s=1}^{n} \sum_{i_1+\cdots+i_n=k-1} c^{k-1}_{i_1\ldots i_n} \sum_{s=1}^{n} \|a_s((a_1)^{i_1} \cdots (a_n)^{i_n} x)\|$$

$$\overset{(4)}{\geq} \sum_{i_1+\cdots+i_n=k-1} c^{k-1}_{i_1\ldots i_n} \|(a_1)^{i_1} \cdots (a_n)^{i_n} x\| = \|w^{k-1}x\|.$$

Hence we have for $k = 1, 2, \ldots$

$$\|w^kx^k\| \geq \|x^k\| \text{ for every } x \in \mathcal{A}.$$

If we denote $\|u\|_s = \lim_{n\to\infty} \|u^n\|^{1/n}$ the spectral norm on $\mathcal{B}$, then by the above inequalities we deduce

$$\|wz\|_s \geq \|z\|_s$$

and

$$\|z\|_s \leq \|wz\|_s = \left\| \sum_{i=1}^{n} za_i t_i \right\|_s \leq \sum_{i=1}^{n} \|a_i z\|_s \tag{5}$$

because $\|t_i\|_s = 1$ for $i = 1, \ldots, n$ ($\|(t_i)^n\| = \|e(t_i)^n\| = 1$ in $\mathcal{B}$). Therefore in the inequality (4) we can replace the norm by the spectral norm.

Let $\mathcal{M}(\mathcal{A})$ be the compact topological space of maximal ideals of $\mathcal{A}$. The Shilov boundary of $\mathcal{A}$ is the uniquely determined closed subset of $\mathcal{M}(\mathcal{A})$ which is minimal among all closed subsets $F \subseteq \mathcal{M}(\mathcal{A})$ with the property that

$$\sup_{M \in F} |x(M)| = \sup_{M \in \mathcal{M}(\mathcal{A})} |x(M)| = \lim_{n \to \infty} \|x^n\|^{1/n}$$

for every $x \in \mathcal{A}$. Recall that $\hat{x}(M) = x(M)$ denotes the Gelfand transform of $x \in \mathcal{B}$ for $M \in \mathcal{M}$, i.e. $\hat{x}(M) = \varphi_M(\tilde{x}_M)$ where φ_M is the isomorphism of $\mathcal{B}/M$ onto $\mathbb{C}$ and $\tilde{x}_M$ is the equivalence class of x in $\mathcal{B}/M$ for $M \in \mathcal{M}$.

If we consider the completion $\overline{\mathcal{A}}$ of the algebra $\widehat{\mathcal{A}}$ of the Gelfand transforms of elements of $\mathcal{A}$ with respect to the sup-norm on $\mathcal{C}(\mathcal{M}(\mathcal{A}))$, then every multiplicative linear functional f on $\mathcal{A}$ can be extended to a multiplicative linear functional on $\overline{\mathcal{A}}$. Hence we can identify $\mathcal{M}(\mathcal{A})$ with $\mathcal{M}(\overline{\mathcal{A}})$. It is well known that under this identification we have $\Gamma(\mathcal{A}) = \Gamma(\overline{\mathcal{A}})$, i.e. $a_1, \ldots, a_n \in M \in \Gamma(\mathcal{A})$ iff $\widehat{a_1}, \ldots, \widehat{a_n} \in \overline{M} \in \Gamma(\overline{\mathcal{A}})$, where $\overline{M}$ corresponds to M under the identification $\Gamma(\mathcal{A}) = \Gamma(\overline{\mathcal{A}})$.

For $\overline{\mathcal{A}}$ we can apply Proposition 1. Then the above formula (5) proves that the elements $\widehat{a_1}, \ldots, \widehat{a_n}$ cannot be in $\overline{M}$ for any $\overline{M} \in \Gamma(\overline{\mathcal{A}})$. Then $a_1, \ldots, a_n$ cannot be in M for any $M \in \Gamma(\mathcal{A})$ and the theorem is proved. □

2. Weights

In the following $\mathcal{X}$ will be a complex Banach space and $\mathcal{B}(\mathcal{X})$ denotes the algebra of all bounded operators on $\mathcal{X}$ as usual. A natural extension of the weight in this case is the following.

Definition 3. (Şt. Frunză) If $\mathcal{L}$ is a Lie subalgebra of $\mathcal{B}(\mathcal{X})$, then a *generalized weight* for $\mathcal{L}$ is a functional $\varphi : \mathcal{L} \to \mathbb{C}$ such that there exists a net $(x_\alpha)_\alpha$ from $\mathcal{X}$ such that $\|x_\alpha\| = 1$ for every α and

$$\lim_{\alpha}(Tx_\alpha - \varphi(T)x_\alpha) = 0$$

for every $T \in \mathcal{L}$. Moreover if there exists $x \in \mathcal{X}$ such that $x_\alpha = x$ for every α, then φ will be called a *classical weight.*

This concept contains the following one given by D. Gurarie and Yu.I. Lyubich.

Definition 4. If $\mathcal{G} \subset \mathcal{B}(\mathcal{X})$ is a separable (with respect to the norm operator topology) group of bounded operators on $\mathcal{X}$, then a scalar function $\chi : \mathcal{G} \to \mathbb{C}$ is called a *weight* for $\mathcal{G}$ if there exists a sequence $(x_n)_{n\in\mathbb{N}}$ from $\mathcal{X}$ such that $\|x_n\| = 1$ for every $n \in \mathbb{N}$ and

$$\lim_{n\to\infty} (Tx_n - \chi(T)x_n) = 0$$

for every $T \in \mathcal{G}$.

Now we proceed to prove the following result.

Theorem 3 (Frunză). *Any solvable Lie algebra of operators on a complex Banach space $\mathcal{X} \neq \{0\}$ has a generalized weight.*

As a corollary of this theorem we have that any solvable group of bounded operators on a complex Banach space has a generalized weight, thus an improvement of the initial result of D. Gurarie and Yu.I. Lyubich which states that *any solvable, separable, locally compact and connected group of bounded operators on $\mathcal{X}$ has a weight.*

The proof of Theorem 3 is an easy consequence of Żelazko's theorem (Theorem 2 above) and of an extension theorem for generalized weights.

By Żelazko's theorem it is possible to prove Theorem 3 in the commutative case.

Proposition 3. *Any commutative Lie subalgebra $\mathcal{L}$ of $\mathcal{B}(\mathcal{X})$ has a generalized weight.*

The extension theorem for generalized weights is the following.

Theorem 4. *Let $\mathcal{L} \subseteq \mathcal{B}(\mathcal{X})$ be a Lie algebra of bounded operators on $\mathcal{X}$. Then any generalized weight for $\mathcal{L}_1 := [\mathcal{L}, \mathcal{L}]$ is identically zero and can be extended to a weight of $\mathcal{L}$.*

Proof of Theorem 3. We have

$$\mathcal{L} = \mathcal{L}_0 \supset \mathcal{L}_1 = [\mathcal{L}_0, \mathcal{L}_0] \supset \cdots \supset \mathcal{L}_n = [\mathcal{L}_{n-1}, \mathcal{L}_{n-1}] = \{0\}$$

for some $n \in \mathbb{N}$ because $\mathcal{L}$ is a solvable Lie algebra. By Proposition 3, $\mathcal{L}_{n-1}$ has a generalized weight which may be extended step by step (by Theorem 4), so we get a generalized weight for the whole algebra $\mathcal{L}$. □

Proof of Proposition 3. We apply Żelazko's theorem (Theorem 2 above) for the commutative Banach algebra spanned by $\mathcal{L}$ and by the identity operator I. Then any maximal ideal M belonging to $\Gamma(\mathcal{A})$ consists of joint topological divisors of zero. Hence there exists a net $(T_\alpha)_\alpha$ from $\mathcal{A}$ such that $\|T_\alpha\| = 1$ for every α and $\lim_\alpha \|T_\alpha T\| = 0$ for every $T \in M$. Let φ_M be the multiplicative linear functional (i.e. character) of $\mathcal{A}$ corresponding to M. Then the restriction $\varphi := \varphi_M|_\mathcal{L}$ is a generalized weight. Obviously we have $T - \varphi(T)I \in Ker\varphi_M = M$ for any $T \in \mathcal{L}$. Hence $\lim_\alpha \|(T - \varphi(T)I)T_\alpha\| = 0$ for any $T \in \mathcal{L}$. On the other hand

$$\|T_\alpha\| = 1 \Longrightarrow \exists x_\alpha \in \mathcal{X},\ \|x_\alpha\| = 1,\ \|T_\alpha x_\alpha\| \geq \frac{1}{2}.$$

Denote $z_\alpha = T_\alpha x_\alpha$ and $y_\alpha = \frac{1}{\|z_\alpha\|} z_\alpha$. Obviously we have $\|z_\alpha\| \geq \frac{1}{2}$, $\|y_\alpha\| = 1$ and we can write

$$\|(T - \varphi(T)I)y_\alpha\| = \frac{1}{\|z_\alpha\|}\|(T - \varphi(T)I)T_\alpha x_\alpha\| \leq 2\|(T - \varphi(T)I)T_\alpha\| \xrightarrow[\alpha]{} 0.$$

Therefore $\|y_\alpha\| = 1$ and $\lim_\alpha \|(T - \varphi(T))y_\alpha\| = 0$. This proves that φ is indeed a generalized weight for $\mathcal{L}$. □

The proof of Theorem 4 requires some auxiliary facts. First, there exists the following characterization of the generalized weights, which is similar to the characterization by means of function δ for the sets of joint topological divisors of zero in a commutative Banach algebra (see Lemma 3 above).

Lemma 4. *If $\mathcal{L}$ is a Lie subalgebra of $\mathcal{B}(\mathcal{X})$ and $\varphi : \mathcal{L} \to \mathbb{C}$ is a scalar function defined on $\mathcal{L}$, then the following assertions are equivalent.*

1. *φ is a generalized weight for $\mathcal{L}$.*
2. *We have*

$$\inf_{\|x\|=1} \{\max_{S\in\mathcal{F}} \|Sx - \varphi(S)x\|\} = 0$$

for every finite subset $\mathcal{F} \subset \mathcal{L}$.

Moreover, if φ is a generalized weight then $\varphi(T) = 0$ for any $T \in \mathcal{L}_1 = [\mathcal{L}, \mathcal{L}]$.

Proof. The implication $1 \Rightarrow 2$ is obvious because, $\mathcal{F}$ being finite, we have

$$\lim_\alpha \max_{S\in\mathcal{F}} \|Sx_\alpha - \varphi(S)x_\alpha\| = 0$$

if φ is a generalized weight with the corresponding net $(x_\alpha)_\alpha$.

$2 \Rightarrow 1$ Indeed, if the assertion 2 holds, then for every finite subset $\mathcal{F} \subset \mathcal{L}$ and every natural number n we can find $x_\alpha \in \mathcal{X}$, $\|x_\alpha\| = 1$, where $\alpha = (\mathcal{F}, n)$, such that

$$\max_{S\in\mathcal{F}} \|Sx_\alpha - \varphi(S)x_\alpha\| < \frac{1}{n}. \tag{6}$$

Let's define $\alpha_1 \leq \alpha_2$ by $\mathcal{F}_1 \subset \mathcal{F}_2$ and $n_1 < n_2$ for any pairs $\alpha_i = (\mathcal{F}_i, n_i)$ $(i = 1, 2)$. Then it is clear that by (6) we can deduce

$$\lim_\alpha \|Sx_\alpha - \varphi(S)x_\alpha\| = 0 \text{ for any } S \in \mathcal{L}$$

and the assertion 1 is proved.

For the last assertion it suffices to consider $T = [S_1, S_2]$ with $S_1, S_2 \in \mathcal{L}$. The following inequality proves that $\lim_\alpha \|[S_1, S_2]x_\alpha\| = 0$

$$\begin{aligned} \|S_1S_2x_\alpha - S_2S_1x_\alpha\| \leq \;& \|S_1(S_2 - \varphi(S_2))x_\alpha\| + \|\varphi(S_2)(S_1 - \varphi(S_1))x_\alpha\| \\ & + \|S_2(\varphi(S_1) - S_1)x_\alpha\|, \end{aligned}$$

and this implies $\lim_\alpha \varphi([S_1, S_2])x_\alpha = 0$. Hence $\varphi([S_1, S_2]) = 0$ since $\|x_\alpha\| = 1$. □

The following step for proving Theorem 4 is the proof of this theorem for classical weights.

Theorem 5. *Let $\mathcal{L} \subseteq \mathcal{B}(\mathcal{X})$ be a Lie algebra of bounded operators on $\mathcal{X}$. Then any classical weight for $\mathcal{L}_1 := [\mathcal{L}, \mathcal{L}]$ is identically zero and can be extended to a weight of $\mathcal{L}$.*

Proof. Let's consider a classical weight φ_0 for $\mathcal{L}_1$. We consider also the following subspace of $\mathcal{X}$:

$$\mathcal{X}_0 = \{x \in \mathcal{X} \mid Tx = \varphi_0(T)x \text{ for any } T \in \mathcal{L}_1\}.$$

We have $\mathcal{X}_0 \neq \{0\}$ because φ_0 is a classical weight. For $S, T \in \mathcal{B}(\mathcal{X})$ recall (see Remark 1 from § 15) that

$$(\exp(z\, ad\, S))(T) = \exp(-zS)T\exp(zS) \qquad \forall z \in \mathbb{C}.$$

Let $\overline{\mathcal{L}}_1$ be the closure of $\mathcal{L}_1$ in the norm operator topology of $\mathcal{B}(\mathcal{X})$. Obviously $(\exp(z\, ad\, S))(T) \in \overline{\mathcal{L}}_1$ whenever $S \in \mathcal{L}$ and $T \in \mathcal{L}_1$. On the other hand it is easy to prove that φ_0 is a continuous linear functional on $\mathcal{L}_1$. Hence there exists a unique extension $\tilde{\varphi}_0$ of φ_0 to $\overline{\mathcal{L}}_1$ and $\tilde{\varphi}_0$ has the following property:

$$Vx = \tilde{\varphi}_0(V)x \text{ for every } x \in \mathcal{X}_0 \text{ and } V \in \overline{\mathcal{L}}_1.$$

Particularly we have, for $S \in \mathcal{L}$ and $T \in \mathcal{L}_1$,

$$\exp(-zS)T\exp(zS)x = ((\exp(z\, ad\, S))(T))x = \tilde{\varphi}_0((\exp(z\, ad\, S))(T))x \tag{7}$$

for any $z \in \mathbb{C}$ and $x \in \mathcal{X}_0$.

We deduce that for any $x \in \mathcal{X}_0$, $S \in \mathcal{L}$ and $T \in \mathcal{L}_1$ we have

$$T\exp(zS)x = \tilde{\varphi}_0((\exp(z\, ad\, S))(T))\exp(zS)x \text{ for every } z \in \mathbb{C}.$$

It results that $\tilde{\varphi}_0((\exp(z\, ad\, S))(T))$ is an eigenvalue of T for any $z \in \mathbb{C}$. Then $z \mapsto \tilde{\varphi}_0((\exp(z\, ad\, S))(T))$ is a bounded function because the spectrum of T is compact. But $z \mapsto \tilde{\varphi}_0((\exp(z\, ad\, S))(T))$ is an analytic function, so it must be constant by Liouville's Theorem.

Therefore, for $S \in \mathcal{L}$ and $T \in \mathcal{L}_1$ we have

$$\tilde{\varphi}_0((\exp(z\, ad\, S))(T)) = \tilde{\varphi}_0((\exp(z\, ad\, S))(T))|_{z=0} = \tilde{\varphi}_0(T) = \varphi_0(T)$$

and by (7) we deduce for $S \in \mathcal{L}$, $T \in \mathcal{L}_1$ and $x \in \mathcal{X}_0$

$$((\exp(z\, ad\, S))(T))x = \varphi_0(T)x \qquad \forall z \in \mathbb{C}.$$

By differentiation in $z = 0$ we obtain

$$((ad\, S)T)x = 0 \text{ for } S \in \mathcal{L},\ T \in \mathcal{L}_1 \text{ and } x \in \mathcal{X}_0.$$

By taking into account the definition of $\mathcal{X}_0$ we can write

$$TSx = \varphi_0(T)x \text{ whenever } x \in \mathcal{X}_0,\ T \in \mathcal{L}_1 \text{ and } S \in \mathcal{L},$$

which means that $\mathcal{X}_0$ is an invariant subspace for every operator $S \in \mathcal{L}$.

We observe that $\mathcal{X}_0 \neq \{0\}$ is a closed subspace of $\mathcal{X}$ and consider the Lie algebras of all restrictions $S|_{\mathcal{X}_0}$ of the operators $S \in \mathcal{L}$. This Lie algebra is commutative. Indeed,

$$[S_1|_{\mathcal{X}_0}, S_2|_{\mathcal{X}_0}]x = \varphi_0([S_1, S_2])x \text{ for } S_1, S_2 \in \mathcal{L} \text{ and } x \in \mathcal{X}_0.$$

Then $[S_1|_{\mathcal{X}_0}, S_2|_{\mathcal{X}_0}]$ is a scalar multiple of the identity on $\mathcal{X}_0$ and by the Kleinecke-Shirokov Theorem (see Remark 1 from § 17) we must have $[S_1|_{\mathcal{X}_0}, S_2|_{\mathcal{X}_0}] = 0$ and consequently $\varphi_0([S_1, S_2]) = 0$ for any $S_1, S_2 \in \mathcal{L}$.

Now we apply Proposition 3 for the commutative Lie algebra

$$\{S|_{\mathcal{X}_0} \mid S \in \mathcal{L}\}.$$

Then there exist $\varphi : \mathcal{L} \to \mathbb{C}$ and a net $(x_\alpha)_\alpha$ from $\mathcal{X}_0$ such that $\|x_\alpha\| = 1$ for every α and $\lim_\alpha \|Sx_\alpha - \varphi(S)x_\alpha\| = 0$ for any $S \in \mathcal{L}$. For $T \in \mathcal{L}_1$ we have $Tx_\alpha = 0$ because $x_\alpha \in \mathcal{X}_0$. Therefore $\varphi(T) = 0 = \varphi_0(T)$. Hence φ extends φ_0 (= the identically zero classical weight on $[\mathcal{L}, \mathcal{L}]$) and Theorem 5 (i.e. Theorem 4 for classical weights) is proved. □

The complete proof of Theorem 4 is the following.

Proof of Theorem 4. Let φ_0 be a generalized weight for the algebra $\mathcal{L}_1 := [\mathcal{L}, \mathcal{L}]$. There exists a net $(x_\alpha)_\alpha$ from $\mathcal{X}$ such that $\|x_\alpha\| = 1$ for every α and $\lim_\alpha \|(T - \varphi_0(T))x_\alpha\| = 0$ for any $T \in \mathcal{L}_1$.

We denote by D the set of indices of the net $(x_\alpha)_\alpha$ and consider the quotient space

$$\widetilde{\mathcal{X}} = l^\infty(D, \mathcal{X})/c_0(D, \mathcal{X})$$

of the space of all bounded functions from D to $\mathcal{X}$ by its subspace of all the bounded functions with limit equal to zero. The generalized weight φ_0 induces in $\widetilde{\mathcal{X}}$ a classical weight as follows: to any operator $S \in \mathcal{L}$ corresponds a bounded linear operator $\widetilde{S}$ on $\widetilde{\mathcal{X}}$ defined by the equality

$$\widetilde{S}((y_\alpha)_{\alpha \in D} + c_0(D, \mathcal{X})) = (Sy_\alpha)_{\alpha \in D} + c_0(D, \mathcal{X}).$$

It is easy to verify that $\widetilde{\mathcal{L}} := \{\widetilde{S} \mid S \in \mathcal{L}\}$ is a Lie algebra of bounded operators on $\widetilde{\mathcal{X}}$ and $S \mapsto \widetilde{S}$ defines a Lie algebra morphism from $\mathcal{L}$ to $\widetilde{\mathcal{L}}$. We have $\xi := (x_\alpha)_{\alpha \in D} + c_0(D, \mathcal{X}) \neq 0$ in $\widetilde{\mathcal{X}}$ because $\|x_\alpha\| = 1$ for every $\alpha \in D$. Moreover $\widetilde{T}\xi = \varphi_0(T)\xi$ for any $T \in \mathcal{L}_1$ because $\lim_\alpha \|(T - \varphi_0(T))x_\alpha\| = 0$ for any $T \in \mathcal{L}_1$.

Hence the scalar function $\tilde{\varphi}_0$ on $\widetilde{\mathcal{L}_1} = \left(\widetilde{\mathcal{L}}\right)_1 = [\widetilde{\mathcal{L}}, \widetilde{\mathcal{L}}]$ defined by $\tilde{\varphi}_0(\widetilde{T}) = \varphi_0(T)$ is a classical weight for $\left(\widetilde{\mathcal{L}}\right)_1$ on the space $\widetilde{\mathcal{X}}$. By Theorem 5, $\tilde{\varphi}_0$ is identically zero; hence also φ_0 is identically zero. By taking into account the last assertion of Lemma 4, if we prove the existence of a generalized weight φ of $\mathcal{L}$ then this generalized weight φ automatically extends φ_0 because $\varphi_0 \equiv 0$ and $\varphi|_{\mathcal{L}_1} \equiv 0$ by Lemma 4.

By Lemma 4 again it suffices to check that $\mathcal{L}$ verifies the condition 2 of Lemma 4 with the same φ. First we observe that the (above defined) classical weight $\tilde{\varphi}_0$ of

$\widetilde{(\mathcal{L})}_1$ extends to a weight $\tilde{\varphi}$ of $\widetilde{\mathcal{L}}$. Hence by Lemma 4 it results that the pair $(\widetilde{\mathcal{L}}, \tilde{\varphi})$ verifies the assertion 2 from Lemma 4. We define $\varphi(S) = \tilde{\varphi}(\widetilde{S})$ for every $\widetilde{S} \in \widetilde{\mathcal{L}}$ $(S \in \mathcal{L})$. It remains to check the following implication:

$$\begin{array}{ccc} (\widetilde{\mathcal{L}}, \tilde{\varphi}) \text{ verifies the} & \Rightarrow & (\mathcal{L}, \varphi) \text{ verifies the} \\ \text{assertion 2 of Lemma 1} & & \text{assertion 2 of Lemma 1.} \end{array}$$

Let $\mathcal{F} = \{S_1, \ldots, S_p\}$ be a finite subset of $\mathcal{L}$. We have

$$\inf_{\tilde{\eta} \in \widetilde{\mathcal{X}}, \|\tilde{\eta}\|=1} \{\max_{S \in \mathcal{F}} \|\widetilde{S}\tilde{\eta} - \tilde{\varphi}(\widetilde{S})\tilde{\eta}\|\} = 0.$$

Then for every $\varepsilon > 0$ there exists $\xi \in \widetilde{\mathcal{X}}$ such that $\|\xi\| = 1$ and

$$\|\widetilde{S}_j\xi - \tilde{\varphi}(\widetilde{S}_j)\xi\| < \varepsilon \text{ for } j = 1, \ldots, p.$$

Suppose that $\xi = (x_\alpha)_{\alpha \in D} + c_0(D, \mathcal{X})$, where $(x_\alpha)_{\alpha \in D} \in l^\infty(D, \mathcal{X})$. By definition of the quotient norm we can find a net $(z_\alpha^j)_{\alpha \in D} \in c_0(D, \mathcal{X})$ such that

$$\sup_\alpha \|S_j x_\alpha - \varphi(S_j)x_\alpha - z_\alpha^j\| < \varepsilon \text{ for } j = 1, \ldots, p.$$

But $(x_\alpha)_{\alpha \in D} \in \xi$ and $\|\xi\| = 1$. Then $\sup_\alpha \|x_\alpha\| \geq 1$ and there exists $\alpha_0 \in D$ with $\|x_{\alpha_0}\| \geq 1/2$ (for instance) and $\|z_{\alpha_0}^j\| < \varepsilon$. If we put $x = \frac{1}{\|x_{\alpha_0}\|} \cdot x_{\alpha_0}$ then $\|x\| = 1$ and for $j = 1, \ldots, p$ we have

$$\begin{aligned} \|S_j x - \varphi(S_j)x\| &\leq 2\|S_j x_\alpha - \varphi(S_j)x_\alpha\| \\ &\leq 2\|S_j x_\alpha - \varphi(S_j)x_\alpha - z_\alpha^j\| + 2\|z_\alpha^j\| \\ &\leq 2\varepsilon + 2\varepsilon = 4\varepsilon. \end{aligned}$$

Therefore $(\mathcal{L}, \varphi)$ verifies the assertion 2 of Lemma 4 and the proof is finished. □

§ 21 Invariant subspaces for LM-decomposable Lie algebras of bounded operators

Definition 1. If $\mathcal{L} \subseteq \mathcal{B}(\mathcal{X})$, then a subspace $\mathcal{X}_0 \subseteq \mathcal{X}$ is called *invariant* for $\mathcal{L}$ if $A\mathcal{X}_0 \subseteq \mathcal{X}_0$ for any $a \in \mathcal{L}$, i.e. if $\mathcal{X}_0$ is invariant for any $A \in \mathcal{L}$.

The following simple fact gives closed invariant subspaces.

Lemma 1. *If $\mathcal{N}$ is a nonzero finite-dimensional ideal of a Lie subalgebra $\mathcal{L}$ of $\mathcal{B}(\mathcal{X})$ and $\mathcal{N}$ contains only nilpotent operators, then the closed subspace*

$$\bigcap_{Q \in \mathcal{N}} Ker Q$$

is invariant for $\mathcal{L}$ and distinct from $\{0\}$ and $\mathcal{X}$.

Proof. We can consider a basis $\{Q_1, \ldots, Q_n\}$ of $\mathcal{N}$ such that for every $X \in \mathcal{N}$ we have

$$(ad X)Q_1 = 0,$$
$$(ad X)Q_2 = 0 \bmod \mathbb{C}Q_1,$$
$$\ldots\ldots\ldots\ldots\ldots\ldots\ldots\ldots\ldots\ldots\ldots\ldots\ldots\ldots$$
$$(ad X)Q_k = 0 \bmod (\mathbb{C}Q_1 + \ldots + \mathbb{C}Q_{k-1}),$$
$$\ldots\ldots\ldots\ldots\ldots\ldots\ldots\ldots\ldots\ldots\ldots\ldots\ldots\ldots$$
$$(ad X)Q_n = 0 \bmod (\mathbb{C}Q_1 + \ldots + \mathbb{C}Q_{n-1}),$$

because $\mathcal{N}$ is a nilpotent finite-dimensional Lie algebra by Rosenblum's Theorem (Theorem 1 from § 13) and Engel's Theorem.

Obviously $Ker Q_1$ is invariant for any $X \in \mathcal{N}$. Particularly it is invariant for Q_2, hence $Ker Q_1 \cap Ker Q_2 \neq \{0\}$ because Q_2 is nilpotent.

Obviously $Ker Q_1 \cap Ker Q_2$ is invariant for any $X \in \mathcal{N}$. Particularly it is invariant for Q_3, hence $Ker Q_1 \cap Ker Q_2 \cap Ker Q_3 \neq \{0\}$ because Q_3 is nilpotent etc. Finally we obtain that

$$Ker Q_1 \cap \cdots \cap Ker Q_n \neq \{0\}$$

and this subspace is distinct from $\mathcal{X}$ because $\mathcal{N} \neq \{0\}$.

We recall that $\mathcal{N}$ was supposed to be an ideal of $\mathcal{L}$. If $X \in \mathcal{L}$ and $x \in \bigcap_{i=1}^{n} Ker Q_i$, then

$$[X, Q_i] = \sum_{j=1}^{n} \lambda_j Q_j \quad \text{for } i = 1, \ldots, n$$

and

$$Q_i X x = X Q_i x + \sum_{j=1}^{n} \lambda_j Q_j x = 0.$$

Then $\bigcap_{i=1}^{n} Ker Q_i$ is a closed invariant subspace for $\mathcal{L}$ and we have obviously

$$\bigcap_{Q \in \mathcal{N}} Ker Q = \bigcap_{i=1}^{n} Ker Q_i.$$

□

The results of § 19 can be written as follows by applying the above lemma.

Theorem 1. *Let $\mathcal{L}$ be a Lie subalgebra of $\mathcal{B}(\mathcal{X})$ and $\mathcal{I}$ an ideal of $\mathcal{L}$ such that $\dim \mathcal{I} < \infty$ and $ad X|_{\mathcal{I}}$ is nilpotent for every $X \in \mathcal{L}$. In these conditions one of the following assertions is true:*

(j) $[Q, \mathcal{I}] = \{0\}$ for every Q which is analytically central in $\mathcal{L}$;

(jj) there exists a closed nontrivial subspace of $\mathcal{X}$ which is invariant to $\mathcal{L}$.

Proof. We apply Theorem 1 from § 19 and observe that $Ker Q$ is invariant to $\mathcal{L}$ if $Q \in Z_{\mathcal{L}}$. □

Theorem 2. *Let $\mathcal{L} \subset \mathcal{B}(\mathcal{X})$ be a Lie algebra of bounded operators. If there exists $N \in \mathbb{N} \cup \{\aleph_0\}$ and*

$$\mathcal{I}_1 \subset \mathcal{I}_2 \subset \cdots \subset \mathcal{I}_n \subset \cdots$$

is an increasing chain of ideals in $\mathcal{L}$ with $\dim \mathcal{I}_n = n$ for $1 \leq n < N$ then one of the following statements holds:

(*j*) *for every positive integer n we have $ad X|_{\mathcal{I}_n}$ nilpotent for any $X \in \mathcal{L}$ and $[Q, \mathcal{I}_n] = \{0\}$ for any analytically central element Q in $\mathcal{L}$;*

(*jj*) *there exists a closed nontrivial subspace of $\mathcal{X}$ which is invariant to $\mathcal{L}$.*

Proof. It suffices to apply Theorem 2 from § 19 and Lemma 1 above. □

Theorem 3. *Let $\mathcal{U}$ be an LM-decomposable Lie subalgebra of $\mathcal{B}(\mathcal{X})$, $\mathcal{U} = \mathcal{R} + \mathcal{G}$ the decomposition of $\mathcal{L}$. One of the following statements holds:*

(I) *there exists a closed nontrivial subspace of $\mathcal{X}$ which is invariant to $\mathcal{U}$;*

(II) *$[\mathcal{G}, \mathcal{R}] = \{0\}$, $\mathcal{R}$ is a quasinilpotent Lie algebra, $\mathcal{R}$ contains no non-zero nilpotent operators and every analytically central element in $\mathcal{R}$ is central.*

Proof. It suffices to apply Theorem 3 from § 19 and Lemma 1 above. □

If some operators from the above Lie algebras have "good spectral properties" (for instance they are compact, decomposable or generalized scalar operators) then the above results can be improved.

We recall that a decomposable operator on a complex Banach space $\mathcal{X}$ possesses a nontrivial spectral maximal subspace iff its spectrum contains at least two distinct points (see Corollary 1 from § 12).

Definition 2. We say that $T \in \mathcal{B}(\mathcal{X})$ is *one-spectral* if there exists $\lambda \in \mathbb{C}$ such that $\sigma(T) = \{\lambda\}$, i.e. $T = \lambda I + Q$ where Q is a quasinilpotent operator. We say that $\mathcal{L} \subset \mathcal{B}(\mathcal{X})$ is *one-spectral* if any $T \in \mathcal{L}$ is one-spectral.

We recall also that a quasinilpotent generalized scalar operator is nilpotent. Hence a one-spectral generalized scalar operator T is of the form $T = \lambda I + Q$ with $\lambda \in \mathbb{C}$ and Q nilpotent.

By the transport theorem of the spectral maximal subspaces (see Theorem 2 from § 13) and by Lemma 1 above, we get the following results.

Proposition 1. *Let $\mathcal{R}$ be a quasinilpotent Lie subalgebra of $\mathcal{B}(\mathcal{X})$. If $\mathcal{R}$ consists only of decomposable operators, then one of the following statements is true:*

(*ms*) *there exists $A \in \mathcal{R}$ possessing a nontrivial spectral maximal subspace invariant to $\mathcal{R}$;*

(*os*) *$\mathcal{R}$ is one-spectral.*

Theorem 4. *Let $\mathcal{U}$ be an LM-decomposable Lie subalgebra of $\mathcal{B}(\mathcal{X})$ with the decomposition $\mathcal{U} = \mathcal{R} + \mathcal{G}$. If $\mathcal{R}$ contains only decomposable operators, then one of the following statements holds:*

(I) *there exists a non-trivial closed subspace of* $\mathcal{X}$ *which is invariant to* $\mathcal{U}$;

(II) $[\mathcal{G},\mathcal{R}] = \{0\}$, $\mathcal{R}$ *is a quasinilpotent Lie algebra,* $\mathcal{R}$ *is one-spectral, contains no non-zero nilpotent operators and every analytically central element in* $\mathcal{R}$ *is central in* $\mathcal{R}$.

The nilpotence property of the generalized scalar operators leads to the following result.

Theorem 5. *Let* $\mathcal{U}$ *be an LM-decomposable Lie subalgebra of* $\mathcal{B}(\mathcal{X})$ *and* $\mathcal{U} = \mathcal{R}+\mathcal{G}$ *the decomposition of* $\mathcal{U}$. *If* $\mathcal{R}$ *consists only of generalized scalar operators, then one of the following assertions is true:*

(I) *there exists a nontrivial closed subspace of* $\mathcal{X}$ *which is invariant to* $\mathcal{U}$;

(II) $\mathcal{R} = \{\lambda I\}_{\lambda\in\mathbb{C}}$.

These results can be improved by using the weights of the solvable Lie algebras (cf. § 20). We begin with some results concerning quasinilpotent Lie algebras.

Lemma 2. *The center of any non-zero quasinilpotent Lie algebra is non-zero.*

Proof. Indeed, if $\mathcal{N}$ is a quasinilpotent Lie algebra, then $\mathcal{N} = \sum_{\alpha\in\Lambda}\mathcal{I}_\alpha$ where $\mathcal{I}_\alpha$ are its finite-dimensional nilpotent ideals. Let's choose $\alpha\in\Lambda$. By Theorem 4 from § 2 there exists a chain

$$\{0\} = \mathcal{I}_0^\alpha \subset \mathcal{I}_1^\alpha \subset \cdots \subset \mathcal{I}_k^\alpha \subset \cdots \subset \mathcal{I}_\alpha$$

such that $[\mathcal{N},\mathcal{I}_k^\alpha] \subseteq \mathcal{I}_{k-1}^\alpha$ and $\dim \mathcal{I}_k^\alpha = k$ for every k. Then $\mathcal{I}_1^\alpha \neq \{0\}$ and $[\mathcal{N},\mathcal{I}_1^\alpha] = \{0\}$. Hence $\{0\} \neq \mathcal{I}_1^\alpha \subseteq Z_\mathcal{N}$. □

The following result gives a relation between the center of a quasinilpotent Lie algebra $\mathcal{N}$ and the whole algebra $\mathcal{N}$ if $\mathcal{N}$ is a Lie algebra of bounded operators.

Theorem 6. *Let* $\mathcal{N}$ *be a quasinilpotent Lie subalgebra of* $\mathcal{B}(\mathcal{X})$. *If the center of* $\mathcal{N}$ *is scalar (i.e.* $Z_\mathcal{N} \subseteq \{\lambda I_\mathcal{X}\}_{\lambda\in\mathbb{C}}$*), then* $\mathcal{N}$ *is scalar.*

Proof. If $Z_\mathcal{N} = \{0\}$ then $\mathcal{N} = \{0\}$ by Lemma 2. Let now $Z_\mathcal{N} = \{\lambda I_\mathcal{X}\}_{\lambda\in\mathbb{C}}$. Obviously, for any finite-dimensional ideal $\mathcal{I}$ of $\mathcal{N}$, $\mathcal{J} := \mathcal{I} + Z_\mathcal{N}$ is a finite-dimensional solvable ideal of $\mathcal{N}$. Then there exists (see Theorem 3 from § 20) a linear form $\chi : \mathcal{J} \to \mathbb{C}$ and a net $\{\xi_n\} \subset \mathcal{X}$ with the property $\|\xi_n\| = 1$ for every n, such that

$$\lim_n (R\xi_n - \chi(R)\xi_n) = 0 \text{ for every } R \in \mathcal{J}. \tag{1}$$

Obviously $\chi|_{[\mathcal{J},\mathcal{J}]} = 0$. It then follows that $\chi|_{Z_\mathcal{N}\cap[\mathcal{J},\mathcal{J}]} = 0$. But $Z_\mathcal{N} = \{\lambda I\}_{\lambda\in\mathbb{C}}$, hence $Z_\mathcal{N}\cap[\mathcal{J},\mathcal{J}] = \{0\}$. We obtain $Z_\mathcal{N}\cap[\mathcal{I},\mathcal{I}] = \{0\}$ for every finite-dimensional ideal $\mathcal{I}$ of $\mathcal{N}$ because $[\mathcal{I},\mathcal{I}] = [\mathcal{J},\mathcal{J}]$. Finally it follows that

$$Z_\mathcal{N} \cap [\mathcal{N},\mathcal{N}] = \{0\}.$$

Now let $\mathcal{N} = \sum_{\alpha\in\Lambda}\mathcal{I}_\alpha$, $\mathcal{I}_\alpha$ finite-dimensional nilpotent ideals of $\mathcal{N}$, the decomposition of the quasinilpotent Lie algebra $\mathcal{N}$ (Remark 3 from § 2). Let $\mathcal{J}_\alpha = \mathcal{I}_\alpha + Z_\mathcal{N}$. Obviously we have

$$Z_\mathcal{N} \cap [\mathcal{J}_\alpha,\mathcal{N}] = \{0\}$$

because $[\mathcal{J}_\alpha,\mathcal{N}] \subseteq [\mathcal{N},\mathcal{N}]$.

Then there exists the following decomposition of $\mathcal{J}_\alpha$ into a direct sum of vector spaces

$$\mathcal{J}_\alpha = Z_{\mathcal{N}} \oplus \mathcal{V}_\alpha \text{ with } \mathcal{J}_\alpha \supset \mathcal{V}_\alpha \supset [\mathcal{J}_\alpha, \mathcal{N}]. \tag{2}$$

The inclusion $\mathcal{V}_\alpha \supset [\mathcal{J}_\alpha, \mathcal{N}]$ shows that $\mathcal{V}_\alpha$ is an ideal of $\mathcal{N}$; the center $Z_{\mathcal{V}_\alpha}$ of $\mathcal{V}_\alpha$ is invariant to any derivation of $\mathcal{V}_\alpha$. But $\mathcal{N}$ is quasinilpotent; hence $ad x|_{\mathcal{V}_\alpha}$ is nilpotent for any $x \in \mathcal{N}$ because $\dim \mathcal{V}_\alpha < \infty$. If $Z_{\mathcal{V}_\alpha} \neq \{0\}$ then, by Engel's Theorem it follows that there exists R_1 such that $0 \neq R_1 \in Z_{\mathcal{V}_\alpha}$ and $(ad X)R_1 = 0$ for any $X \in \mathcal{N}$. By (2) we deduce $R_1 \notin Z_{\mathcal{N}}$; so we have at the same time $R_1 \notin Z_{\mathcal{N}}$ and $0 \neq R_1 \in Z_{\mathcal{N}}$, which is a contradiction. Hence $Z_{\mathcal{V}_\alpha} = \{0\}$, so that $\mathcal{V}_\alpha$ is zero by Lemma 2 because $\mathcal{V}_\alpha$ is a finite-dimensional nilpotent Lie algebra. Then by (2) we deduce $\mathcal{J}_\alpha = Z_{\mathcal{N}}$ for every $\alpha \in \Lambda$, so that $\mathcal{N} = Z_{\mathcal{N}}$ and the proof is finished. □

Using this result we can improve the result concerning the invariant subspaces of an LM-decomposable Lie algebra of bounded operators which contains generalized scalar operators. We shall prove that we can obtain the same conclusion as in Theorem 5 with a weaker hypothesis.

Theorem 7. *Let $\mathcal{U}$ be an LM-decomposable Lie subalgebra of $\mathcal{B}(\mathcal{X})$ with the decomposition $\mathcal{U} = \mathcal{R} + \mathcal{G}$ (see Definition 3 from § 4). If the center of $\mathcal{R}$ contains only generalized scalar operators, then one of the following assertions is true:*

(I) *there exists a non-zero commutative finite-dimensional ideal of $\mathcal{U}$ consisting of nilpotent operators;*

(II_1) *there exists $A \in Z_{\mathcal{R}}$ with nontrivial spectral maximal subspaces which are invariant to $\mathcal{U}$;*

(II_2) *$\mathcal{R}$ is scalar (i.e. $\mathcal{R} = \{\lambda(R)I\}_{R\in\mathcal{R}}$, $\lambda(R) \in \mathbb{C}$).*

Proof. By Theorem 4 from § 18 it remains to study the case when the assertion (II) is true. We have either $\{0\} \neq Z_{\mathcal{R}} \subseteq Z_{\mathcal{U}}$ or $\mathcal{R} = \{0\}$ because $\mathcal{R}$ is quasinilpotent (see the Lemma 2) and $[\mathcal{G}, \mathcal{R}] = \{0\}$. If $\mathcal{R} \neq \{0\}$, then there exist two possibilities:

a) there exists $R \in Z_{\mathcal{R}}$ such that the spectrum of R contains at least two distinct points;

b) for any $R \in Z_{\mathcal{R}}$ we have $\sigma(R) = \{\lambda(R)\}$ with $\lambda(R) \in \mathbb{C}$.

In the case *a*) we have (II_1) because R possesses nontrivial spectral maximal subspaces (cf. Corollary 1 from § 14) which are invariant to $\mathcal{U}$ because R commutes with $\mathcal{U}$ (see e.g. Remark 4 from § 12). Hence in this case we can put $A := R$ and (II_1) holds.

In the case *b*) we can write

$$R = \lambda(R)I + Q_R \text{ with } \lambda(R) \in \mathbb{C},\ Q_R \text{ nilpotent, for any } R \in Z_{\mathcal{R}}.$$

If there exists $R \in Z_{\mathcal{R}}$ with $Q_R \neq 0$ then we have (I). Hence only the case $Z_{\mathcal{R}} = \{\lambda(R)I\}_{R\in Z_{\mathcal{R}}}$ remains to be studied. But in this case the above Theorem 6 shows that $\mathcal{R} = \{\lambda(R)I\}_{R\in\mathcal{R}}$ where $\lambda(R) \in \mathbb{C}$, hence we have ($\mathrm{II}_2$). □

Corollary 1. *Let $\mathcal{U} \subset \mathcal{B}(\mathcal{X})$ be an LM-decomposable Lie algebra of bounded operators with the decomposition $\mathcal{U} = \mathcal{R} + \mathcal{G}$. If $Z_{\mathcal{R}}$, the center of $\mathcal{R}$, consists of generalized scalar operators, then we have either*

(i) *there exists a nontrivial closed subspace of $\mathcal{X}$ which is invariant to $\mathcal{U}$,*

or

(s) *either $\mathcal{U} = \mathcal{G}$ or $\mathcal{U} = \mathbb{C}I + \mathcal{G}$, where $\mathcal{G}$ is a Lie algebra such that all its ideals are primitive.*

Finally we shall study the case when an LM-decomposable Lie subalgebra $\mathcal{U}$ of $\mathcal{B}(\mathcal{X})$ contains a non-zero compact operator. Hence let us suppose that

$$\mathcal{K}_{\mathcal{U}} := \{K \in \mathcal{U} \mid K \text{ compact operator}\} \neq \{0\}.$$

We shall analyse the case when the radical $\mathcal{R}$ of $\mathcal{U}$ is not scalar.

By Theorem 4 from § 18 we have either (I) or (II). If we have (II), then by Theorem 6 above we deduce that $Z_{\mathcal{R}}$ is not scalar. But $[\mathcal{G}, \mathcal{R}] = \{0\}$; hence $Z_{\mathcal{R}} \subseteq Z_{\mathcal{U}}$ and it results that there exists an $A_0 \in Z_{\mathcal{U}}$ which is not scalar. Particularly A_0 commutes with a non-zero compact operator because $\{0\} \neq \mathcal{K}_{\mathcal{U}} \subseteq \mathcal{U}$. We have proved the following theorem.

Theorem 8. *Let $\mathcal{U}$ be an LM-decomposable Lie subalgebra of $\mathcal{B}(\mathcal{X})$ with the decomposition $\mathcal{U} = \mathcal{R} + \mathcal{G}$. If $\mathcal{U}$ contains a non-zero compact operator, then one of the following assertions is true:*

(1) *either $\mathcal{U} = \mathcal{G}$ or $\mathcal{U} = \mathbb{C}I + \mathcal{G}$, where $\mathcal{G}$ is a Lie algebra such that all its ideals are primitive ;*

(2) *$\{0\} \neq \mathcal{R} \neq \mathbb{C}I$ and one of the following statements holds:*

(2_1) *there exists a non-scalar operator $A_0 \in Z_{\mathcal{U}}$ such that A_0 commutes with a non-zero compact operator;*

(2_2) *there exists a commutative non-zero finite-dimensional ideal of $\mathcal{U}$ consisting of nilpotent operators.*

Corollary 2. *Let $\mathcal{U} \subset \mathcal{B}(\mathcal{X})$ be an LM-decomposable Lie algebra of bounded operators with the decomposition $\mathcal{U} = \mathcal{R} + \mathcal{G}$. If $\mathcal{U}$ contains non-zero compact operators then one of the following statements holds:*

(1) *either $\mathcal{U} = \mathcal{G}$ or $\mathcal{U} = \mathbb{C}I + \mathcal{G}$, where $\mathcal{G}$ is a Lie algebra such that all its ideals are primitive ;*

(2) *$\mathcal{U}$ has a nontrivial closed invariant subspace.*

Proof. We apply the above theorem, Lomonosov's Theorem and Lemma 1 from the beginning of the present paragraph. □

Corollary 3. *Either every quasisolvable Lie subalgebra of $\mathcal{B}(\mathcal{X})$ which contains a non-zero compact operator has a nontrivial closed invariant subspace or $\dim \mathcal{X} \leq 1$.*

Corollary 4. *Either every finite-dimensional Lie subalgebra of $\mathcal{B}(\mathcal{X})$ which contains a non-zero compact operator has a nontrivial closed invariant subspace or* $\dim \mathcal{X} < \infty$.

The proof of Corollary 3 is obvious.

The proof of Corollary 4 results by the "unitary trick" of Weyl. We consider $\mathcal{G}_u$ a compact form of the complex semisimple Lie algebra $\mathcal{G}$. If $\mathcal{G}$ has no nontrivial closed invariant subspaces, then $\mathcal{G}_u$ has no nontrivial closed invariant subspaces. The operators

$$\exp G_u \quad (G_u \in \mathcal{G}_u)$$

generate a compact, connected and simply connected Lie group K which results being topologically irreducible. It is well known that in this case $\dim \mathcal{X} < \infty$ because K is compact by Remark 1 from § 6.

Finally we shall prove some consequences of the asymptotic formulas for commutators, concerning invariant subspaces.

Theorem 9. *If $Q \in \mathcal{B}(\mathcal{X})$ and F is a spectral set of Q, then $\mathcal{X}_Q(F)$ is an invariant subspace for the associative operator algebra*

$$\mathcal{A}_Q = \{S \in \mathcal{B}(\mathcal{X}) \mid \lim_{m\to\infty} \|(adQ)^m S\|^{\frac{1}{m}} = 0\}.$$

Proof. Choose $f \in \mathcal{O}(\sigma(Q))$ such that $f \equiv 1$ on a neighbourhood of F and $f \equiv 0$ on a neighbourhood of $\sigma(Q)\setminus F$. By Corollary 4 from § 15, we get $[f(Q), S] = 0$, hence $f(Q)\mathcal{X} = \mathcal{X}_Q(F)$ is an invariant subspace for every operator $S \in \mathcal{A}_Q$. □

Corollary 5. *Let $\mathcal{U}$ be an LM-decomposable Lie subalgebra of $\mathcal{B}(\mathcal{X})$, $\mathcal{U} = \mathcal{R} + \mathcal{G}$ the decomposition of $\mathcal{U}$. If $\mathcal{R}$ contains an operator Q whose spectrum is non-connected, then there exists a nontrivial closed invariant subspace for $\mathcal{U}$.*

Proof. By Theorem 3 from § 19 and the above Lemma 1, only one case remains to be studied, namely the case when the assertion (II) of Theorem 3 from § 19 holds. Particularly,

$$[\mathcal{G}, \mathcal{R}] = \{0\} \text{ and } \mathcal{R} \text{ is a quasinilpotent Lie algebra.}$$

Let $Q \in \mathcal{R}$ with non-connected spectrum. Then there exists a spectral set F of Q, $\emptyset \neq F \neq \sigma(Q)$. So $\mathcal{X}_Q(F)$ is a non-trivial invariant subspace for $\mathcal{A}_Q$ by the above Theorem 9. But $\mathcal{R} \subseteq \mathcal{A}_Q$ because $\mathcal{R}$ is quasinilpotent. On the other hand, $\mathcal{X}_Q(F)$ is invariant to $\mathcal{G}$ because $[\mathcal{G}, Q] = \{0\}$. So $\mathcal{X}_Q(F)$ is a non-trivial invariant subspace for $\mathcal{U}$ and the proof is finished. □

Corollary 6. *Let $\mathcal{R}$ be a quasisolvable Lie subalgebra of $\mathcal{B}(\mathcal{X})$. If $\mathcal{R}$ contains an operator with non-connected spectrum, then there exists a nontrivial closed invariant subspace for $\mathcal{R}$.*

§ 22 The irreducible representations of an LM-decomposable Lie algebra. Infinite-dimensional variant of Lie's Theorem on a complex Banach space

We will consider the algebra $\mathcal{B}(\mathcal{X})$ of all bounded linear operators on some complex Banach space $\mathcal{X}$. We refer only to the following types of irreducibility: operatorially irreducibility, completely irreducibility, topologically irreducibility (or transitivity) and paratransitivity.

Definition 1. A subset $\mathcal{S}$ of $\mathcal{B}(\mathcal{X})$ is called *operatorially irreducible* if its commutant $\mathcal{S}'$ in $\mathcal{B}(\mathcal{X})$ consists of scalar operators.

Definition 2. A subset $\mathcal{S} \subseteq \mathcal{B}(\mathcal{X})$ is called *completely irreducible* if its associative hull is weakly dense in $\mathcal{B}(\mathcal{X})$.

Recall that a *paraclosed subspace* of $\mathcal{X}$ means the range of a bounded linear operator from an arbitrary Banach space to $\mathcal{X}$.

Notation 1. For $\mathcal{S} \subseteq \mathcal{B}(\mathcal{X})$ we shall denote by

$$Lat\, \mathcal{S}$$

the lattice of all closed subspaces of $\mathcal{X}$ which are invariant to $\mathcal{S}$. Moreover we shall denote by

$$Lat^{1/2}\, \mathcal{S}$$

the lattice of all paraclosed subspaces of $\mathcal{X}$ which are invariant to $\mathcal{S}$.

Definition 3. A subset $\mathcal{S} \subseteq \mathcal{B}(\mathcal{X})$ is *topologically irreducible* (or *transitive*) respectively *paratransitive* if $Lat\, \mathcal{S} = \{\{0\}, \mathcal{X}\}$, respectively $Lat^{1/2}\, \mathcal{S} = \{\{0\}, \mathcal{X}\}$.

Definition 4. If $\mathcal{L}$ is a Lie algebra, then a Banach space representation $\rho : \mathcal{L} \to \mathcal{B}(\mathcal{X})$ is called *operatorially, completely* or *topologically irreducible* if $\rho(\mathcal{L})$ is respectively operatorially, completely or topologically irreducible. The representation ρ is called *paratransitive* if $\rho(\mathcal{L})$ is paratransitive.

Obviously the results from § 21 have irreducible variants as follows.

As a corollary of Theorem 3 from § 21 we obtain:

Theorem 1. *Let $\mathcal{L}$ be an LM-decomposable Lie subalgebra of $\mathcal{B}(\mathcal{X})$ and $\mathcal{L} = \mathcal{R} + \mathcal{G}$ its decomposition. If $\mathcal{L}$ is topologically irreducible, then $[\mathcal{G}, \mathcal{R}] = \{0\}$, $\mathcal{R}$ is a quasinilpotent Lie algebra, $\mathcal{R}$ contains no non-zero nilpotent operators and every analytically central element in $\mathcal{R}$ is central in $\mathcal{R}$.*

As a corollary of Theorem 4 from § 21 we obtain:

Theorem 2. *Let $\mathcal{L}$ be an LM-decomposable Lie algebra and $\mathcal{L} = \mathcal{R} + \mathcal{G}$ its decomposition. If $\rho : \mathcal{L} \to \mathcal{B}(\mathcal{X})$ is a topologically irreducible representation with the range* *consisting of decomposable operators, then* $\rho([\mathcal{R},\mathcal{G}]) = \{0\}$, $\rho(\mathcal{R})$ *is a quasinilpotent Lie algebra which is one-spectral, contains no non-zero nilpotent operators and every analytically central element in* $\rho(\mathcal{R})$ *is central in* $\rho(\mathcal{R})$.

By Corollary 1 from § 21 we deduce the following result:

Theorem 3. *Let* $\mathcal{L}$ *be an LM-decomposable Lie algebra and* $\mathcal{L} = \mathcal{R} + \mathcal{G}$ *its decomposition. If* $\rho : \mathcal{L} \to \mathcal{B}(\mathcal{X})$ *is a topologically irreducible representation such that the center of* $\rho(\mathcal{R})$ *consists of generalized scalar operators, then either* $\rho(\mathcal{L}) = \rho(\mathcal{G})$ *or* $\rho(\mathcal{L}) = \mathbb{C}I + \rho(\mathcal{G})$, *where* $\rho(\mathcal{G})$ *is a Lie algebra such that all its ideals are primitive.*

Corollary 2 from § 21 gives an analogous result:

Theorem 4. *If* $\mathcal{L}$ *is an LM-decomposable Lie algebra and* $\rho : \mathcal{L} \to \mathcal{B}(\mathcal{X})$ *is a topologically irreducible representation which contains a compact operator in its range, then* $\rho(\mathcal{L})$ *is a Lie algebra such that all its ideals are primitive.*

We can describe also some corollaries of the above results, which are infinite-dimensional variants of Lie's classical theorem which says that for every irreducible finite-dimensional representation of a solvable Lie algebra in a complex vector space $\mathcal{V}$ we have $\dim \mathcal{V} \leq 1$.

Corollary 1. *Any quasisolvable topologically irreducible Lie subalgebra of* $\mathcal{B}(\mathcal{X})$ *is a quasinilpotent Lie algebra whose elements are only operators with connected spectra.*

Corollary 2. *Every topologically irreducible representation of a quasisolvable Lie algebra with the range consisting of decomposable operators on some complex Banach space is one-spectral.*

Corollary 3. *If the center of a topologically irreducible representation on a Banach space of a quasisolvable Lie algebra consists only of generalized scalar operators, then the representation is scalar (i.e. of dimension* ≤ 1*).*

Corollary 4. *Every topologically irreducible generalized scalar representation of a finite-dimensional Lie algebra in a complex Banach space* $\mathcal{X}$ *is finite-dimensional (i.e.* $\dim \mathcal{X} < \infty$*).*

Proof. See the "unitary trick" of Weyl as in the proof of Corollary 4 from § 21. □

Corollary 5. *If* ρ *is a topologically irreducible representation of a quasisolvable Lie algebra on the complex Banach space* $\mathcal{X}$ *and the range of* ρ *contains a compact operator, then* $\dim \mathcal{X} \leq 1$.

Now we discuss the operatorially irreducibility. As a corollary of Theorem 6 from § 21 we obtain:

Theorem 5. *Each operatorially irreducible representation* ρ *of a quasinilpotent Lie algebra by bounded operators on a complex Banach space* $\mathcal{X}$ *is scalar (i.e.* $\dim \mathcal{X} \leq 1$*).*

Proof. It suffices to observe that the range $\mathcal{N}$ of ρ is a quasinilpotent Lie algebra. Then its center $Z_{\mathcal{N}}$ is scalar because $\mathcal{N}$ is operatorially irreducible. By Theorem 6 from § 21 we deduce that $\mathcal{N}$ is scalar and, since it is operatorially irreducible, we must have $\dim \mathcal{X} \leq 1$. □

For an arbitrary representation, Theorem 1 above gives in the case of the operatorially irreducibility the same conclusion as in Theorem 3 above.

Theorem 6. *If $\mathcal{U}$ is an LM-decomposable, topologically and operatorially irreducible Lie subalgebra of $\mathcal{B}(\mathcal{X})$ then either $\mathcal{U} = \mathcal{G}$ or $\mathcal{U} = \mathbb{C}I + \mathcal{G}$, where $\mathcal{G}$ is a Lie algebra such that all its ideals are primitive. If in addition $\mathcal{G}$ is finite-dimensional (particularly if $\mathcal{U}$ is finite-dimensional), then* $\dim \mathcal{X} < \infty$.

Proof. Let $\mathcal{U} = \mathcal{R} + \mathcal{G}$ be a decomposition of $\mathcal{U}$ and apply Theorem 1 above. We deduce that $[\mathcal{G}, \mathcal{R}] = \{0\}$ and that $\mathcal{R}$ is a quasinilpotent Lie algebra. Obviously $Z_{\mathcal{R}} \subseteq Z_{\mathcal{U}}$. But $\mathcal{U}$ is operatorially irreducible, hence either $Z_{\mathcal{R}} = \{0\}$ or $Z_{\mathcal{R}} = \mathbb{C}I$. By Theorem 6 from § 21 we obtain either $\mathcal{R} = \{0\}$ or $\mathcal{R} = \mathbb{C}I$ because $\mathcal{R}$ is a quasinilpotent Lie algebra. The last part of the conclusion can be obtained by the "unitary trick" (see the proof of Corollary 4 from § 21). □

Remark 1. The above theorem is true for quasisolvable Lie algebras ($\mathcal{G} = \{0\}$) and for finite-dimensional Lie algebras. So *every topologically and operatorially irreducible representation of a finite-dimensional Lie algebra in a complex Banach space is finite-dimensional.*

Now let $\mathcal{A}(= \mathcal{R} + \mathcal{G})$ be an LM-decomposable Lie subalgebra in $\mathcal{B}(\mathcal{X})$ and let $\widetilde{\mathcal{A}}$ be the weakly closed associative subalgebra of $\mathcal{B}(\mathcal{X})$ generated by $\mathcal{A}$ and I. We suppose that $\dim \mathcal{X} > 1$. If $\dim \mathcal{A} \leq 1$, then we have $\widetilde{\mathcal{A}} \neq \mathcal{B}(\mathcal{X})$ because $\dim \mathcal{X} > 1$.

Next let us suppose that $\dim \mathcal{A} > 1$. In this case we have by Theorem 3 from § 19 one of the following statements:

(I) There exists a finite-dimensional commutative ideal $\mathcal{N} \neq \{0\}$ of $\mathcal{A}$ consisting of nilpotent operators.

(II) $\mathcal{R}$ is a quasinilpotent Lie algebra, $[\mathcal{R}, \mathcal{G}] = \{0\}$, $\mathcal{R}$ contains no non-zero nilpotent operators and every analytically central element in $\mathcal{R}$ is central in $\mathcal{R}$.

If we have (I), then the closed subspace $\bigcap_{Q \in \mathcal{N}} Ker Q$ is different from $\{0\}, \mathcal{X}$ and invariant to $\mathcal{A}$, hence to $\widetilde{\mathcal{A}}$. In this case we have obviously $\widetilde{\mathcal{A}} \neq \mathcal{B}(\mathcal{X})$.

In case (II) we can have either

(II_a) $\dim \mathcal{R} \leq 1$

or

(II_b) $\dim \mathcal{R} > 1$.

In the situation (II_b) let $\mathcal{I}_1 \subset \mathcal{I}_2$ be two ideals of $\mathcal{R}$ (hence of $\mathcal{A} = \mathcal{R} + \mathcal{G}$ since $[\mathcal{R}, \mathcal{G}] = \{0\}$), $\dim \mathcal{I}_1 = 1$, $\dim \mathcal{I}_2 = 2$. Let $\{A, B\}$ be a basis of $\mathcal{I}_2$ with $A \in \mathcal{I}_1$. We remark that $[T, A] = 0$ for every $T \in \mathcal{R}$, hence also for every $T \in \mathcal{A}$.

We can have one of the following cases:

(*i*) $A \neq \xi I$ for every $\xi \in \mathbb{C}$.

(*ii*) $A = \xi I$ for some $\xi \in \mathbb{C}$.

If we have (ii), then

$$[T, B] = \lambda(T)B + \mu(T)I \text{ for every } T \in \mathcal{A},$$

where λ, μ are certain linear functionals on $\mathcal{A}$. By the Wintner-Wielandt Theorem we have: $\lambda(T) = 0 \Rightarrow \mu(T) = 0$ hence $\mu(\cdot) = c\lambda(\cdot)$ for some $c \in \mathbb{C}\backslash\{0\}$. Therefore for $X := B + cI$ we have $[T, X] = \lambda(T)X$ for every $T \in \mathcal{A}$; moreover $X \neq \xi I$ for every $\xi \in \mathbb{C}$ because $\dim \mathcal{I}_2 = 2$.

Hence in any case (i.e., either (i) or (ii)) there exists $X \in \mathcal{R}$ such that

$$[T, X] = \lambda(T)X,\ \lambda(T) \in \mathbb{C} \quad \forall T \in \mathcal{A} \quad \text{and} \quad X \neq \xi I \quad \forall \xi \in \mathbb{C}.$$

But $\mathcal{R}$ contains no non-zero nilpotent operators. Hence $\lambda(T) = 0$ for every $T \in \mathcal{A}$. Hence it results that there exists $X \in \mathcal{R}, X \neq \xi I$ for each $\xi \in \mathbb{C}$, such that $[T, X] = 0$ for every $T \in \mathcal{A}$ hence for every $T \in \widetilde{\mathcal{A}}$. We deduce that ($\mathrm{II}_b$) (i.e. $\dim \mathcal{R} > 1$) implies $\widetilde{\mathcal{A}} \neq \mathcal{B}(\mathcal{X})$.

It remains to study the case when either $\mathcal{A} = \mathbb{C}I + \mathcal{G}$ or $\mathcal{A} = \mathcal{G}$, where $\mathcal{G}$ is a Lie algebra such that every ideal of $\mathcal{G}$ is primitive. So we have proved the following theorem.

Theorem 7. *Let $\mathcal{A}$ be an LM-decomposable Lie subalgebra of $\mathcal{B}(\mathcal{X})$. Let $\widetilde{\mathcal{A}}$ be the weakly closed associative subalgebra of $\mathcal{B}(\mathcal{X})$ generated by $\mathcal{A}$ and I (the unit element of $\mathcal{B}(\mathcal{X})$). If $\dim \mathcal{X} > 1$, then we have*

$$\text{either } \widetilde{\mathcal{A}} \neq \mathcal{B}(\mathcal{X}) \text{ or } \mathcal{A} = \mathbb{C}I + \mathcal{G} \text{ or } \mathcal{A} = \mathcal{G},$$

where $\mathcal{G} \neq \{0\}$ is a Lie algebra such that every ideal of $\mathcal{G}$ is primitive (particularly $\mathcal{G}$ is semisimple).

Corollary 6. *Let $\mathcal{R}$ be a quasisolvable Lie subalgebra of $\mathcal{B}(\mathcal{X})$ and $\widetilde{\mathcal{R}}$ be the weakly closed associative subalgebra of $\mathcal{B}(\mathcal{X})$ generated by $\mathcal{R}$ and I (the identity operator on $\mathcal{X}$). If $\dim \mathcal{X} > 1$, then we have $\widetilde{\mathcal{R}} \neq \mathcal{B}(\mathcal{X})$.*

The following result is also an infinite-dimensional variant of the classical Lie's Theorem and it is an easy consequence of Theorem 7.

Corollary 7. *Let $\mathcal{R}$ be a quasisolvable completely irreducible Lie subalgebra of $\mathcal{B}(\mathcal{X})$. Then $\dim \mathcal{X} \leq 1$.*

Corollary 8. *Let $\mathcal{A}$ be a finite-dimensional Lie subalgebra of $\mathcal{B}(\mathcal{X})$. If we have $\dim \mathcal{X} > 1$, then one of the following assertions is true:*

(I) $\widetilde{\mathcal{A}} \neq \mathcal{B}(\mathcal{X})$;

(II) $\dim \mathcal{X} < \infty$ *and either* $\mathcal{A} = \mathbb{C}I + \mathcal{G}$ *or* $\mathcal{A} = \mathcal{G}$ *where* $\mathcal{G}$ *is semisimple.*

Proof. By the above Theorem 7 only the case when either $\mathcal{A} = \mathcal{G} + \mathbb{C}I$ or $\mathcal{A} = \mathcal{G}$ remains to be studied, where $\mathcal{G}$ is a semisimple finite-dimensional Lie algebra. But in this case $\widetilde{\mathcal{A}}$ is also of finite dimension (see Theorem 2 from § 30). If $\widetilde{\mathcal{A}} = \mathcal{B}(\mathcal{X})$, then $\dim \mathcal{X} < \infty$.

Hence either $\widetilde{\mathcal{A}} \neq \mathcal{B}(\mathcal{X})$, or $\dim \mathcal{X} < \infty$ and either $\mathcal{A} = \mathbb{C}I + \mathcal{G}$ or $\mathcal{A} = \mathcal{G}$ for a certain semisimple Lie algebra $\mathcal{G}$. □

Then we deduce also the following infinite-dimensional variant of Lie's Theorem.

Theorem 8. *Let $\mathcal{A}$ be a finite-dimensional completely irreducible Lie subalgebra of $\mathcal{B}(\mathcal{X})$. Then $\dim \mathcal{X} < \infty$; if $\dim \mathcal{X} > 1$, then the solvable radical of $\mathcal{A}$ is scalar.*

The proof of the above Theorem 7 remains valid for a class of Lie algebras which contains the class of the quasisolvable Lie algebras. The following definition of this class is suggested by the above proof.

Definition 5. We denote by $\mathcal{R}_p$ the class of all Lie algebras $\mathcal{A}$ such that there exists a chain

$$\mathcal{I}_1 \subset \mathcal{I}_2 \subset \cdots \subset \mathcal{I}_p$$

of ideals of $\mathcal{A}$ such that $\dim \mathcal{I}_k = k$ for $1 \leq k \leq p$.

Remark 2. If $\mathcal{A} \in \mathcal{R}_p$, then $\dim \mathcal{R} \geq p$, where $\mathcal{R}$ is the solvable radical of $\mathcal{A}$.

Remark 3. If $\mathcal{A}$ is an ideally finite Lie algebra over an algebraically closed field and $[\mathcal{R}, \mathcal{G}] = \{0\}$, where $\mathcal{A} = \mathcal{R} + \mathcal{G}$ is a LM-decomposition of $\mathcal{A}$, $\mathcal{R}$ is the solvable radical of $\mathcal{A}$, then $\mathcal{A} \in \mathcal{R}_p$ iff $\dim \mathcal{R} \geq p$.

Remark 4. If $p \geq q$, then $\mathcal{R}_p \subset \mathcal{R}_q$.

Remark 5. If $\dim \mathcal{A} < \infty$, then $\mathcal{A} \in \mathcal{R}_{\dim \mathcal{A}}$ means that $\mathcal{A}$ is solvable.

Theorem 9. *Let $\mathcal{A}$ be a Lie subalgebra of $\mathcal{B}(\mathcal{X})$. If $\mathcal{A} \in \mathcal{R}_p$ for some $p \geq 2$ and $\dim \mathcal{X} > 1$, then $\widetilde{\mathcal{A}} \neq \mathcal{B}(\mathcal{X})$.*

Proof. As in the proof of the above Theorem 7 we have $\mathcal{I}_1 \subset \mathcal{I}_2 \subset \mathcal{A}$ two ideals of $\mathcal{A}$ with $\dim \mathcal{I}_1 = 1$, $\dim \mathcal{I}_2 = 2$, a basis $\{A, B\}$ of $\mathcal{I}_2$ with $A \in \mathcal{I}_1$ ... and in the same way we deduce that $\widetilde{\mathcal{A}} \neq \mathcal{B}(\mathcal{X})$. □

In an obvious manner we can deduce the following "irreducible" variant of Theorem 9.

Corollary 9. *If $\mathcal{A} \subseteq \mathcal{B}(\mathcal{X})$ is a completely irreducible Lie subalgebra of class $\mathcal{R}_p$ for some $p \geq 2$ then $\dim \mathcal{X} \leq 1$.*

It's also possible to give another "irreducible" variant of the above results when "irreducible" means paratransitive. This possibility is given by the following Burnside type theorem (see C. Foiaş [1]).

Theorem 10. *If $\mathcal{S}$ is a paratransitive weakly closed associative subalgebra of $\mathcal{B}(\mathcal{X})$, then $\mathcal{S} = \mathcal{B}(\mathcal{X})$.*

Corollary 10. *Let $\mathcal{A}$ be a Lie subalgebra of $\mathcal{B}(\mathcal{X})$ and $\mathcal{A} \in \mathcal{R}_p$ for some $p \geq 2$. If $\mathcal{A}$ is paratransitive, then $\dim \mathcal{X} \leq 1$.*

Proof. If $\dim \mathcal{X} > 1$, then $\widetilde{\mathcal{A}} \neq \mathcal{B}(\mathcal{X})$ by Theorem 9. By the above Burnside type theorem we get $Lat^{1/2}\widetilde{\mathcal{A}} \neq \{\{0\}, \mathcal{X}\}$. But $Lat^{1/2}\mathcal{A} = Lat^{1/2}\widetilde{\mathcal{A}}$, which means that $\mathcal{A}$ cannot be paratransitive. □

§ 23 The associative envelope of a Lie algebra of quasinilpotent operators

Now we shall establish some variants of Engel's Theorem holding on an infinite-dimensional complex Banach space $\mathcal{X}$, respectively Hilbert space $\mathcal{H}$ (see Theorems 1–3 below). First let's introduce some useful notation: for a subset $\mathcal{G}$ of $\mathcal{B}(\mathcal{X})$ we denote by $\mathcal{A}_0(\mathcal{G})$ the associative complex subalgebra of $\mathcal{B}(\mathcal{X})$ generated by $\mathcal{G}$. Moreover we denote by $\overline{\mathcal{A}_0(\mathcal{G})}$ the closure of $\mathcal{A}_0(\mathcal{G})$ with respect to the norm operator topology.

Theorem 1. *Let $\mathcal{G}$ be a finite-dimensional complex Lie subalgebra of $\mathcal{B}(\mathcal{X})$ consisting only of quasinilpotent operators. Then $\mathcal{G}$ is a nilpotent Lie algebra and $\mathcal{A}_0(\mathcal{G})$ contains only quasinilpotent operators. Moreover, if $\mathcal{G}$ contains only nilpotent operators, then $\mathcal{A}_0(\mathcal{G})$ also contains only nilpotent operators.*

We postpone the proof of Theorem 1 until after Corollary 2 from § 26 because it needs the spectral theory which will be developed in Chapter IV.

In the following we establish variants of Theorem 1 by strengthening the conditions on operators from the Lie algebra in exchange for weakening the condition on the algebra itself. A first variant is the following.

Theorem 2. *Let $\mathcal{G}$ be a complex Lie subalgebra consisting only of compact operators. Assume that $\mathcal{G}$ has a dense Lie subalgebra which is locally finite and contains only quasinilpotent operators. Then $\overline{\mathcal{A}_0(\mathcal{G})}$ contains only quasinilpotent operators.*

Proof. It is well known that the set of all compact quasinilpotent operators is closed with respect to the norm operator topology. Consequently it suffices to prove that every element of $\mathcal{A}_0(\mathcal{G})$ is quasinilpotent. To this end let $A \in \mathcal{A}_0(\mathcal{G})$. Then there exist $G_1, \dots, G_n \in \mathcal{G}$ and a polynomial $p \in \mathbb{C}\langle X_1, \dots, X_n\rangle$ in n non-commuting variables such that $p(0) = 0$ and $p(G_1, \dots, G_n) = A$.

Now let $\mathcal{L}$ be the dense Lie subalgebra of $\mathcal{G}$ which is locally finite and contains only quasinilpotent operators. Then for every $i \in \{1, \dots, n\}$ there exists a sequence $\{G_i^{(j)}\}_{j\geq 1}$ from $\mathcal{L}$ such that $\lim_{j\to\infty} \|G_i^{(j)} - G_i\| = 0$. Particularly

$$\lim_{j\to\infty} p(G_1^{(j)}, \dots, G_n^{(j)}) = p(G_1, \dots, G_n) \tag{1}$$

(with respect to the norm operator topology on $\mathcal{B}(\mathcal{X})$). Now fix $j \geq 1$ for the moment. We have $G_1^{(j)}, \dots, G_n^{(j)} \in \mathcal{L}$ and $\mathcal{L}$ is locally finite, hence there exists a finite-dimensional Lie subalgebra $\mathcal{L}_j$ of $\mathcal{L}$ such that $G_1^{(j)}, \dots, G_n^{(j)} \in \mathcal{L}_j$. But $\mathcal{L}_j$ contains only quasinilpotent operators (since $\mathcal{L}_j \subseteq \mathcal{L}$) hence by Theorem 1 we deduce that $p(G_1^{(j)}, \dots, G_n^{(j)})$ is quasinilpotent. But this operator is also compact, hence by (1) it follows that $p(G_1, \dots, G_n)(= A)$ is also quasinilpotent. □

In the next variant of Theorem 1 (see Theorem 3 below) we allow the Lie algebra to be arbitrary but we have to work only with Hilbert-Schmidt operators (of course on a Hilbert space). We shall need the following characterization of the quasinilpotent operators, some power of which belongs to the trace-class.

Lemma 1. *Let $A \in \mathcal{B}(\mathcal{H})$ be such that A^{n_0} belongs to the trace-class for some $n_0 \geq 1$. Then A^n belongs to the trace-class for every $n \geq n_0$ and the following assertions are equivalent:*
(i) *A is quasinilpotent;*
(ii) *$Tr(A^n) = 0$ for every $n \geq n_0$.*

Proof. Since the trace-class is an ideal of the associative algebra $\mathcal{B}(\mathcal{H})$ and A^{n_0} belongs to this ideal, it follows that $A^{n_0} \cdot A^{n-n_0} = A^n$ also belongs to it for every $n \geq n_0$.

Now, for proving the desired equivalence, first recall Lidskiĭ's Theorem asserting that for a trace-class operator $T \in \mathcal{B}(\mathcal{H})$ we have

$$Tr\, T = \sum_{\lambda \in \sigma(T)} \lambda \tag{2}$$

where each $\lambda \in \sigma(T) \setminus \{0\}$ occurs a number of times equal to its spectral multiplicity and the series from (2) is absolutely convergent. Now let's come back to the proof. If $\sigma(A) = \{0\}$, then $\sigma(A^n) = \{0\}$ for every $n \geq n_0$ hence (ii) follows by (2).

Now assume that (ii) holds. For proving that A is quasinilpotent it suffices to check that $\sigma(A^{n_0}) = \{0\}$. Hence by replacing A by A^{n_0} we may assume $n_0 = 1$, i.e. A belongs to the trace-class. Suppose that A is not quasinilpotent and let $\{\lambda_i\}_{i \geq 1}$ be a sequence of complex numbers with the properties:

1° $|\lambda_1| \geq |\lambda_2| \geq \cdots$;
2° $\sigma(A) = \{\lambda_i \mid i \geq 1\}$;
3° every $\lambda \in \sigma(A) \setminus \{0\}$ occurs in the sequence $\{\lambda_i\}_{i \geq 1}$ a number of times equal to its spectral multiplicity.

Then by (2) and by the hypothesis (ii) we get

$$\sum_{i=1}^{\infty} (\lambda_i)^n = 0 \text{ for every } n \geq 1. \tag{3}$$

Since $\lim_{i \to \infty} \lambda_i = 0$ and not all the λ_i's equal zero (by our supposition), by 1° we may assume without loss of generality that for a certain $k \geq 1$ we have

$$1 = |\lambda_1| = \cdots = |\lambda_k| > |\lambda_{k+1}| \geq \cdots .$$

Then by the Lebesgue Dominated Convergence Theorem we easily deduce

$$\lim_{n \to \infty} \left(\sum_{i=k+1}^{\infty} (\lambda_i)^n \right) = 0.$$

Hence by (3) there exists $n_1 \geq 1$ such that

$$\left| \sum_{i=1}^{k} (\lambda_i)^n \right| < \frac{1}{3} \text{ for every } n \geq n_1. \tag{4}$$

On the other hand consider the point $w = (\lambda_1, \dots, \lambda_k)$ of the k-dimensional torus $\mathbf{T}^k$ and denote by G the closed subgroup of $\mathbf{T}^k$ generated by w. It is well known that G is either finite or isomorphic to a torus $\mathbf{T}^l$ (with $l \leq k$) and in both cases the set

$$\{w^m = ((\lambda_1)^m, \dots, (\lambda_k)^m) | m \geq 1\}$$

is dense in G. Particularly there exists $m \geq n_1$ such that $|(\lambda_i)^m - 1| < 1/3k$ for $i = 1, \dots, k$. Hence

$$\left|\sum_{i=1}^{k}(\lambda_i)^m\right| = \left|k - \sum_{i=1}^{k}(1-(\lambda_i)^m)\right| \geq k - \sum_{i=1}^{k}|1-(\lambda_i)^m| > k - \frac{1}{3} \geq \frac{2}{3},$$

which contradicts (4). So the assumption that (i) does not hold is not true. □

We need also the following description of the associative hull of a Lie subalgebra of $\mathcal{B}(\mathcal{X})$.

Lemma 2. *If $\mathcal{G}$ is a complex Lie subalgebra of $\mathcal{B}(\mathcal{X})$, then the set*

$$\left\{\sum_{i=1}^{m}\alpha_i B_i^{k_i} | B_1, \dots, B_m \in \mathcal{G}; \alpha_1, \dots, \alpha_m \in \mathbb{C}; k_1, \dots, k_m \geq 1; m \geq 1\right\}$$

coincides with $\mathcal{A}_0(\mathcal{G})$.

Proof. Denote by $\mathcal{M}$ the set referred to in the statement. Obviously $\mathcal{M}$ is a linear subspace of $\mathcal{A}_0(\mathcal{G})$. Hence for proving that $\mathcal{M} = \mathcal{A}_0(\mathcal{G})$ it suffices to check that if $f : \mathcal{B}(\mathcal{X}) \to \mathbb{C}$ is a linear functional such that $f|_{\mathcal{M}} = 0$, then $f|_{\mathcal{A}_0(\mathcal{G})} = 0$. To this end we prove by induction on n that the following implication holds:

$$A_1, \dots, A_n \in \mathcal{G} \Longrightarrow f(A_1 \dots A_n) = 0. \tag{5}$$

Since $\mathcal{G} \subseteq \mathcal{M} \subseteq Ker f$, the above implication holds for $n = 1$. Next assume that it holds for $n-1$ and take $A_1, \dots, A_n \in \mathcal{G}$. For every $t_1, \dots, t_n \in \mathbb{C}$ we have $(t_1A_1 + \dots + t_nA_n)^n \in \mathcal{M}$, so $f((t_1A_1 + \dots + t_nA_n)^n) = 0$. Since $t_1, \dots, t_n \in \mathbb{C}$ are arbitrary we easily deduce

$$f(\sum_{\sigma \in S(n)} A_{\sigma(1)} \dots A_{\sigma(n)}) = 0, \tag{6}$$

where $S(n)$ denotes the group of all permutations of the set $\{1, \dots, n\}$. But for every $\sigma \in S(n)$ we have

$$A_{\sigma(1)} \dots A_{\sigma(n)} = A_1 \dots A_n + \sum_j \theta_j, \tag{7}$$

where each θ_j is a product of at most $n-1$ factors from $\mathcal{G}$. (This fact is obvious if σ is a transposition of the type $(i, i+1)$; generally every permutation can be written as a product of such transpositions.) Then (7) together with the induction hypothesis imply $f(A_{\sigma(1)} \dots A_{\sigma(n)}) = f(A_1 \dots A_n)$ for every $\sigma \in S(n)$. Hence by (6) we get $f(A_1 \dots A_n) = 0$ and (5) is proved. □

Now we can prove:

Theorem 3. *Let $\mathcal{G}$ be a complex Lie subalgebra of $\mathcal{B}(\mathcal{H})$ consisting only of quasinilpotent Hilbert-Schmidt operators. Then $\overline{\mathcal{A}_0(\mathcal{G})}$ contains only quasinilpotent operators.*

Proof. As in the proof of Theorem 2 it suffices to prove that every element of $\mathcal{A}_0(\mathcal{G})$ is quasinilpotent. To this end let $A \in \mathcal{A}_0(\mathcal{G})$. Then A is a Hilbert-Schmidt operator, hence A^n belongs to the trace-class for every $n \geq 2$. Then by Lemma 1 it suffices to check that $Tr(A^n) = 0$ for each $n \geq 2$. To this end let's fix $n \geq 2$. Remark that $A^n \in \mathcal{A}_0(\mathcal{G})$, hence by Lemma 2 there exist $m \geq 1$; $\alpha_1, \dots, \alpha_m \in \mathbb{C}\backslash\{0\}$; $k_1, \dots, k_m \geq 1$; $B_1, \dots, B_m \in \mathcal{G}$ such that

$$A^n = \alpha_1 B_1^{k_1} + \cdots + \alpha_m B_m^{k_m}. \tag{8}$$

Obviously we may assume without loss of generality that $k_1 = 1$ and $k_i \geq 2$ for $i = 2, \dots, m$. Since B_i is a Hilbert-Schmidt operator, it then follows that $B_i^{k_i}$ belongs to the trace-class for $i = 2, \dots, m$. Moreover, since A^n also belongs to the trace-class, by (8) one gets that $B_1^{k_1}$ is also a trace-class operator if $\alpha_1 \neq 0$. Now, since each B_i is a quasinilpotent operator, by Lemma 1 ($(i) \Rightarrow (ii)$) and by (8) it follows $Tr(A^n) = 0$. □

§ 24 Commutativity modulo the Jacobson radical

The aim of the present paragraph is to expose some results concerning the Jacobson radical of the closed associative subalgebra of a Banach (associative) algebra. We shall use the notation introduced at the beginning of § 23.

Proposition 1. *Let $\mathcal{L}$ be a Lie subalgebra of the complex unital Banach algebra $\mathcal{A}$. If $\mathcal{I}$ is a finite-dimensional solvable ideal of $\mathcal{L}$ and we denote*

$$\mathcal{N}_{\mathcal{I}} := \{a \in \mathcal{I} \mid a \text{ is nilpotent}\},$$

then $\mathcal{N}_{\mathcal{I}}$ is also a Lie ideal of $\mathcal{L}$ and $\mathcal{N}_{\mathcal{I}} \subseteq rad\,\overline{\mathcal{A}(\mathcal{L})}$.

We postpone the proof of Proposition 1 until after Lemma 5 from § 28.

Lemma 1. *Let $\mathcal{B}$ be a complex unital Banach algebra with $rad\mathcal{B} = \{0\}$ (i.e., $\mathcal{B}$ is semisimple) and let $\mathcal{L}$ be a Lie subalgebra of $\mathcal{B}$ such that the closed associative subalgebra generated by $\mathcal{L}$ equals $\mathcal{B}$. If $\mathcal{I}$ is a finite-dimensional ideal of $\mathcal{L}$ such that $ad\,b|_{\mathcal{I}}$ is a nilpotent map for every $b \in \mathcal{L}$, then $[\mathcal{I}, \mathcal{L}] = \{0\}$.*

Proof. The Lie algebra $\{ad\,b|_{\mathcal{I}} : \mathcal{I} \to \mathcal{I} \mid b \in \mathcal{L}\}$ is nilpotent by Engel's Theorem since $\dim \mathcal{I} < \infty$. Consequently there exists $m \geq 1$ such that

$$(ad\,b_m) \cdots (ad\,b_1) b_0 = 0 \quad \forall b_1, \dots, b_m \in \mathcal{L}\, \forall b_0 \in \mathcal{I}. \tag{1}$$

If $m = 1$, then (1) expresses just the desired conclusion. Now assume that (1) holds for some $m > 1$ and prove that it holds also for $m - 1$. To this end take $b_1, \ldots, b_{m-1} \in \mathcal{L}$ and $b_0 \in \mathcal{I}$ arbitrary and denote

$$y := (ad\, b_{m-1}) \cdots (ad\, b_1) b_0,$$

and

$$u := \begin{cases} (ad\, b_{m-2}) \cdots (ad\, b_1) b_0 & \text{if } m > 2 \\ b_0 & \text{if } m = 2. \end{cases}$$

We must prove that $y = 0$. If we apply (1) for $b_m := y$ and $b_{m-1}, \ldots, b_0$ already chosen, then we deduce $[y, u] = 0$. But $y = [b_{m-1}, u]$, hence $[[b_{m-1}, u], u] = 0$. So $y = [b_{m-1}, u]$ is quasinilpotent by the Kleinecke-Sirokov theorem (see Remark 1 from § 17). But $[b_m, y] = 0$ for every $b_m \in \mathcal{L}$ by (1). Since $\mathcal{B}$ is the closed associative subalgebra generated by $\mathcal{L}$, it then follows that $[\mathcal{B}, y] = \{0\}$. Since y is quasinilpotent, this implies that by is quasinilpotent for every $b \in \mathcal{B}$, i.e. $y \in rad\, \mathcal{B}$. Hence $y = 0$ by the hypothesis $rad\, \mathcal{B} = \{0\}$. □

Theorem 1. *Let $\mathcal{A}$ be a complex unital Banach algebra, $\mathcal{L}$ be a Lie subalgebra of $\mathcal{A}$ and $\mathcal{A}(\mathcal{L})$ be the associative unital subalgebra generated by $\mathcal{L}$. If $\mathcal{I}$ is a finite-dimensional solvable ideal of $\mathcal{L}$ then $[\mathcal{I}, \mathcal{L}] \subseteq rad\, \overline{\mathcal{A}(\mathcal{L})}$.*

Proof. Denote $\mathcal{B} := \overline{\mathcal{A}(\mathcal{L})} / rad\, \overline{\mathcal{A}(\mathcal{L})}$ and let $\pi : \overline{\mathcal{A}(\mathcal{L})} \to \mathcal{B}$ be the canonical projection. We must prove that $[\pi(\mathcal{I}), \pi(\mathcal{L})] = \{0\}$. To this end we show that the conditions of Lemma 1 hold for the ideal $\pi(\mathcal{I})$ of the Lie algebra $\pi(\mathcal{L}) \subseteq \mathcal{B}$. Obviously we have $rad\, \mathcal{B} = \{0\}$.

Next suppose that for some $b \in \pi(\mathcal{L})$ the map $ad\, b|_{\pi(\mathcal{I})}$ is not nilpotent. Then there exist $\lambda \in \mathbb{C} \backslash \{0\}$ and $a \in \pi(\mathcal{I}) \backslash \{0\}$ such that $[b, a] = \lambda a$. Then $a \in \mathcal{N}_{\pi(\mathcal{I})}$, i.e. a is a nilpotent element of $\pi(\mathcal{I})$ (see e.g. Theorem 3 from § 17). But by Proposition 1 it follows that

$$\mathcal{N}_{\pi(\mathcal{I})} \subseteq rad\, \overline{\mathcal{A}(\pi(\mathcal{L}))} = rad\, \mathcal{B} = \{0\}.$$

Hence $a = 0$, which is a contradiction with the choice of a.

Consequently $ad\, b|_{\pi(\mathcal{I})}$ is a nilpotent map for every $b \in \pi(\mathcal{L})$, and Lemma 1 can be applied. □

Corollary 1. *Let $\mathcal{A}$ be a complex unital Banach algebra. If $\mathcal{L}$ is a quasisolvable Lie subalgebra of $\mathcal{A}$ and $\overline{\mathcal{A}(\mathcal{L})}$ is the closed associative unital subalgebra generated by $\mathcal{L}$, then $[\overline{\mathcal{A}(\mathcal{L})}, \overline{\mathcal{A}(\mathcal{L})}] \subseteq rad\, \overline{\mathcal{A}(\mathcal{L})}$. That is, $\overline{\mathcal{A}(\mathcal{L})}$ is commutative modulo its Jacobson radical.*

Proof. Let $\{\mathcal{I}_\alpha\}_{\alpha \in \Lambda}$ be the family of the finite-dimensional solvable ideals of $\mathcal{L}$ and let $\pi : \overline{\mathcal{A}(\mathcal{L})} \to \mathcal{B}$ be the canonical projection, where $\mathcal{B} := \overline{\mathcal{A}(\mathcal{L})} / rad\, \overline{\mathcal{A}(\mathcal{L})}$. Then $[\pi(\mathcal{I}_\alpha), \mathcal{B}] = \{0\}$ for every $\alpha \in \Lambda$ by Theorem 1. But obviously $\mathcal{B}$ equals the closed associative unital subalgebra generated by $\cup_{\alpha \in \Lambda} \pi(\mathcal{I}_\alpha) (= \pi(\mathcal{L}))$, hence we deduce $[\mathcal{B}, \mathcal{B}] = \{0\}$. □

Notes

The first result concerning the existence of weights for solvable Lie algebras of bounded operators was obtained by D. Gurarie and Yu.I. Lyubich [1] for finite-dimensional algebras as a consequence of an analogous result for separable solvable groups when the weight is defined by a countable sequence of unit vectors. Şt. Frunză [2] proved the existence of weights for arbitrary solvable Lie algebras of bounded operators when the weight is defined by an arbitrary net of unit vectors (see Theorem 3 from § 20).

The problem of extension of Lie's Theorem for solvable Lie algebras of generalized scalar operators was raised by C. Foiaş. The first result concerning the existence of invariant subspaces for a solvable Lie algebra of bounded operators was obtained in M. Şabac [1] when the Lie algebra consists of generalized scalar operators. This result was improved for solvable or LM-decomposable Lie algebras of decomposable operators in M. Şabac [2], [3], [4], [5], [6], [7], [10]. One uses the existence of nilpotents (nilpotence criteria), the invariant subspace associated to an ideal of nilpotents (see Lemma 1 from § 21), special structure of invariant subspaces for special classes of operators, particularly Lomonosov's theorem. All these results are described in § 21. Theorem 9 and Corollary 6 were proved in D. Beltiţă and M. Şabac [1]. Corollary 5 appears here for the first time.

The extensions of Lie's Theorem (irreducible variants) were obtained as follows:

- the results for topological irreducibility in M. Şabac [1], [2], [3], [4], [5], [6], [7];
- the characterization of paratransitive or operatorially (i.e. completely) irreducible solvable Lie algebras of bounded operators in C. Foiaş and M. Şabac [1];
- the characterization of irreducible and operatorially irreducible finite-dimensional Lie algebras of bounded operators in D. Gurarie [1].

All these results are presented in § 22. Theorem 7 from § 22 is due to M. Şabac, appears here for the first time and contains the results of C. Foiaş and M. Şabac [1]. Also Corollary 8 and Theorem 8 appear here for the first time and are due to M. Şabac.

The first infinite-dimensional variant of Engel's Theorem was proved in W. Wojtynski [1]; see also W. Wojtynski [2], [3] for related discussions and some open problems. For discussion of the continuity of spectrum on compact operators (used in the proof of Theorem 2 from § 23) see N. Dunford and J.T. Schwartz [1]; see also P.R. Halmos [1]. For the properties of trace-class operators see N. Dunford and J.T. Schwartz [2]. The main result of § 23, i.e. Theorem 3, was proved by W. Wojtynski [1]. Theorem 1 seems to appear here for the first time. We note that V.S. Shul'man and Yu.V. Turovskiĭ [1], [2] have proved recently that, if $\mathcal{G}$ is an arbitrary Lie algebra of compact quasinilpotent operators on a Banach space, then $\overline{\mathcal{A}_0(\mathcal{G})}$ also consists of compact quasinilpotent operators. (Particularly one can remove the hypothesis of locally finiteness in Theorem 2 from § 23.) The result is announced also in the end of Yu.V. Turovskiĭ [6].

Theorem 1 and Corollary 1 from § 24 were proved by Yu.V. Turovskiĭ [5]. We give here essentially the original proof but Proposition 1 allows us to make it "shorter" (of course, modulo the spectral theory developed in Chapter IV). Moreover we notice that our Lemma 1 is a variant of Theorem 1 of C.R. Miers [6]. The commutativity modulo the Jacobson radical has many applications to invariant subspace problems; see Yu.V. Turovskiĭ [2], [4], [5], [6], V.S. Shul'man [3], V.S. Shul'man and Yu.V. Turovskiĭ [1], [2], B.A. Barnes and A. Katavolos [1], [2], A. Katavolos and C. Stamatopoulos [1].

Chapter IV

Spectral Theory for Solvable Lie Algebras of Operators

§ 25 Spectral theory for representations of Lie algebras

As usual we shall denote by $\mathcal{X}$ a complex Banach space.

Definition 1. Let $\rho: \mathcal{E} \to \mathcal{B}(\mathcal{X})$ be a representation of the complex finite-dimensional Lie algebra $\mathcal{E}$. The *spectrum* of ρ, denoted $\sigma(\rho)$, is the set defined in the following way

$$\sigma(\rho) = \{\lambda \in \widehat{\mathcal{E}} \mid Kos\,(\rho - \lambda) \text{ is not exact}\}.$$

(We recall that $\widehat{\mathcal{E}}$ denotes the set of *characters* of $\mathcal{E}$, that is $\lambda \in \widehat{\mathcal{E}}$ iff $\lambda : \mathcal{E} \to \mathbb{C}$ is a Lie morphism. Moreover, for $\lambda \in \widehat{\mathcal{E}}$ we denote by $\rho - \lambda : \mathcal{E} \to \mathcal{B}(\mathcal{X})$ the representation defined by $e \mapsto \rho(e) - \lambda(e) I_{\mathcal{X}}$.)

Now let's collect a few simple facts concerning the above definition.

Remark 1. With the notation of Definition 1, assume moreover that $\dim \mathcal{E} = 1$ and choose $e \in \mathcal{E} \backslash \{0\}$. So $\mathcal{E} = \mathbb{C} \cdot e$ and each $\widetilde{\lambda} \in \widehat{\mathcal{E}}$ is uniquely determined by its value $\widetilde{\lambda}(e) =: \lambda \in \mathbb{C}$. If we denote $T := \rho(e) \in \mathcal{B}(\mathcal{X})$, then we have (see Example 1 in § 10)

$$Kos\,(\rho - \widetilde{\lambda}) : \qquad 0 \leftarrow \mathcal{X} \xleftarrow{T-\lambda} \mathcal{X} \leftarrow 0.$$

Consequently $\sigma(\rho)$ can be identified with the (usual) spectrum $\sigma(T)$ of the bounded linear operator T by means of the bijective map $\widehat{\mathcal{E}} \to \mathbb{C}$, $\widetilde{\lambda} \mapsto \widetilde{\lambda}(e)$. (For the connection between Definition 1 and the Taylor spectrum of a commuting tuple of operators see Theorem 3 (4°) from § 27 and the remarks following Definition 1 from § 27.)

Remark 2. With the notation of Definition 1, assume moreover that the Lie algebra $\mathcal{E}$ is nilpotent. If $\widetilde{\lambda} \in \widehat{\mathcal{E}}$ is a character such that there exists a sequence $\{x_j\}_{j \geq 1}$

consisting of unit vectors from $\mathcal{X}$ such that

$$\lim_{j\to\infty}(\rho(e)-\widetilde{\lambda}(e))x_j = 0 \text{ for every } e \in \mathcal{E},$$

then $\widetilde{\lambda} \in \sigma(\rho)$. For proving this fact denote $\dim \mathcal{E} = n$ and choose a basis $\{e_1, \dots, e_n\}$ in $\mathcal{E}$ such that

$$[e_j, e_i] = \sum_{k<i} c_{ji}^k e_k \quad (1 \le i < j \le n).$$

Now denote the Koszul complex of the representation $\rho - \widetilde{\lambda} : \mathcal{E} \to \mathcal{B}(\mathcal{X})$ in the following way:

$$Kos\,(\rho-\widetilde{\lambda}) : 0 \leftarrow \mathcal{X} \xleftarrow{\alpha_1} \mathcal{X}\otimes\mathcal{E} \xleftarrow{\alpha_2} \cdots \xleftarrow{\alpha_n} \mathcal{X}\otimes\Lambda^n\mathcal{E} \leftarrow 0.$$

Now, since the structure constants have the property $c_{ij}^j = 0$ for every i, j, by Example 2 from § 10 we obtain

$$\alpha_n(x\otimes \underline{e}) = \sum_{i=1}^{n}(-1)^{i-1}(\rho-\widetilde{\lambda})(e_i)x \otimes \overset{i}{\widehat{\underline{e}}} \text{ for each } x \in \mathcal{X},$$

where $\underline{e} = e_1 \wedge \cdots \wedge e_n \in \Lambda^n\mathcal{E}$. So we have

$$\lim_{j\to\infty} \alpha_n(x_j \otimes \underline{e}) = 0.$$

Then the operator $\alpha_n : \mathcal{X}\otimes\Lambda^n\mathcal{E} \to \mathcal{X}\otimes\Lambda^{n-1}\mathcal{E}$ either is not one-to-one or its range is not closed. Particularly either $Ker\,\alpha_n \neq \{0\}$ or $Ran\,\alpha_n \neq Ker\,\alpha_{n-1}$, so one of the homology spaces

$$\mathbf{H}_n(\rho-\widetilde{\lambda}),\ \mathbf{H}_{n-1}(\rho-\widetilde{\lambda})$$

must be different from $\{0\}$. So $Kos\,(\rho-\widetilde{\lambda})$ is not exact and $\widetilde{\lambda} \in \sigma(\rho)$ as desired.

Remark 3. Now assume that we have a short exact sequence

$$0 \to \mathcal{G} \hookrightarrow \mathcal{F} \xrightarrow{h} \mathcal{E} \to 0,$$

where $\mathcal{G}, \mathcal{F}, \mathcal{E}$ are finite-dimensional Lie algebras and h is a Lie algebra morphism. Let $\rho : \mathcal{E} \to \mathcal{B}(\mathcal{X})$ be a representation. If either $[\mathcal{G}, \mathcal{F}] = \{0\}$ (i.e. $\mathcal{G}$ is contained in the center of $\mathcal{F}$) or $\mathcal{F}$ is nilpotent then we have a bijective map

$$\sigma(\rho) \to \sigma(\rho\circ h),\ \widetilde{\lambda} \mapsto \widetilde{\lambda}\circ h,$$

by Propositions 2 and 3 from § 10.

Our first aim in the following is to show that, in fairly general conditions, the above defined spectrum of a Banach space representation of a Lie algebra possesses some basic properties of the spectrum of an operator (such as compactness and nonemptyness).

Theorem 1. *The spectrum of every representation $\rho : \mathcal{E} \to \mathcal{B}(\mathcal{X})$ of a finite-dimensional Lie algebra $\mathcal{E}$ is a compact subset of the finite-dimensional vector space $\widehat{\mathcal{E}}$.*

Proof. It is easily seen that one can apply Theorem 2 from § 9 to deduce that $\sigma(\rho)$ is a closed subset of $\widehat{\mathcal{E}}$.

Next we prove by induction on $n = \dim \mathcal{E}$ that for every representation $\rho : \mathcal{E} \to \mathcal{B}(\mathcal{X})$ the spectrum $\sigma(\rho)$ is a bounded subset of $\widehat{\mathcal{E}}$. For $n = 1$ see Remark 1 above. Now assume that the assertion holds for representations of Lie algebras with dimensions strictly less than n. If $\mathcal{E} = [\mathcal{E}, \mathcal{E}]$ then $\widehat{\mathcal{E}} = \{0\}$ and the conclusion is trivial since $\sigma(\rho) \subseteq \widehat{\mathcal{E}}$. Hence we may assume that

$$\dim \mathcal{E} - \dim[\mathcal{E}, \mathcal{E}] =: m > 0.$$

Choose a basis $\{e_1, \dots, e_n\}$ in $\mathcal{E}$ such that $\{e_{m+1}, \dots, e_n\}$ is a basis in $[\mathcal{E}, \mathcal{E}]$ and denote

$$\mathcal{I} := \bigoplus_{i=2}^{m} \mathbb{C}e_i \oplus [\mathcal{E}, \mathcal{E}].$$

Hence denoting $e := e_1$ we have $\mathcal{E} = \mathcal{I} \oplus \mathbb{C}e$, where $\mathcal{I}$ is an ideal of $\mathcal{E}$. So by Proposition 4 from § 10 we obtain

$$Kos\,(\rho - \lambda) = Con\,(Kos\,(\rho|_{\mathcal{I}} - \lambda|_{\mathcal{I}}), \theta^{\cdot}_{e} - \lambda(e)) \qquad (\lambda \in \widehat{\mathcal{E}}) \tag{1}$$

(where we used the obvious fact that $\theta^{\cdot}_{\rho-\lambda,e} = \theta^{\cdot}_{\rho,e} - \lambda(e)$, cf. the definition of $\theta^{\cdot}_{\rho,e}$ before Proposition 4 from § 10). Hence for $\lambda \in \widehat{\mathcal{E}}$ the complex $Kos\,(\rho - \lambda)$ coincides with the totalization of the following bicomplex,

$$\begin{array}{ccccccc}
 & & 0 & & 0 & & \\
 & & \downarrow & & \downarrow & & \\
0 & \leftarrow & \mathcal{X} \otimes \Lambda^{n-1}\mathcal{I} & \stackrel{\theta^{n-1}_{e} - \lambda(e)}{\longleftarrow} & \mathcal{X} \otimes \Lambda^{n-1}\mathcal{I} & \leftarrow & 0 \\
 & & \downarrow & & \downarrow & & \\
\cdots & & \cdots & & \cdots & & \cdots \\
 & & \downarrow & & \downarrow & & \\
0 & \leftarrow & \mathcal{X} \otimes \Lambda^{p}\mathcal{I} & \stackrel{\theta^{p}_{e} - \lambda(e)}{\longleftarrow} & \mathcal{X} \otimes \Lambda^{p}\mathcal{I} & \leftarrow & 0 \\
 & & \downarrow & & \downarrow & & \\
\cdots & & \cdots & & \cdots & & \cdots \\
 & & \downarrow & & \downarrow & & \\
0 & \leftarrow & \mathcal{X} \otimes \mathcal{I} & \stackrel{\theta^{1}_{e} - \lambda(e)}{\longleftarrow} & \mathcal{X} \otimes \mathcal{I} & \leftarrow & 0 \\
 & & \downarrow & & \downarrow & & \\
0 & \leftarrow & \mathcal{X} & \stackrel{\theta^{0}_{e} - \lambda(e)}{\longleftarrow} & \mathcal{X} & \leftarrow & 0 \\
 & & \downarrow & & \downarrow & & \\
 & & 0 & & 0 & &
\end{array}$$

where both columns coincide with $Kos(\rho|_{\mathcal{I}} - \lambda|_{\mathcal{I}})$. So, by Corollary 1 ((a′) and (b′)) from § 8, the following implication holds

$$\lambda \in \sigma(\rho) \Rightarrow \lambda|_{\mathcal{I}} \in \sigma(\rho|_{\mathcal{I}}) \text{ and } \lambda(e) \in \sigma(\theta^p_e) \text{ for some } p \in \{0, \dots, n-1\}. \tag{2}$$

But $\sigma(\rho|_{\mathcal{I}})$ is compact by the induction hypothesis and θ^p_e is a bounded linear operator on the Banach space $\mathcal{X} \otimes \Lambda^p\mathcal{I}$, hence its spectrum $\sigma(\theta^p_e)$ is compact for $p = 0, \dots, n-1$. So by (2) we deduce that $\sigma(\rho)$ is a bounded set. Since we have already proved that it is also closed, the proof ends. □

Now, for later use we notice the following consequence of the preceding proof.

Remark 4. Let $\rho : \mathcal{E} \to \mathcal{B}(\mathcal{X})$ be a representation of the finite-dimensional Lie algebra $\mathcal{E}$. There exists a bounded subset $\mathcal{M}$ of $\widehat{\mathcal{E}}$ such that for every invertible operator $S \in \mathcal{B}(\mathcal{X})$ we have

$$\sigma((Ad\,S) \circ \rho) \subset \mathcal{M},$$

where $Ad\,S : \mathcal{B}(\mathcal{X}) \to \mathcal{B}(\mathcal{X})$ is defined by $T \mapsto S^{-1}TS$.

Indeed, if $S \in \mathcal{B}(\mathcal{X})$ is an invertible operator, then

$$S \otimes I_{\Lambda \cdot \mathcal{I}} : Kos(\rho - \lambda) \to Kos((Ad\,S) \circ \rho - \lambda)$$

defines an isomorphism of complexes. This implies

$$\sigma(\rho) = \sigma((Ad\,S) \circ \rho) \text{ for every } S \in \mathcal{B}(\mathcal{X}). \tag{3}$$

Hence one can choose the bounded set $\mathcal{M} := \sigma(\rho)$ by Theorem 1.

In the following we are concerned mainly with representations of finite-dimensional solvable Lie algebras. The following result is a variant of the projection property of the Taylor spectrum of a commuting tuple of operators (see also Corollary 2 from § 26 below). It is useful to introduce the notation

$$\mathcal{F}|_A := \{f|_A \mid f \in \mathcal{F}\} \quad (\text{"the projection of } \mathcal{F} \text{ on } A\text{"}),$$

where $\mathcal{F}$ is a set of functions defined on a certain set B containing A. Now we can state:

Theorem 2. *Let $\rho : \mathcal{E} \to \mathcal{B}(\mathcal{X})$ be a representation of the finite-dimensional solvable Lie algebra $\mathcal{E}$. Then the spectrum $\sigma(\rho)$ has the projection property on every ideal, in the sense that for every ideal $\mathcal{I}$ of $\mathcal{E}$ we have*

$$\sigma(\rho|_{\mathcal{I}}) = \sigma(\rho)|_{\mathcal{I}}.$$

Proof. Since $\mathcal{E}$ is solvable, it obviously suffices to prove the desired equality in the assumption $\dim(\mathcal{E}/\mathcal{I}) = 1$. Then take $e \in \mathcal{E}$ such that $\mathcal{E} = \mathcal{I} \oplus \mathbb{C}e$. By the implication (2) from the proof of Theorem 1 we deduce

$$\sigma(\rho)|_{\mathcal{I}} \subseteq \sigma(\rho|_{\mathcal{I}}).$$

To prove the converse inclusion take $\widetilde{\lambda}_0 \in \sigma(\rho|_{\mathcal{I}})$. By Proposition 4 from § 10 it easily follows that we have the morphism of complexes

$$\theta^{\cdot}_e : Kos\,(\rho|_{\mathcal{I}} - \widetilde{\lambda}_0) \to Kos\,(\rho|_{\mathcal{I}} - \widetilde{\lambda}_0).$$

But $Kos\,(\rho|_{\mathcal{I}} - \widetilde{\lambda}_0)$ is not exact (since $\widetilde{\lambda}_0 \in \sigma(\rho|_{\mathcal{I}})$). Hence by Słodkowski's Lemma (Theorem 3 from § 9) there exists $\lambda \in \mathbb{C}$ such that the complex

$$Con\,(Kos\,(\rho|_{\mathcal{I}} - \widetilde{\lambda}_0), \theta^{\cdot}_e - \lambda)$$

is not exact. Then the linear functional $\widetilde{\lambda} : \mathcal{E} \to \mathbb{C}$ defined by $\widetilde{\lambda}|_{\mathcal{I}} := \widetilde{\lambda}_0$ and $\widetilde{\lambda}(e) := \lambda$ satisfies $\widetilde{\lambda}|_{[\mathcal{E},\mathcal{E}]} = 0$ by Lemma 1 below. Consequently $\widetilde{\lambda} \in \widehat{\mathcal{E}}$. Moreover by the choice of λ and by (1) (see the proof of Theorem 1 above) we deduce that $Kos\,(\rho - \widetilde{\lambda})$ is not exact. Hence $\widetilde{\lambda} \in \sigma(\rho)$ and $\widetilde{\lambda}|_{\mathcal{I}} = \widetilde{\lambda}_0$, that is $\widetilde{\lambda}_0 \in \sigma(\rho)|_{\mathcal{I}}$. □

Lemma 1. *Let $\mathcal{E}$ be a finite-dimensional Lie algebra, $\mathcal{I}$ be an ideal of $\mathcal{E}$ and $\rho : \mathcal{E} \to \mathcal{B}(\mathcal{X})$ be a representation. Then for every $\varphi \in \sigma(\rho|_{\mathcal{I}})$ we have $\varphi|_{[\mathcal{E},\mathcal{I}]} = 0$. Particularly, if $\dim(\mathcal{E}/\mathcal{I}) = 1$, then $\varphi|_{[\mathcal{E},\mathcal{E}]} = 0$.*

Proof. Take $a \in \mathcal{E}$ arbitrary. We have $(ad\,a)\mathcal{I} \subseteq \mathcal{I}$ and we must prove that $\varphi([a,\mathcal{I}]) = \{0\}$, i.e. $\varphi \circ ((ad\,a)|_{\mathcal{I}}) = 0$. This is obviously equivalent to

$$\varphi \circ \exp(t \cdot ad\,a|_{\mathcal{I}}) = \varphi \text{ for each } t \in \mathbb{C}, \tag{4}$$

where $\exp(t \cdot ad\,a) : \mathcal{E} \to \mathcal{E}$ is the automorphism defined by

$$\exp(t \cdot ad\,a) = \sum_{n=0}^{\infty} \frac{t^n}{n!} (ad\,a)^n.$$

Since $(ad\,a)\mathcal{I} \subseteq \mathcal{I}$ it is obvious that $\mathcal{I}$ is an invariant subspace for the invertible map $\exp(t \cdot ad\,a)$. Consequently $(\exp(t \cdot ad\,a))|_{\mathcal{I}} : \mathcal{I} \to \mathcal{I}$ is also an automorphism. By Proposition 2 from § 10 it then follows that

$$\varphi \circ (\exp(t \cdot ad\,a)|_{\mathcal{I}}) \in \sigma(\rho|_{\mathcal{I}} \circ (\exp(t \cdot ad\,a)|_{\mathcal{I}})) \text{ for each } t \in \mathbb{C}. \tag{5}$$

On the other hand, since $\rho : \mathcal{E} \to \mathcal{B}(\mathcal{X})$ is a Lie algebra morphism we have

$$\rho \circ (ad\,b) = (ad\,(\rho(b))) \circ \rho \text{ for each } b \in \mathcal{E}.$$

This implies

$$\rho \circ \exp(ad\,b) = \exp(ad\,(\rho(b))) \circ \rho \qquad (b \in \mathcal{E}).$$

But by Rosenblum's formula (see Remark 1 from § 15) we have $\exp(ad\,(\rho(b))) = Ad\,(e^{\rho(b)})$, so

$$\rho \circ \exp(ad\,b) = (Ad\,(e^{\rho(b)})) \circ \rho \qquad (b \in \mathcal{E}).$$

Consequently (5) implies

$$\varphi \circ (\exp(t \cdot ad\,a)|_{\mathcal{I}}) \in \sigma(Ad\,(e^{t \cdot \rho(a)}) \circ \rho|_{\mathcal{I}}) \text{ for each } t \in \mathbb{C}.$$

Hence by Remark 4 (applied to the representation $\rho|_{\mathcal{I}} : \mathcal{I} \to \mathcal{B}(\mathcal{X})$) we obtain that there exists a bounded set $\mathcal{M} \subseteq \widehat{\mathcal{I}}$ such that

$$\varphi \circ (\exp(t \cdot ad\, a)|_{\mathcal{I}}) \in \mathcal{M} \text{ for each } t \in \mathbb{C}.$$

Consequently the holomorphic function

$$\mathbb{C} \to \widehat{\mathcal{I}},\ t \mapsto \varphi \circ (\exp(t \cdot ad\, a)|_{\mathcal{I}}),$$

is constant by Liouville's Theorem. This implies (4).

For the second assertion use the fact that $[\mathcal{E}, \mathcal{E}] = [\mathcal{E}, \mathcal{I}]$ whenever $\dim(\mathcal{E}/\mathcal{I}) = 1$. □

Next we are going to apply Theorem 2 for proving that, for representations of nilpotent Lie algebras, the spectrum has the projection property on every Lie subalgebra (see Corollary 1 below). To this end we need the following well-known fact.

Lemma 2. *Let $\mathcal{L}$ be a subalgebra of the finite-dimensional nilpotent Lie algebra $\mathcal{E}$. If $\dim(\mathcal{E}/\mathcal{L}) = 1$, then $\mathcal{L}$ is an ideal of $\mathcal{E}$. More generally, if $\dim(\mathcal{E}/\mathcal{L}) = k$, then there exists a sequence of subalgebras of $\mathcal{E}$,*

$$\mathcal{L} = \mathcal{L}_0 \subset \mathcal{L}_1 \subset \cdots \subset \mathcal{L}_k = \mathcal{E},$$

such that $\mathcal{L}_j$ is an ideal of $\mathcal{L}_{j+1}$ and $\dim(\mathcal{L}_{j+1}/\mathcal{L}_j) = 1$ for $j = 0, \ldots, k-1$.

Corollary 1. *Let $\rho : \mathcal{E} \to \mathcal{B}(\mathcal{X})$ be a representation of the finite-dimensional nilpotent Lie algebra $\mathcal{E}$. Then the spectrum $\sigma(\rho)$ has the projection property on every subalgebra, i.e. for each subalgebra $\mathcal{L}$ of $\mathcal{E}$ we have*

$$\sigma(\rho|_{\mathcal{L}}) = \sigma(\rho)|_{\mathcal{L}}.$$

Proof. Let $\mathcal{L}$ be a subalgebra of $\mathcal{E}$. Denote $\dim(\mathcal{E}/\mathcal{L}) =: k$ and choose a chain

$$\mathcal{L} = \mathcal{L}_0 \subset \mathcal{L}_1 \subset \cdots \subset \mathcal{L}_k = \mathcal{E}$$

as in Lemma 2. Since $\mathcal{L}_j$ is an ideal of $\mathcal{L}_{j+1}$ for $j = 0, \ldots, k-1$, by Theorem 2 we obtain

$$\sigma(\rho)|_{\mathcal{L}} = (\sigma(\rho)|_{\mathcal{L}_{k-1}})|_{\mathcal{L}} = \sigma(\rho|_{\mathcal{L}_{k-1}})|_{\mathcal{L}} = \cdots = \sigma(\rho|_{\mathcal{L}_1})|_{\mathcal{L}} = \sigma(\rho|_{\mathcal{L}}),$$

and the proof is finished. □

Particularly we have the following useful result holding for representations of finite-dimensional nilpotent Lie algebras. It allows us to compute the spectra of the operators from the image of the representation by means of the spectrum of the representation itself.

Corollary 2. *Let $\rho : \mathcal{E} \to \mathcal{B}(\mathcal{X})$ be a representation of the finite-dimensional nilpotent Lie algebra $\mathcal{E}$. Then for every $e \in \mathcal{E}$ we have*

$$\sigma(\rho(e)) = \{\lambda(e) | \lambda \in \sigma(\rho)\}.$$

Proof. Let $e \in \mathcal{E}$. First one applies Corollary 1 to the subalgebra $\mathcal{L} := \mathbb{C}e$ and get

$$\sigma(\rho|_{\mathcal{L}}) = \sigma(\rho)|_{\mathcal{L}} = \{\lambda|_{\mathcal{L}} \,|\lambda \in \sigma(\rho)\}.$$

Then one applies Remark 1. □

Now we establish a basic fact.

Theorem 3. *The spectrum of every representation of a finite-dimensional solvable Lie algebra $\mathcal{E}$ in a non-zero Banach space is a compact nonempty subset of $\widehat{\mathcal{E}}$.*

Proof. Let $\rho : \mathcal{E} \to \mathcal{B}(\mathcal{X})$ be a representation. The set $\sigma(\rho)(\subset \widehat{\mathcal{E}})$ is compact by Theorem 2. For proving the nonemptyness, choose an ideal $\mathcal{I}$ of $\mathcal{E}$ such that $\dim \mathcal{I} = 1$. (This is possible since $\mathcal{E}$ is solvable.) Let $e \in \mathcal{I} \setminus \{0\}$, so $\mathcal{I} = \mathbb{C}e$ and denote $T = \rho(e) \in \mathcal{B}(\mathcal{X})$. Next we proceed as in the proof of Corollary 2 but applying Theorem 2 instead of Corollary 1. Namely by Remark 1 we have

$$\sigma(T) = \{\lambda_0(e) | \lambda_0 \in \sigma(\rho|_{\mathcal{I}})\}.$$

Particularly $\sigma(\rho|_{\mathcal{I}}) \neq \emptyset$. But $\sigma(\rho|_{\mathcal{I}}) = \sigma(\rho)|_{\mathcal{I}}$ by Theorem 1, so $\sigma(\rho)$ is also nonempty. □

The following result can be viewed as another infinite-dimensional variant of Lie's Theorem.

Theorem 4. *If $\mathcal{G}$ is a finite-dimensional solvable Lie subalgebra of $\mathcal{B}(\mathcal{X})$, then every element of $[\mathcal{G}, \mathcal{G}]$ is a quasinilpotent operator.*

Proof. Consider the identity representation $\rho : \mathcal{G} \to \mathcal{B}(\mathcal{X})$, $\rho(T) = T$ for every $T \in \mathcal{G}$. By Theorem 2 we obtain

$$\sigma(\rho|_{[\mathcal{G},\mathcal{G}]}) = \sigma(\rho)|_{[\mathcal{G},\mathcal{G}]}.$$

But each element of $\sigma(\rho) \subset \widehat{\mathcal{G}}$ is a functional vanishing on $[\mathcal{G}, \mathcal{G}]$, so $\sigma(\rho|_{[\mathcal{G},\mathcal{G}]}) = \{0\}$. Then, since $[\mathcal{G}, \mathcal{G}]$ is a nilpotent Lie algebra, by Corollary 2 we obtain $\sigma(T) = \{0\}$ for each $T \in [\mathcal{G}, \mathcal{G}]$. □

Our next aim is to show that the property expressed in Corollary 1 (namely the projection property on subalgebras which are not necessarily ideals) characterizes the nilpotent Lie algebras. To this end we establish below certain facts which are converses to Corollary 1 (see Theorems 5 and 6 below). We shall need the following estimate of the projection of the spectrum on certain special subalgebras.

Proposition 1. *Let $\mathcal{E}$ be a finite-dimensional Lie algebra possessing a hyperplane subalgebra $\mathcal{B}$ (i.e. $\mathcal{B}$ is a Lie subalgebra of $\mathcal{E}$ and $\dim\mathcal{B} = \dim\mathcal{E} - 1$) and an ideal $\mathcal{I}$ such that $\dim\mathcal{I} = 1$ and $\mathcal{E} = \mathcal{I} + \mathcal{B}$. Let $\gamma \in \widehat{\mathcal{B}}$ be such that*

$$[b, a] = \gamma(b)a \text{ for each } a \in \mathcal{I} \text{ and } b \in \mathcal{B}. \tag{6}$$

Then for every representation $\rho : \mathcal{E} \to \mathcal{B}(\mathcal{X})$ the following estimate of $\sigma(\rho)|_{\mathcal{B}}$ ("the projection of $\sigma(\rho)$ on $\mathcal{B}$") holds

$$\sigma(\rho|_{\mathcal{B}})\Delta(\gamma + \sigma(\rho|_{\mathcal{B}})) \subseteq \sigma(\rho)|_{\mathcal{B}} \subseteq \sigma(\rho|_{\mathcal{B}}) \cup (\gamma + \sigma(\rho|_{\mathcal{B}})), \tag{7}$$

where Δ denotes the usual symmetric difference of sets. If moreover $\mathcal{I} \subseteq Ker\,\rho$, then

$$\sigma(\rho)|_{\mathcal{B}} = \sigma(\rho|_{\mathcal{B}}) \cup (\gamma + \sigma(\rho|_{\mathcal{B}})). \tag{8}$$

Proof. We begin with a preliminary observation. Let's choose $a \in \mathcal{I}\backslash\{0\}$ and for $\lambda \in \widehat{\mathcal{E}}$ denote by $X_{\rho-\lambda}$ the following bicomplex

$$\begin{array}{ccccccc}
 & & \cdots & \cdots & \cdots & & \\
 & & \beta_{p+1} \downarrow & & \beta'_{p+1} \downarrow & & \\
0 & \leftarrow & \mathcal{X} \otimes \Lambda^p\mathcal{B} & \xleftarrow{\rho(a)\otimes I_{\Lambda^p\mathcal{B}}} & \mathcal{X} \otimes \Lambda^p\mathcal{B} & \leftarrow & 0 \\
 & & \beta_p \downarrow & & \beta'_p \downarrow & & \\
 & & \cdots & \cdots & \cdots & & \\
 & & \beta_2 \downarrow & & \beta'_2 \downarrow & & \\
0 & \leftarrow & \mathcal{X} \otimes \mathcal{I} & \xleftarrow{\rho(a)\otimes I_{\mathcal{B}}} & \mathcal{X} \otimes \mathcal{I} & \leftarrow & 0 \\
 & & \beta_1 \downarrow & & \beta'_1 \downarrow & & \\
0 & \leftarrow & \mathcal{X} & \xleftarrow{\rho(a)} & \mathcal{X} & \leftarrow & 0 \\
 & & \downarrow & & \downarrow & & \\
 & & 0 & & 0 & &
\end{array} \tag{9}$$

where the right column is $Kos\,(\rho|_{\mathcal{B}} - \lambda|_{\mathcal{B}} + \gamma)$ and the left column is $Kos\,(\rho|_{\mathcal{B}} - \lambda|_{\mathcal{B}})$. By (6) we obtain

$$\rho(a)(\rho(b) - \lambda(b) + \gamma(b)) = (\rho(b) - \lambda(b))\rho(a) \text{ for each } b \in \mathcal{B},$$

and this easily implies that (9) is a commutative diagram. Next we remark that $\Lambda^p\mathcal{E} = (\Lambda^p\mathcal{B}) \oplus (a \wedge (\Lambda^{p-1}\mathcal{B})) \cong (\Lambda^p\mathcal{B}) \oplus (\Lambda^{p-1}\mathcal{B})$ (since $\mathcal{E} = \mathcal{B} \oplus \mathbb{C}a$) and then a straightforward computation shows that the totalization of $X_{\rho-\lambda}$ is isomorphic to $Kos\,(\rho - \lambda)$.

Now let's come back to the proof of (7). Let $\lambda_0 \in \sigma(\rho|_{\mathcal{B}})\Delta(\gamma + \sigma(\rho|_{\mathcal{B}}))$. If we define a linear functional $\lambda : \mathcal{E} \to \mathbb{C}$ by $\lambda|_{\mathcal{B}} := \lambda_0$ and $\lambda|_{\mathcal{I}} = 0$, then we have $\lambda \in \widehat{\mathcal{E}}$ and $0 \in \sigma(\rho|_{\mathcal{B}} - \lambda|_{\mathcal{B}})\Delta(\gamma + \sigma(\rho|_{\mathcal{B}} - \lambda|_{\mathcal{B}}))$. Consequently one column of (9) is exact while the other is not exact. Hence $Tot\,(X_{\rho-\lambda})$ is not exact by Proposition 3 from §8. Then $Kos\,(\rho - \lambda)$ is not exact by the preliminary observation, so $\lambda \in \sigma(\rho)$. Particularly $\lambda_0 = \lambda|_{\mathcal{B}} \in \sigma(\rho)|_{\mathcal{B}}$.

For proving the second inclusion from (7) start with $\lambda \in \sigma(\rho)$. Then $Kos(\rho-\lambda)$ is not exact, so $Tot(X_{\rho-\lambda})$ is not exact by the preliminary observation. Consequently at least one column of (9) is not exact (see Corollary 1 (a') from § 8). This implies $0 \in \sigma(\rho|_{\mathcal{B}} - \lambda|_{\mathcal{B}}) \cup (\gamma + \sigma(\rho|_{\mathcal{B}} - \lambda|_{\mathcal{B}}))$ hence $\lambda|_{\mathcal{B}} \in \sigma(\rho|_{\mathcal{B}}) \cup (\gamma + \sigma(\rho|_{\mathcal{B}}))$ and (7) is completely proved.

Now assume $\mathcal{I} \subseteq Ker\,\rho$, i.e. $\rho(a) = 0$. In view of (7) we must prove only the inclusion "$\supseteq$" from (8). Let λ_0 be arbitrary in the right-hand member of (8). We define a linear functional $\lambda : \mathcal{E} \to \mathbb{C}$ by $\lambda|_{\mathcal{B}} := \lambda_0$ and $\lambda|_{\mathcal{I}} = 0$. Then $\lambda \in \widehat{\mathcal{E}}$ and at least one column of (9) is not exact. Hence by Proposition 2 from § 8 the complex $Tot(X_{\rho-\lambda})$ is not exact. Then the preliminary observation implies that $Kos(\rho-\lambda)$ is not exact, so $\lambda \in \sigma(\rho)$. Consequently $\lambda_0 = \lambda|_{\mathcal{B}} \in \sigma(\rho)|_{\mathcal{B}}$ and (8) is proved. □

Before establishing the announced converses of Corollary 1 we are going to show another application of Proposition 1. Namely we shall compute the spectra of certain concrete representations of Lie algebras. To this end we first particularize Proposition 1 in the following way (using Remark 1):

Corollary 3. *Let $\mathcal{E}$ be a Lie algebra of dimension 2 with a basis $\{a, b\}$ such that $[b, a] = \gamma a$ for some $\gamma \in \mathbb{C}$. If $\rho : \mathcal{E} \to \mathcal{B}(\mathcal{X})$ is a representation and we denote $T := \rho(e) \in \mathcal{B}(\mathcal{X})$, then $\sigma_1 \subseteq \sigma(\rho) \subseteq \sigma_2$ where*

$$\sigma_1 := \{\varphi : \mathcal{E} \to \mathbb{C} \text{ linear } \mid \varphi(a) = 0,\ \varphi(b) \in \sigma(T)\Delta(\gamma + \sigma(T))\},$$

$$\sigma_2 := \{\varphi : \mathcal{E} \to \mathbb{C} \text{ linear } \mid \varphi(a) = 0,\ \varphi(b) \in \sigma(T) \cup (\gamma + \sigma(T))\}.$$

Example 1. Let $\mathcal{E}$ be a Lie algebra of dimension 2 with a basis $\{a, b\}$ such that $[b, a] = \gamma a$ for some $\gamma \in \mathbb{C}\backslash\{0\}$. For some fixed $\lambda \in \mathbb{C}$ we define a representation $\rho : \mathcal{E} \to \mathcal{B}(\mathbb{C}^2)$ by

$$\rho(b) = \begin{pmatrix} \lambda+\gamma & 0 \\ 0 & \lambda \end{pmatrix} =: T,\ \rho(a) = \begin{pmatrix} 0 & 1 \\ 0 & 0 \end{pmatrix} =: N.$$

(We have $[T, N] = \gamma N$ so ρ is indeed a representation.) With the notation of Corollary 3 we have

$$\sigma_1 = \{\varphi : \mathcal{E} \to \mathbb{C} \text{ linear } \mid \varphi(a) = 0,\ \varphi(b) \in \{\lambda, \lambda + 2\gamma\}\},$$

$$\sigma_2 = \{\varphi : \mathcal{E} \to \mathbb{C} \text{ linear } \mid \varphi(a) = 0,\ \varphi(b) \in \{\lambda, \lambda + \gamma, \lambda + 2\gamma\}\}.$$

and $\sigma_1 \subseteq \sigma(\rho) \subseteq \sigma_2$. Hence for a precise computation of $\sigma(\rho)$ we have to decide whether or not $\varphi \in \sigma(\rho)$, where $\varphi : \mathcal{E} \to \mathbb{C}$ is a linear functional such that $\varphi(a) = 0$ and $\varphi(b) = \lambda + \gamma$. We have obviously

$$Ker\,N \cap Ker\,(T - \varphi(b) + \gamma) = Ker\,N \cap Ker\,(T - \lambda) = \{0\}$$

and

$$Ran\,N + Ran\,(T - \varphi(b)) = Ran\,N + Ran\,(T - \lambda - \gamma) = \mathbb{C}^2,$$

so the complex $Kos(\rho - \varphi)$ is exact in view of Example 3 from § 10. Consequently we have $\varphi \notin \sigma(\rho)$ and it follows that

$$\sigma(\rho) = \sigma_1 = \{\varphi : \mathcal{E} \to \mathbb{C} \text{ linear } \mid \varphi(a) = 0,\ \varphi(b) \in \{\lambda, \lambda + 2\gamma\}\}.$$

Now let's come back to the converses of Corollary 1 above.

Theorem 5. *Let $\mathcal{E}$ be a finite-dimensional Lie algebra. If there exists a representation $\rho : \mathcal{E} \to \mathcal{B}(\mathcal{X})$ such that $\mathcal{X} \neq \{0\}$ and for every subalgebra $\mathcal{L}$ of $\mathcal{E}$ with $\dim \mathcal{L} \leq 2$ we have*

$$\sigma(\rho|_{\mathcal{L}}) = \sigma(\rho)|_{\mathcal{L}}$$

then the Lie algebra $\mathcal{E}$ is nilpotent.

Proof. Suppose that $\mathcal{E}$ is not nilpotent. Then by Engel's Theorem there exists $b \in \mathcal{E} \backslash \{0\}$ such that $ad\, b$ is not nilpotent, so we can find $a \in \mathcal{E}\backslash\{0\}$ and $\gamma \in \mathbb{C}\backslash\{0\}$ such that $[b, a] = \gamma a$. Denote $\mathcal{I} := \mathbb{C}a$, $\mathcal{B} := \mathbb{C}b$. Then $\mathcal{I} + \mathcal{B}$ is a subalgebra of $\mathcal{E}$ and the hypothesis implies

$$\sigma(\rho|_{\mathcal{B}}) = \sigma(\rho)|_{\mathcal{B}} = (\sigma(\rho)|_{\mathcal{I}+\mathcal{B}})|_{\mathcal{B}} = \sigma(\rho|_{\mathcal{I}+\mathcal{B}})|_{\mathcal{B}}.$$

But if one applies Proposition 1 for the ideal $\mathcal{I}$ and the subalgebra $\mathcal{B}$ in the Lie algebra $\mathcal{I} + \mathcal{B}$, then the first inclusion from (7) shows that $\sigma(\rho|_{\mathcal{I}+\mathcal{B}})|_{\mathcal{B}}$ contains elements which do not belong to $\sigma(\rho|_{\mathcal{B}})$ (since $\gamma \neq 0$ and $\sigma(\rho|_{\mathcal{B}})$ is compact and nonempty; see also Corollary 3). This fact contradicts the above equalities. □

Next we are going to prove a variant of Theorem 5, where the projection property is assumed for the "large" subalgebras (instead of the "small" subalgebras as in Theorem 5). To this end we shall use Proposition 1 on the basis of the following auxiliary results.

Lemma 3. *Let $\mathcal{E}$ be a finite-dimensional Lie algebra.*

(a) *If $\mathcal{I}$ is an ideal of $\mathcal{E}$ such that $\dim \mathcal{I} = 1$ and $[\mathcal{I}, \mathcal{E}] \neq \{0\}$ then there exists a hyperplane subalgebra $\mathcal{B}$ of $\mathcal{E}$ such that $\mathcal{E} = \mathcal{I} + \mathcal{B}$.*

(b) *Let $\mathcal{J}$ be an ideal of $\mathcal{E}$ such that $\dim(\mathcal{E}/\mathcal{J}) = 1$. If there exists a one-dimensional ideal of $\mathcal{J}$ which does not commute with $\mathcal{J}$, then there exists a one-dimensional ideal of $\mathcal{E}$ which does not commute with $\mathcal{E}$.*

Proof. The assertion (a) is obtained as a by-product in the proof of Levi's Theorem (Theorem 1 from § 4). See e.g. the proof of theorem 1.6.9 from the book of J. Dixmier [1].

Now we prove the assertion (b). By hypothesis there exists $a_1^0 \in \mathcal{J}\backslash\{0\}$ and a character $\tilde{\lambda} \in \widehat{\mathcal{J}}\backslash\{0\}$ such that $[a, a_1^0] = \tilde{\lambda}(a)a_1^0$ for every $a \in \mathcal{J}$. Denote

$$\mathcal{J}_1 = \{a_1 \in \mathcal{J} \mid [a, a_1] = \tilde{\lambda}(a)a_1 \text{ for every } a \in \mathcal{J}\}.$$

For $x \in \mathcal{E}$, $a \in \mathcal{J}$ and $a_1 \in \mathcal{J}_1$ we have

$$[a, [x, a_1]] = [[a, x], a_1] + [x, [a, a_1]] = \tilde{\lambda}([a, x])a_1 + [x, [a, a_1]].$$

Hence by Lemma from § 5 of Chapter V from the book of J.P. Serre [1] we get $\tilde{\lambda}([a,x]) = 0$, hence

$$[a,[x,a_1]] = [x,[a,a_1]] = \tilde{\lambda}(a)[x,a_1] \text{ for } x \in \mathcal{E},\ a \in \mathcal{J} \text{ and } a_1 \in \mathcal{J}_1.$$

Particularly $[\mathcal{E}, \mathcal{J}_1] \subset \mathcal{J}_1$. Let $x_0 \in \mathcal{E} \backslash \mathcal{J}$, so $\mathcal{E} = \mathcal{J} \oplus \mathbb{C}x_0$. There exist $\mu \in \mathbb{C}$ and $a_1 \in \mathcal{J}_1 \backslash \{0\}$ such that $(adx_0)a_1 = \mu a_1$. Since $a_1 \in \mathcal{J}_1$, we have

$$[a,a_1] = \tilde{\lambda}(a)a_1 \text{ for all } a \in \mathcal{J}.$$

Since $\mathcal{E} = \mathcal{J} \oplus \mathbb{C}x_0$, it follows that $\mathbb{C}a_1$ is a one-dimensional ideal of $\mathcal{E}$. But $\tilde{\lambda} \not\equiv 0$, so this ideal does not commute with $\mathcal{E}$ (because, if $a_0 \in \mathcal{E}$ and $\tilde{\lambda}(a_0) \neq 0$ then $[a_0,a_1] = \tilde{\lambda}(a_0)a_1 \neq 0$). □

Corollary 4. *Let $\mathcal{E}$ be a finite-dimensional solvable Lie algebra. If $\mathcal{E}$ has a representation $\rho : \mathcal{E} \to \mathcal{B}(\mathcal{X})$ such that $\mathcal{X} \neq \{0\}$ and for every hyperplane subalgebra $\mathcal{L}$ of $\mathcal{E}$ we have*

$$\sigma(\rho)|_{\mathcal{L}} \subseteq \sigma(\rho|_{\mathcal{L}}),$$

then every one-dimensional ideal of $\mathcal{E}$ commutes with $\mathcal{E}$.

Proof. Suppose that we can find an ideal $\mathcal{I}$ of $\mathcal{E}$ such that $\dim \mathcal{I} = 1$ and $[\mathcal{I}, \mathcal{E}] \neq \{0\}$. By Lemma 3 (a) there exists a hyperplane subalgebra $\mathcal{B}$ of $\mathcal{E}$ such that $\mathcal{E} = \mathcal{I} + \mathcal{B}$. Since $\mathcal{I}$ is a one-dimensional ideal, there exists $\gamma \in \widehat{\mathcal{B}} \backslash \{0\}$ such that $[b,a] = \gamma(b)a$ for every $b \in \mathcal{B}$ and $a \in \mathcal{I}$. Then by Proposition 1 we obtain

$$(\sigma(\rho|_{\mathcal{B}}) + \gamma) \backslash \sigma(\rho|_{\mathcal{B}}) \subset \sigma(\rho)|_{\mathcal{B}}.$$

Since $\sigma(\rho|_{\mathcal{B}})$ is a compact nonempty set (by Theorem 3) and $\gamma \neq 0$, the left-hand side of the above inclusion is a nonempty set of elements which do not belong to $\sigma(\rho|_{\mathcal{B}})$. Particularly the above inclusion implies $\sigma(\rho)|_{\mathcal{B}} \not\subseteq \sigma(\rho|_{\mathcal{B}})$, a contradiction with the hypothesis. □

For stating the next result we need the following definition.

Definition 2. Let $\mathcal{E}$ be a finite-dimensional Lie algebra. We say that $\mathcal{E}$ is a *metabelian Lie algebra* if $[[\mathcal{E},\mathcal{E}],[\mathcal{E},\mathcal{E}]] = \{0\}$.

For stating the following results we recall the S^p classes of solvable Lie algebras from § 3.

The following lemma is concerned with some classes of solvable Lie algebras to which we shall apply Corollary 4.

Lemma 4. *Let $\mathcal{E}$ be a finite-dimensional Lie algebra so that each one-dimensional ideal of $\mathcal{E}$ commutes with $\mathcal{E}$. If moreover $\mathcal{E}$ has one of the following properties:*

(i) *$\mathcal{E}$ is solvable and* $\dim \mathcal{E} \leq 3$*;*

(ii) *$\mathcal{E}$ is metabelian;*

(iii) *$\mathcal{E}$ is belongs to the class S^2;*

then $\mathcal{E}$ is nilpotent.

Proof. (*i*) Since $\mathcal{E}$ is solvable, $[\mathcal{E},\mathcal{E}]$ is nilpotent and moreover $\mathcal{E} \neq [\mathcal{E},\mathcal{E}]$. Therefore $\dim[\mathcal{E},\mathcal{E}] \leq 2$ and $[\mathcal{E},\mathcal{E}]$ is abelian. Consequently $\mathcal{E}$ is metabelian and this case is treated below at (*ii*).

(*ii*) Assume that $\mathcal{E}$ is metabelian but it is not nilpotent. Let $\mathcal{H}$ be a Cartan subalgebra of $\mathcal{E}$ and use notation analogous to Proposition 4 from § 5, i.e. we denote by $\{\mathcal{E}^\alpha\}_{\alpha\in R}$ the family of root spaces of $\mathcal{E}$ with respect to $\mathcal{H}$. Since $\mathcal{E}$ is not nilpotent it must have at least one non-zero root. Furthermore we have $\oplus_{0\neq\alpha\in R}\mathcal{E}^\alpha \subseteq [\mathcal{E},\mathcal{E}]$ (see Proposition 4 (*c*) from § 5), hence the hypothesis that $\mathcal{E}$ is metabelian implies $[\mathcal{E}^\alpha,\mathcal{E}^\beta] = \{0\}$ for every $\alpha,\beta \in R\backslash\{0\}$. Now choose $\alpha \in R\backslash\{0\}$ and (by Engel's Theorem) $a \in \mathcal{E}^\alpha\backslash\{0\}$ such that $[h,a] = \alpha(h)a$ for each $h \in \mathcal{H}$. Since $[\mathcal{E}^\beta, a] = \{0\}$ for each $\beta \in R\backslash\{0\}$ and $\mathcal{E} = \oplus_{\beta\in R}\mathcal{E}^\beta$, it then follows that $[\mathcal{E},\mathbb{C}a] = \mathbb{C}a$. Hence $\mathbb{C}a$ is an ideal of $\mathcal{E}$ which does not commute with $\mathcal{E}$, a contradiction with the hypothesis.

(*iii*) We proceed by induction on $n = \dim\mathcal{E}$. The case $n \leq 3$ was discussed at (*i*). Now suppose that $n \geq 4$ and the conclusion holds for Lie algebras belonging to S^2 and having dimension strictly less than n. Let $\mathcal{J}$ be a proper ideal of $\mathcal{E}$. If $\mathcal{J}$ has a one-dimensional ideal which does not commute with $\mathcal{J}$ then also $\mathcal{E}$ posseses a one-dimensional ideal which does not commute with $\mathcal{E}$ (we can apply Lemma 3 (*b*) step by step because $\mathcal{E}$ belongs to S^2, so it is solvable), which is a contradiction with the hypothesis. Hence every one-dimensional ideal of $\mathcal{J}$ commutes with $\mathcal{J}$. Consequently, since $\mathcal{J}$ in turn belongs to S^2, the induction hypothesis implies that $\mathcal{J}$ is nilpotent. So every proper ideal of $\mathcal{E}$ is nilpotent.

Now let $e \in \mathcal{E}$, arbitrary. Since $\mathcal{E}$ belongs to the class S^2, there exists an ideal $\mathcal{J}$ of $\mathcal{E}$ such that $e \in \mathcal{J} \neq \mathcal{E}$. We have just seen that $\mathcal{J}$ is nilpotent so there exists $m \geq 1$ such that $(ad\,e|_{\mathcal{J}})^m = 0$, that is $(ad\,e)^m\mathcal{J} = \{0\}$. But $\mathcal{J}$ is an ideal of $\mathcal{E}$ and $a \in \mathcal{J}$, so $(ad\,e)\mathcal{E} \subseteq \mathcal{J}$. Hence $(ad\,e)^{m+1} = 0$ on $\mathcal{E}$. Since $e \in \mathcal{E}$ was arbitrary, by Engel's Theorem we obtain that $\mathcal{E}$ is nilpotent. □

Now by Corollary 4 and Lemma 4 we obtain the following variant of Theorem 5.

Theorem 6. *Let $\mathcal{E}$ be a finite-dimensional Lie algebra which has one of the following properties:*
(i) *$\mathcal{E}$ is solvable and $\dim\mathcal{E} \leq 3$;*
(ii) *$\mathcal{E}$ is metabelian;*
(iii) *$\mathcal{E}$ belongs to the class S^2.*
If there exists a representation $\rho : \mathcal{E} \to \mathcal{B}(\mathcal{X})$ such that $\mathcal{X} \neq \{0\}$ and for every hyperplane subalgebra $\mathcal{L}$ of $\mathcal{E}$ we have

$$\sigma(\rho)|_{\mathcal{L}} \subseteq \sigma(\rho|_{\mathcal{L}})$$

then the Lie algebra $\mathcal{E}$ is nilpotent.

As another aspect of the spectral theory of representations, now we study how small the diameters of spectra of representations of a normed solvable Lie algebra can be. We shall see (cf. Theorem 7 below) that these diameters are bounded from below by some constant which is strictly positive if the algebra is not nilpotent. Particularly the spectrum of every representation of a solvable Lie algebra which is not nilpotent contains at least two distinct points (see Corollary 5 below). To obtain these results we need the following elementary fact.

Lemma 5. *Let* $\gamma_1, \ldots, \gamma_s \in \mathbb{C}\backslash\{0\}$. *For every nonempty compact subset* K *of* $\mathbb{C}$ *denote* $K_0 = K$ *and, for* $p = 1, \ldots, s$,

$$K_p = \bigcup_{1 \le i_1 < \cdots < i_p \le s} (K + \gamma_{i_1} + \cdots + \gamma_{i_p}).$$

Then there exists a positive number ω *depending on* $\gamma_1, \ldots, \gamma_s$ *and independent of* K, *such that the following set*

$$M = \{z \in \mathbb{C} \mid \textit{There exists a unique } p \textit{ in } \{0, \ldots, s\} \textit{such that } z \in K_p\}$$

has diameter at least ω. *In particular,* M *has at least two distinct elements.*

Proof. After a suitable rotation of the complex plane we may suppose that $x_i = Re\,\gamma_i \neq 0$ for $1 \le i \le s$. Only one of the following situations can appear:

1° Not all the numbers $x_1, \ldots, x_s$ have the same sign.

2° All the numbers $x_1, \ldots, x_s$ have the same sign.

We shall study only the first possibility since the second one can be treated similarly. In the case 1° we necessarily have $s \ge 2$. After a suitable renumbering we may suppose that there exists $q \in \{1, \ldots, s-1\}$ such that all the numbers $x_1, \ldots, x_q$ are negative and all the numbers $x_{q+1}, \ldots, x_s$ are positive. We define

$$\omega := -x_1 - \cdots - x_q + x_{q+1} + \cdots + x_s = \sum_{i=1}^{s} |x_i|.$$

Let $\zeta^-, \zeta^+ \in K$ be such that

$$Re\,\zeta^- = \inf\{Re\,\lambda \mid \lambda \in K\}, \; Re\,\zeta^+ = \sup\{Re\,\lambda \mid \lambda \in K\}.$$

We shall verify that the following complex numbers

$$\lambda^- := \zeta^- + \sum_{j=1}^{q} \gamma_j, \; \lambda^+ := \zeta^+ + \sum_{j=q+1}^{s} \gamma_j$$

are two elements of M with $|\lambda^- - \lambda^+| \ge \omega$. First observe that

$$|\lambda^- - \lambda^+| \ge Re\,(\lambda^+ - \lambda^-) = Re(\zeta^+ - \zeta^-) + \omega \ge \omega.$$

Moreover, if $p \in \{1, \ldots, s\}\backslash\{q\}$ and $1 \le i_1 < \cdots < i_p \le s$ then $x_1 + \cdots + x_q < x_{i_1} + \cdots + x_{i_p}$ (because $x_1, \ldots, x_q$ are *all* the negative terms of the sequence $x_1, \ldots, x_s$) and $x_1 + \cdots + x_q < 0$. Hence for any $\lambda \in K$ we have

$$Re\,\lambda^- \le Re\,(\lambda + \gamma_1 + \cdots + \gamma_q) < Re\,(\lambda + \gamma_{i_1} + \cdots + \gamma_{i_p})$$

and

$$Re\,\lambda^- < Re\,\zeta^- \le Re\,\lambda.$$

Hence $\lambda^- \notin K_p$ for $p \in \{0,\dots,s\}\backslash\{q\}$; but $\lambda^- \in K_q$, so $\lambda^- \in M$. One checks similarly that $\lambda^+ \notin K_p$ for $p \in \{0,\dots,s\}\backslash\{s-q\}$; but $\lambda^+ \in K_{s-q}$, so we have also $\lambda^+ \in M$. □

In the next statement, we consider a finite-dimensional Lie algebra endowed with a vector space norm. Then the norm of every linear functional on that algebra makes sense. Particularly we can speak about the diameter (with respect to this dual norm) of the spectrum of a Banach space representation, since this is a compact set of characters (see Theorem 1; by Theorem 3 it is even nonempty if our Lie algebra is solvable).

Theorem 7. *Let $\mathcal{E}$ be a finite-dimensional solvable Lie algebra which is endowed with a vector space norm $\|\cdot\|$. Then $\mathcal{E}$ possesses Banach space representations with spectra of arbitrarily small diameters iff it is nilpotent.*

Proof. If $\mathcal{E}$ is nilpotent then we consider the trivial representation of $\mathcal{E}$ in an arbitrary non-zero Banach space,

$$\rho : \mathcal{E} \to \mathcal{B}(\mathcal{X}),\ \rho(e) = 0 \text{ for every } e \in \mathcal{E}.$$

Then by Corollary 2 we have $\lambda(e) = 0$ for every $\lambda \in \sigma(\rho)$ and $e \in \mathcal{E}$. So $\sigma(\rho) = \{0\}$ and we obtain a representation ρ with $diam\,(\sigma(\rho)) = 0$.

For the converse assertion we prove an equivalent statement. Namely, if the normed solvable finite-dimensional Lie algebra $\mathcal{E}$ is not nilpotent, then there exists a number $\omega > 0$ such that for every representation $\rho : E \to \mathcal{B}(\mathcal{X})$ we have

$$diam\,(\sigma(\rho)) \ge \omega.$$

To this end we proceed by induction on $n = \dim \mathcal{E}$.

If $n = 1$ the assertion is trivial. Next assume that $n \ge 2$ and that the above assertion holds for normed solvable Lie algebras of dimension strictly less than n. Endow any subalgebra of $\mathcal{E}$ with the norm inherited from $\mathcal{E}$.

If $\mathcal{E}$ has a proper Lie ideal $\mathcal{J}$ which is not nilpotent, then in view of the induction hypothesis we can choose $\omega > 0$ which is not greater than the diameter of the spectrum of an arbitrary Banach space representation of $\mathcal{J}$. Particularly, by projection property on ideals (Theorem 2 above) we obtain

$$diam\,(\sigma(\rho)) \ge diam\,(\sigma(\rho)|_{\mathcal{J}}) = diam\,(\sigma(\rho|_{\mathcal{J}})) \ge \omega$$

for every Banach space representation ρ of $\mathcal{E}$.

Next assume that every proper ideal of $\mathcal{E}$ is nilpotent. Since $\mathcal{E}$ is solvable we can choose a basis $\{e_1,\dots,e_n\}$ of $\mathcal{E}$ such that $\|e_k\| = 1\,(1 \le k \le n)$,

$$[e_j, e_i] = \sum_{k=1}^{i} c_{ji}^k e_k \qquad (1 \le i < j \le n) \tag{10}$$

and $\{e_k | 1 \leq k \leq m\}$ is a basis of $[E, E]$ for $m = \dim[\mathcal{E}, \mathcal{E}]$. Denote by $\mathcal{I}$ the ideal of $\mathcal{E}$ spanned by $e_1, \ldots, e_{n-1}$. Then $\mathcal{I}$ is nilpotent by our assumption. So $c^i_{ji} = 0$ for $1 \leq i < j \leq n-1$. Further we study the two possible cases:

a) There exists an i_0 in $\{1, \ldots, n-1\}$ such that $c^{i_0}_{n\,i_0} = 0$. Then the determinant

$$\begin{vmatrix} c^{n-1}_{n,n-1} & \cdots & c^1_{n,n-1} \\ 0 & \ddots & \vdots \\ 0 & 0 & c^1_{n,1} \end{vmatrix}$$

vanishes. Hence by (10) the vectors

$$[e_n, e_1], \ldots, [e_n, e_{n-1}]$$

are linearly dependent. Consequently, since we already know that $c^i_{ji} = 0$ for $1 \leq i < j \leq n-1$, we easily deduce as above that any $n-1$ vectors from the set

$$\{[e_j, e_i] \mid 1 \leq i < j \leq n\}$$

are linearly dependent. So we have

$$m = \dim[E, E] \leq n - 2.$$

Then $\mathcal{J}_1 := \mathbb{C}e_n + [E, E]$ and $\mathcal{J}_2 := \mathbb{C}e_{n-1} + \cdots + \mathbb{C}e_{m+1} + [E, E]$ are proper ideals of $\mathcal{E}$. Hence they are nilpotent by our assumption. Since $\mathcal{E} = \mathcal{J}_1 + \mathcal{J}_2$ we deduce that E is nilpotent, a contradiction with the hypothesis. Hence the present case cannot actually appear.

b) We have $\gamma_i := c^i_{n,i} \neq 0$ for every $i \in \{1, \ldots, n-1\}$. In this case we can apply Lemma 5 to the numbers $\gamma_1, \ldots, \gamma_{n-1}$; denote by ω the corresponding positive number, as in Lemma 5. We show that if $\rho : \mathcal{E} \to \mathcal{B}(\mathcal{X})$ is an arbitrary representation then we have $diam(\sigma(\rho)) \geq \omega$. Let's denote

$$T := \rho(e_n) \in \mathcal{B}(\mathcal{X}).$$

For each $p \geq 0$, for the operator

$$\theta^p_{e_n} : \mathcal{X} \otimes \Lambda^p \mathcal{I} \to \mathcal{X} \otimes \Lambda^p \mathcal{I}$$

we obtain (by Proposition 5 from § 10) that for every $\lambda \in \mathbb{C}$ we have

$$\theta^p_{e_n} - \lambda \text{ is not invertible} \Longleftrightarrow \lambda \in \bigcup_{1 \leq i_1 < \cdots < i_p \leq n-1} (\sigma(T) + \gamma_{i_1} + \cdots + \gamma_{i_p}). \quad (11)$$

Next we consider the commutative diagram

$$
\begin{array}{ccccccc}
 & & 0 & & 0 & & \\
 & & \downarrow & & \downarrow & & \\
0 & \leftarrow & \mathcal{X}\otimes\Lambda^{n-1}\mathcal{I} & \longleftarrow & \mathcal{X}\otimes\Lambda^{n-1}\mathcal{I} & \leftarrow & 0 \\
 & & \downarrow & & \downarrow & & \\
\cdots & & \cdots & & \cdots & & \cdots \\
 & & \downarrow & & \downarrow & & \\
0 & \leftarrow & \mathcal{X}\otimes\Lambda^{p}\mathcal{I} & \longleftarrow & \mathcal{X}\otimes\Lambda^{p}\mathcal{I} & \leftarrow & 0 \\
 & & \downarrow & & \downarrow & & \\
\cdots & & \cdots & & \cdots & & \cdots \\
 & & \downarrow & & \downarrow & & \\
0 & \leftarrow & \mathcal{X}\otimes\mathcal{I} & \longleftarrow & \mathcal{X}\otimes\mathcal{I} & \leftarrow & 0 \\
 & & \downarrow & & \downarrow & & \\
0 & \leftarrow & \mathcal{X} & \longleftarrow & \mathcal{X} & \leftarrow & 0 \\
 & & \downarrow & & \downarrow & & \\
 & & 0 & & 0 & &
\end{array}
\tag{12}
$$

whose columns coincide with $Kos(\rho|_{\mathcal{I}})$ and whose p-th row is $\theta^{p}_{e_n} - \lambda$ (see Proposition 4 from § 10). In view of the choice of ω (see (11) and Lemma 5 applied for $K = \sigma(T)$ and $s = n-1$), we obtain two numbers $\lambda_1, \lambda_2 \in \mathbb{C}$ with $|\lambda_1 - \lambda_2| \geq \omega(> 0)$ such that for $\lambda \in \{\lambda_1, \lambda_2\}$ precisely $n-1$ rows of (12) are exact. Then for such λ the totalization of (12) is not exact by Proposition 3 from § 8 (applied for rows instead of columns). Now define a linear functional $\widetilde{\lambda}_j : \mathcal{E} \to \mathbb{C}$ by $\widetilde{\lambda}_j|_{\mathcal{I}} = 0$ and $\widetilde{\lambda}_j(e_n) = \lambda_j$ for $j = 1, 2$. (Recall that $\mathcal{I}$ is the ideal spanned by $e_1, \ldots, e_{n-1}$; we have $\mathcal{E} = \mathcal{I} \oplus \mathbb{C}e_n$, so $[\mathcal{E}, \mathcal{E}] \subseteq \mathcal{I}$.) Then $\widetilde{\lambda}_j$ vanishes on $[\mathcal{E}, \mathcal{E}]$, so $\widetilde{\lambda}_j \in \widehat{\mathcal{E}}$. Moreover $Kos(\rho - \widetilde{\lambda}_j)$ is not exact since it is the totalization of (12) for $\lambda = \lambda_j$ (see Proposition 4 (*b*) from § 10). Consequently $\widetilde{\lambda}_1, \widetilde{\lambda}_2 \in \sigma(\rho)$. But $\|\widetilde{\lambda}_1 - \widetilde{\lambda}_2\| \geq |\widetilde{\lambda}_1(e_n) - \widetilde{\lambda}_2(e_n)| = |\lambda_1 - \lambda_2| \geq \omega$, so $diam(\sigma(\rho)) \geq \omega$. This finishes the proof. □

Corollary 5. *Let $\rho : \mathcal{E} \to \mathcal{B}(\mathcal{X})$ be a representation of the finite-dimensional solvable Lie algebra $\mathcal{E}$. If $\sigma(\rho)$ contains only one element, say $\widetilde{\lambda}$, then $\mathcal{E}$ is a nilpotent Lie algebra and for any $e \in E$ the spectrum of the operator $\rho(e) \in \mathcal{B}(\mathcal{X})$ is $\{\widetilde{\lambda}(e)\}$. Particularly, if $\sigma(\rho) = \{0\}$ then ρ is a representation by quasinilpotent operators.*

Proof. Endow $\mathcal{E}$ with an arbitrary vector space norm and then apply Theorem 7 to obtain the nilpotency of E. Then use Corollary 2 above. □

Another aspect of the spectral theory of Banach space representations of finite-dimensional solvable Lie algebras consists of the duality properties which we are going to establish now. These are variants of the well-known fact that the spectrum of a bounded linear operator on a Banach space coincides with the spectrum of the dual operator acting on the dual Banach space. In these results we make use of the character $\chi_{\mathcal{E}}$ of a solvable Lie algebra $\mathcal{E}$ that was introduced just before Theorem 1 in § 11.

Theorem 8. *Let $\rho : \mathcal{E} \to \mathcal{B}(\mathcal{X})$ be a representation of the finite-dimensional solvable Lie algebra $\mathcal{E}$. If $\rho^* : \mathcal{E}^{op} \to \mathcal{B}(\mathcal{X}^*)$ is its dual representation then*

$$\sigma(\rho^*) = \sigma(\rho) + \chi_{\mathcal{E}}.$$

Proof. First remark that $\widehat{\mathcal{E}} = \widehat{\mathcal{E}^{op}}$ by the definition of the opposite Lie algebra $\mathcal{E}^{op}$. Now take $\lambda \in \widehat{\mathcal{E}} = \widehat{\mathcal{E}^{op}}$ arbitrary. By applying Theorem 1 from § 9 and Theorem 1 from § 11 we obtain successively:

$$\lambda \notin \sigma(\rho) \Leftrightarrow Kos(\rho - \lambda) \text{ is exact } \Leftrightarrow (Kos(\rho - \lambda))^* \text{ is exact}$$

$$\Leftrightarrow Kos(\rho^* - \lambda + \chi_{\mathcal{E}}) \text{ is exact } \Leftrightarrow \lambda - \chi_{\mathcal{E}} \notin \sigma(\rho^*).$$

Consequently $\sigma(\rho^*) = \sigma(\rho) + \chi_{\mathcal{E}}$. □

Corollary 6. *If $\mathcal{E}$ is a finite-dimensional nilpotent Lie algebra then for every representation $\rho : \mathcal{E} \to \mathcal{B}(\mathcal{X})$ we have $\sigma(\rho^*) = \sigma(\rho)$.*

Proof. Since $\mathcal{E}$ is nilpotent we have $\chi_{\mathcal{E}} = 0$. Then apply Theorem 8. □

Finally we show how one can compute the spectrum of a tensor product of Hilbert space representations of finite-dimensional Lie algebras.

Theorem 9. *Let $\rho_j : \mathcal{E}_j \to \mathcal{B}(\mathcal{H}_j)$ be a Hilbert space representation of the finite-dimensional Lie algebra $\mathcal{E}_j$ $(j = 1, 2)$. Then the spectrum of the (Hilbert) tensor product*

$$\rho_1 \overline{\otimes} \rho_2 : \mathcal{E}_1 \times \mathcal{E}_2 \to \mathcal{B}(\mathcal{H}_1 \overline{\otimes} \mathcal{H}_2)$$

can be computed in the following way

$$\sigma(\rho_1 \overline{\otimes} \rho_2) = \sigma(\rho_1) \otimes \sigma(\rho_2) := \{\lambda_1 \otimes \lambda_2 \mid \lambda_1 \in \sigma(\rho_1),\ \lambda_2 \in \sigma(\rho_2)\}.$$

Proof. First recall that if $\lambda_j \in \sigma(\rho_j) \subset \widehat{\mathcal{E}_j}$, then $\lambda_j : \mathcal{E}_j \to \mathbb{C}$ is a Lie algebra morphism hence it can be viewed as a representation in a one-dimensional vector space $(j = 1, 2)$. Consequently it makes sense to consider the tensor product representation $\lambda_1 \otimes \lambda_2$; it is also a representation (of the Lie algebra $\mathcal{E}_1 \times \mathcal{E}_2$) in a one-dimensional vector space. Hence $\lambda_1 \otimes \lambda_2$ can be viewed as a character of $\mathcal{E}_1 \times \mathcal{E}_2$ acting in the following way:

$$\lambda_1 \otimes \lambda_2 : \mathcal{E}_1 \times \mathcal{E}_2 \to \mathbb{C},\ (e_1, e_2) \mapsto \lambda_1(e_1) + \lambda_2(e_2).$$

In fact it is easily seen that the following map is bijective

$$\widehat{\mathcal{E}_1} \times \widehat{\mathcal{E}_2} \to \widehat{\mathcal{E}_1 \times \mathcal{E}_2},\ (\lambda_1, \lambda_2) \mapsto \lambda_1 \otimes \lambda_2. \tag{13}$$

(This map should not be confused with the bilinear map canonically associated to the tensor product of the vector spaces $\widehat{\mathcal{E}_1}$ and $\widehat{\mathcal{E}_2}$.)

Now let's come back to the proof. Consider $\lambda_j \in \widehat{\mathcal{E}_j}$ arbitrary ($j = 1, 2$). For every $e_j \in \mathcal{E}_j$ ($j = 1, 2$) we have

$$((\rho_1 - \lambda_1)\overline{\otimes}(\rho_2 - \lambda_2))(e_1, e_2) = (\rho_1 - \lambda_1)(e_1)\overline{\otimes}I_{\mathcal{H}_2} + I_{\mathcal{H}_1}\overline{\otimes}(\rho_2 - \lambda_2)(e_2)$$

$$= (\rho_1(e_1)\overline{\otimes}I_{\mathcal{H}_2} + I_{\mathcal{H}_1}\overline{\otimes}\rho_2(e_2)) - (\lambda_1(e_1) + \lambda_2(e_2))I_{\mathcal{H}_1}\overline{\otimes}I_{\mathcal{H}_2}$$

$$= (\rho_1\overline{\otimes}\rho_2)(e_1, e_2) - (\lambda_1 \otimes \lambda_2)(e_1, e_2) \cdot I_{\mathcal{H}_1}\overline{\otimes}I_{\mathcal{H}_2}$$

$$= (\rho_1\overline{\otimes}\rho_2 - \lambda_1 \otimes \lambda_2)(e_1, e_2).$$

Consequently,

$$(\rho_1 - \lambda_1)\overline{\otimes}(\rho_2 - \lambda_2) = \rho_1\overline{\otimes}\rho_2 - \lambda_1 \otimes \lambda_2 \quad (\lambda_1 \in \widehat{\mathcal{E}_1}, \lambda_2 \in \widehat{\mathcal{E}_2}).$$

Hence, by Theorem 2 from § 11 we obtain

$$Kos\,(\rho_1\overline{\otimes}\rho_2 - \lambda_1 \otimes \lambda_2) = Kos\,(\rho_1 - \lambda_1)\overline{\otimes}Kos\,(\rho_2 - \lambda_2).$$

Then, by Theorem 4 from § 9, we deduce

$$\lambda_1 \otimes \lambda_2 \in \sigma(\rho_1\overline{\otimes}\rho_2) \Leftrightarrow \lambda_1 \in \sigma(\rho_1) \text{ and } \lambda_2 \in \sigma(\rho_2).$$

This finishes the proof in view of the bijectivity of the map (13). □

§ 26 Spectral theory for systems of operators generating nilpotent Lie algebras

In the present paragraph we apply the spectral theory of Banach space representations of Lie algebras (cf. § 25) for obtaining a noncommutative variant of Taylor joint spectrum and some corresponding variants of the properties of the Taylor joint spectrum of commuting tuples of operators. These variants hold for tuples of operators generating nilpotent Lie algebras. First of all let's introduce some useful notation. As usual, $\mathcal{X}$ will denote a complex Banach space. Furthermore, for an arbitrary n-tuple $T = (T_1, \ldots, T_n) \in \mathcal{B}(\mathcal{X})^n$ we denote:

a) for every $\lambda = (\lambda_1, \ldots, \lambda_n) \in \mathbb{C}^n$, $T - \lambda := (T_1 - \lambda_1, \ldots, T_n - \lambda_n) \in \mathcal{B}(\mathcal{X})^n$;

b) $\mathcal{E}(T)$ the Lie subalgebra of $\mathcal{B}(\mathcal{X})$ generated by $T_1, \ldots, T_n$;

c) $Kos\,(T, \mathcal{X})$ the Koszul complex of the identical representation

$$id_{\mathcal{E}(T)} : \mathcal{E}(T) \to \mathcal{E}(T),\ id_{\mathcal{E}(T)}(S) = S \text{ for every } S \in \mathcal{E}(T),$$

whenever $\dim \mathcal{E}(T) < \infty$.

For the following definition note that, if $\dim \mathcal{E}(T) < \infty$, then we have also $\dim \mathcal{E}(T - \lambda) < \infty$ for every $\lambda \in \mathbb{C}^n$.

Definition 1. Let $T = (T_1, \ldots, T_n) \in \mathcal{B}(\mathcal{X})^n$ such that $\dim \mathcal{E}(T) < \infty$. The *Taylor (joint) spectrum* of T is a subset of $\mathbb{C}^n$ denoted $\sigma(T)$ (or $\sigma(T, \mathcal{X})$) and defined in the following way:

$$\sigma(T) = \sigma(T, \mathcal{X}) = \{\lambda \in \mathbb{C}^n \mid \mathit{Kos}\,(T - \lambda, \mathcal{X}) \text{ is not exact}\}.$$

Generally, if $\mathcal{Y}$ is a joint invariant (closed) subspace for the operators $T_1, \ldots, T_n$, then we denote by $\sigma(T, \mathcal{Y})$ the Taylor spectrum of the n-tuple $(T_1|_{\mathcal{Y}}, \ldots, T_n|_{\mathcal{Y}})$.

Note that for $n = 1$ the Taylor spectrum reduces to the usual spectrum of an operator (see Remark 1 from § 25).

Next, for obtaining the main properties of the Taylor spectrum of an n-tuple $T \in \mathcal{B}(\mathcal{X})^n$ which generates a nilpotent Lie subalgebra of $\mathcal{B}(\mathcal{X})$ (recall that every nilpotent, finitely-generated Lie algebra is finite-dimensional), first we establish the connection with the spectral theory of Banach space representations of Lie algebras. To this end we need the following fact.

Lemma 1. *Let $\mathcal{E}$ be a finite-dimensional Lie algebra generated by $e_1, \ldots, e_n (\in \mathcal{E})$. Consider a representation $\rho : \mathcal{E} \to \mathcal{B}(\mathcal{X})$ and denote $T_i := \rho(e_i) \in \mathcal{B}(\mathcal{X})$ for $i = 1, \ldots, n$ and set $T := (T_1, \ldots, T_n) \in \mathcal{B}(\mathcal{X})^n$. Then for each $\lambda = (\lambda_1, \ldots, \lambda_n) \in \sigma(T)$ there exists a character $\widetilde{\lambda} \in \widehat{\mathcal{E}}$ such that $\widetilde{\lambda}(e_i) = \lambda_i$ for $i = 1, \ldots, n$.*

Proof. Assume that there exists no character $\widetilde{\lambda} \in \widehat{\mathcal{E}}$ with the desired property. As is easily seen, this assumption is equivalent to the existence of certain numbers $\alpha_1, \ldots, \alpha_n \in \mathbb{C}$ with the properties $\alpha_1 e_1 + \cdots + \alpha_n e_n \in [\mathcal{E}, \mathcal{E}]$ and $\alpha_1\lambda_1 + \cdots + \alpha_n\lambda_n \neq 0$. (Compare the negated assertions and use the fact that $\widetilde{\lambda} : \mathcal{E} \to \mathbb{C}$ would be linear with $\widetilde{\lambda}|_{[\mathcal{E},\mathcal{E}]} = 0$.) Then we may assume $\alpha_1\lambda_1 + \cdots + \alpha_n\lambda_n = 1$ and setting $c := \alpha_1 e_1 + \cdots + \alpha_n e_n \in [\mathcal{E}, \mathcal{E}]$ we deduce

$$\sum_{i=1}^{n} \alpha_i T_i = \rho(c) \in \rho([\mathcal{E}, \mathcal{E}]) = [\rho(\mathcal{E}), \rho(\mathcal{E})] = [\mathcal{E}(T), \mathcal{E}(T)] = [\mathcal{E}(T - \lambda), \mathcal{E}(T - \lambda)].$$

Consequently

$$\left(\sum_{i=1}^{n} \alpha_i\lambda_i\right) I_{\mathcal{X}} = \rho(c) - \sum_{i=1}^{n} \alpha_i(T_i - \lambda_i) \subset [\mathcal{E}(T - \lambda), \mathcal{E}(T-\lambda)] + \mathcal{E}(T-\lambda) - \mathcal{E}(T-\lambda).$$

Thus $e := I_{\mathcal{X}} \in \mathcal{E}(T - \lambda)$. With the notation of Proposition 4 from § 10 it easily follows that $\theta_e^{\cdot} : \mathit{Kos}\,(T - \lambda, \mathcal{X}) \to \mathit{Kos}\,(T - \lambda, \mathcal{X})$ is the identity morphism, so $\mathbf{H}_p(\theta_e^{\cdot}) = id_{\mathbf{H}_p(\mathit{Kos}\,(T-\lambda,\mathcal{X}))}$ for every $p \geq 0$. Hence by Proposition 4 (c) from § 10 we obtain $\mathbf{H}_p(\mathit{Kos}\,(T - \lambda, \mathcal{X})) = \{0\}$ for $p \geq 0$, that is the complex $\mathit{Kos}\,(T - \lambda, \mathcal{X})$ is exact. But this is a contradiction with the hypothesis $\lambda \in \sigma(T)$. □

The connection between the Taylor spectrum of a tuple of operators and the spectrum of a representation of a nilpotent Lie algebra is, in a certain sense, "faithful". More precisely, if the image of this representation contains all the operators of the tuple, then the Taylor spectrum of the tuple can be computed by means of the spectrum of the representation:

Theorem 1. *Let $T = (T_1, \ldots, T_n) \in \mathcal{B}(\mathcal{X})^n$ and $\rho : \mathcal{E} \to \mathcal{B}(\mathcal{X})$ be a representation of the finite-dimensional nilpotent Lie algebra $\mathcal{E}$ such that $T_1, \ldots, T_n \in \rho(\mathcal{E})$. If $e_1, \ldots, e_n \in \mathcal{E}$ have the properties $T_i := \rho(e_i) \in \mathcal{B}(\mathcal{X})$ for $i = 1, \ldots, n$ then*

$$\sigma(T) = \{(\widetilde{\lambda}(e_1), \ldots, \widetilde{\lambda}(e_n)) \in \mathbb{C}^n \mid \widetilde{\lambda} \in \sigma(\rho)\}.$$

If moreover $e_1, \ldots, e_n$ generate the Lie algebra $\mathcal{E}$, then the map

$$\sigma(\rho) \to \sigma(T),\ \widetilde{\lambda} \mapsto (\widetilde{\lambda}(e_1), \ldots, \widetilde{\lambda}(e_n)) \tag{1}$$

is bijective.

Proof. By considering the subalgebra of $\mathcal{E}$ generated by $e_1, \ldots, e_n$ and applying Corollary 1 from § 25 one easily deduces that we may assume without loss of generality that $\mathcal{E}$ itself is generated by $e_1, \ldots, e_n$. Assuming this fact we obtain $\rho(\mathcal{E}) = \mathcal{E}(T)$.

Now let $\widetilde{\lambda} \in \widehat{\mathcal{E}}$, $\widetilde{\lambda}(e_i) = \lambda_i$ for $i = 1, \ldots, n$ and set $\lambda = (\lambda_1, \ldots, \lambda_n) \in \mathbb{C}^n$. Then the Lie morphism $\rho - \widetilde{\lambda} : \mathcal{E} \to \mathcal{E}(T - \lambda)$ is onto. Hence Proposition 3 from § 10 implies that the following equivalence holds

$$\lambda \in \sigma(T) \Leftrightarrow \widetilde{\lambda} \in \sigma(\rho). \tag{2}$$

Particularly the map (1) is well defined, i.e. its values belong indeed to $\sigma(T)$. The fact that (1) is one-to-one follows from the assumption that $e_1, \ldots, e_n$ generate the Lie algebra $\mathcal{E}$. Finally, the map (1) is onto by (2) and Lemma 1. □

Remark 1.(a) Theorem 1 is a generalization of Corollary 2 from § 25.

(b) Several results of the spectral theory of representations of nilpotent Lie algebras can be translated in terms of Taylor spectrum of tuples generating nilpotent Lie algebras by means of Theorem 1 above. For example one can prove that, if $T = (T_1, \ldots, T_n) \in \mathcal{B}(\mathcal{X})^n$ generates a nilpotent Lie algebra and the point $\lambda = (\lambda_1, \ldots, \lambda_n) \in \mathbb{C}^n$ has the property that

$$\lim_{j \to \infty} (T_k - \lambda_k) x_j = 0 \qquad (1 \le k \le n)$$

for a certain sequence of unit vectors $\{x_j\}_{j \ge 1}$ from $\mathcal{X}$, then $\lambda \in \sigma(T)$. (It suffices to apply Remark 2 from § 25 and Theorem 1 above for $\mathcal{E} = \mathcal{E}(T)$, $\rho = id_{\mathcal{E}(T)} : \mathcal{E}(T) \to \mathcal{B}(\mathcal{X})$ and $\widetilde{\lambda} : \mathcal{E}(T) \to \mathbb{C}$ linear, defined by $\widetilde{\lambda}(T_k) = \lambda_k$ for $k = 1, \ldots, n$. It is easily seen that $\widetilde{\lambda}$ is well defined; moreover it is a character of $\mathcal{E}(T)$ in view of Theorem 4 from § 25.)

Another consequence of Theorem 1 is the following result which will be used in the proof of the fundamental Theorem 4 below.

Corollary 1. *Let $\rho : \mathcal{E} \to \mathcal{B}(\mathcal{X})$ be a representation of the finite-dimensional nilpotent Lie algebra $\mathcal{E}$, $\{e_1, \ldots, e_n\}$ a basis in $\mathcal{E}$ and $\varphi_1, \ldots, \varphi_m \in \mathbb{C}[X_1, \ldots, X_n]$ linear polynomials (i.e. the total degree of φ_i is 1 and $\varphi_i(0) = 0$ for $i = 1, \ldots, m$).*

Denote $T_i := \rho(e_i) \in \mathcal{B}(\mathcal{X})$ $(1 \le i \le n)$, $T = (T_1, \ldots, T_n) \in \mathcal{B}(\mathcal{X})^n$ and $\varphi := (\varphi_1, \ldots, \varphi_m) \in (\mathbb{C}[X_1, \ldots, X_n])^m$. Then the m-tuple $\varphi(T) \in \mathcal{B}(\mathcal{X})^m$ generates a nilpotent Lie algebra and

$$\sigma(\varphi(T)) = \varphi(\sigma(T)).$$

Proof. The Lie algebra $\mathcal{E}(\varphi(T))$ is nilpotent since it is a subalgebra of the nilpotent Lie algebra $\mathcal{E}(T)$. Next, if $\mathcal{L}$ denotes the Lie subalgebra of $\mathcal{E}$ generated by $\varphi_1(e_1, \ldots, e_n), \ldots, \varphi_m(e_1, \ldots, e_n) \in \mathcal{E}$ then by Corollary 1 from § 25 we obtain

$$\sigma(\rho|_{\mathcal{L}}) = \sigma(\rho)|_{\mathcal{L}}.$$

Hence in view of the fact that $\rho(\mathcal{L}) = \mathcal{E}(\varphi(T))$ we obtain (setting $e = (e_1, \ldots, e_n) \in \mathcal{E}^n$ and $\lambda = (\widetilde{\lambda}(e_1), \ldots, \widetilde{\lambda}(e_n)) \in \mathbb{C}^n$)

$$\begin{aligned} \sigma(\varphi(T)) &= \{(\widetilde{\lambda}(\varphi_1(e)), \ldots, \widetilde{\lambda}(\varphi_m(e))) \mid \widetilde{\lambda} \in \sigma(\rho|_{\mathcal{L}})\} \\ &= \{(\widetilde{\lambda}(\varphi_1(e)), \ldots, \widetilde{\lambda}(\varphi_m(e))) \mid \widetilde{\lambda} \in \sigma(\rho)\} \\ &= \{(\varphi_1(\lambda), \ldots, \varphi_m(\lambda)) \mid \widetilde{\lambda} \in \sigma(\rho)\} \\ &= \{\varphi(\lambda) \mid \widetilde{\lambda} \in \sigma(\rho)\}. \end{aligned}$$

Hence again by Theorem 1 we obtain $\sigma(\varphi(T)) = \varphi(\sigma(T))$. □

As another consequence of Theorem 1, we state the following fact obtained by means of Theorem 3 from § 25.

Theorem 2. *If $\mathcal{X} \neq \{0\}$ and the n-tuple $T \in \mathcal{B}(\mathcal{X})^n$ generates a nilpotent Lie algebra, then its Taylor spectrum $\sigma(T)$ is a compact non-empty subset of $\mathbb{C}^n$.*

The following duality property of the Taylor spectrum follows from Corollary 6 from § 25 by means of Theorem 1 above.

Theorem 3. *If the n-tuple $T = (T_1, \ldots, T_n) \in \mathcal{B}(\mathcal{X})^n$ generates a nilpotent Lie algebra, then the dual n-tuple $T^* = (T_1^*, \ldots, T_n^*) \in \mathcal{B}(\mathcal{X}^*)$ also generates a nilpotent Lie algebra and $\sigma(T^*) = \sigma(T)$.*

Now we prove the polynomial spectral mapping theorem for the Taylor spectrum of tuples generating nilpotent Lie algebras.

Theorem 4. *Let $T \in \mathcal{B}(\mathcal{X})^n$ be an n-tuple generating a nilpotent Lie subalgebra $\mathcal{E}$ of $\mathcal{B}(\mathcal{X})$. If $p \in (\mathbb{C}\langle X_1, \ldots, X_n\rangle)^m$ is an m-tuple of polynomials (in the noncommuting indeterminates $X_1, \ldots, X_n$) whose images in the enveloping algebra $U(\mathcal{E})$ generate a finite-dimensional Lie algebra $\widetilde{\mathcal{F}}$, then the Lie algebra generated by the m-tuple $p(T) \in \mathcal{B}(\mathcal{X})^m$ is nilpotent and $\sigma(p(T)) = p(\sigma(T))$.*

Proof. Let $T = (T_1, \ldots, T_n)$. Denote by $\rho : U(\mathcal{E}) \to \mathcal{B}(\mathcal{X})$ the canonical extension of the identical representation $id_{\mathcal{E}} : \mathcal{E} \to \mathcal{B}(\mathcal{X})$. Moreover, denote $e_i := T_i$ ("T_i as element of the abstract algebra $U(\mathcal{E})$") for $i = 1, \ldots, n$. As indicated in the statement of the theorem, we denote by $\widetilde{\mathcal{F}}$ the Lie subalgebra of $U(\mathcal{E})$ generated by the elements

$$p_1(e_1, \ldots, e_n), \ldots, p_m(e_1, \ldots, e_n) \in U(\mathcal{E}),$$

where $(p_1, \dots, p_m) := p \in (\mathbb{C}\langle X_1, \dots, X_n\rangle)^m$. Then we can write

$$\rho(\widetilde{\mathcal{F}}) = \mathcal{E}(p(T)) =: \mathcal{F}_0(\subset \mathcal{B}(\mathcal{X})).$$

Furthermore consider the Lie algebra

$$\theta(\widetilde{\mathcal{F}}) =: \mathcal{F} \subset \mathcal{B}(\mathcal{X} \otimes \Lambda\mathcal{E}),$$

constructed by means of the canonical extension $\theta : U(\mathcal{E}) \to \mathcal{B}(\mathcal{X} \otimes \Lambda\mathcal{E})$ of the representation $\theta : \mathcal{E} \to \mathcal{B}(\mathcal{X} \otimes \Lambda\mathcal{E})$ from Proposition 4 from § 10 (applied for $\mathcal{I} = \mathcal{E}$). We shall use throughout the present proof the notation and facts exposed in the last part of § 10 (more precisely, after the proof of Proposition 5 from § 10).

Now let's come back to the proof of the desired conclusion.

First we prove the inclusion "$\supseteq$". To this end it suffices to verify that, if $0 \in \sigma(T)$ and $p(0) = 0$, then $0 \in \sigma(p(T))$. Consider the bicomplex $\mathcal{B}(\mathcal{E}, \mathcal{F}, \mathcal{X})$. Since $p(0) = 0$, the free term of each $\widetilde{f} \in \widetilde{\mathcal{F}}$ is 0. Hence by Remark 4 from § 10 we obtain

$$\tau_q \equiv 0 \text{ for every } q \geq 0.$$

Consequently by Remark 3 from § 10 (see also Remark 6 from § 8) we get

$$\mathbf{H}_q(\alpha_{1\cdot}) = 0 \text{ for every } q \geq 0. \tag{3}$$

On the other hand $0 \in \sigma(T)$ implies that $Kos\,(\rho)$ is not exact (by Theorem 1 above), so the 0-th row of $\mathcal{B}(\mathcal{E}, \mathcal{F}, \mathcal{X})$ is not exact. In other words,

$$\mathcal{Y}_{0\cdot} \text{ is not exact.} \tag{4}$$

(Recall that the bicomplex $\mathcal{B}(\mathcal{E}, \mathcal{F}, \mathcal{X})$ was denoted $(\mathcal{Y}_{\cdot\cdot}, \alpha_{\cdot\cdot}, \beta_{\cdot\cdot})$, see § 10.) Finally, since $\mathbf{H}_q(Kos\,(\rho) \otimes \Lambda^p\mathcal{E}) = \mathbf{H}_q(Kos\,(\rho)) \otimes \Lambda^p\mathcal{E}$, we deduce that for each $q \geq 0$,

$$\text{if } \mathbf{H}_q(\mathcal{Y}_{0\cdot}) = \{0\}, \text{ then } \mathbf{H}_q(\mathcal{Y}_{p\cdot}) = \{0\} \text{ for every } p \geq 0. \tag{5}$$

Now, in view of (3), (4) and (5), by Proposition 4 from § 8 we deduce that at least one column of the bicomplex $\mathcal{B}(\mathcal{E}, \mathcal{F}, \mathcal{X})$ is not exact. Consequently $Kos\,(\rho_q)$ is not exact for a certain $q \geq 0$ (cf. the definition of the columns of $\mathcal{B}(\mathcal{E}, \mathcal{F}, \mathcal{X})$ in § 10). Hence $Kos\,(\mathcal{F}_0, \mathcal{X})(= Kos\,(id_{\mathcal{F}_0}))$ is not exact (by Proposition 9 from § 10). But we denoted $\mathcal{E}(p(T))$ by $\mathcal{F}_0$, so we have in fact $0 \in \sigma(p(T))$, as desired. The inclusion "$\supseteq$" is proved.

Now we prove the inclusion "$\subseteq$". To this end it suffices to verify that if $0 \in \sigma(p(T))$ then there exists $\lambda \in \sigma(T)$ such that $p(\lambda) = 0$. The proof of this fact has two stages.

1° First suppose that $e_1, \dots, e_n$ constitute a basis in $\mathcal{E}$ such that

$$\{e_1, \dots, e_m\} \text{ is a basis in } [\mathcal{E}, \mathcal{E}] \text{ and } [e_j, e_i] = \sum_{k=1}^{i-1} c_{ji}^k e_k \; (1 \leq i < j \leq n), \tag{6}$$

where $m := \dim[\mathcal{E}, \mathcal{E}]$. Moreover consider the ideals of $\mathcal{E}$,

$$\mathcal{I}_k := \mathbb{C}e_1 + \cdots + \mathbb{C}e_k \qquad (1 \leq k \leq n).$$

We have $0 \in \sigma(p(T))$ hence $Kos\,(id_{\mathcal{F}_0})(= Kos\,(\mathcal{F}_0, \mathcal{X}))$ is not exact. Then by Propositions 9 and 7 from § 10 we deduce that $Kos\,(\rho_0)$ is not exact.

On the other hand we have obviously $Tot(\{0\}, \mathcal{F}, \mathcal{X}) = Kos\,(\rho_0)$, so by Proposition 6 from § 10 (applied for $\mathcal{I} = \{0\}$, $\mathcal{J} = \mathcal{I}_1$) we deduce

$$Tot(\mathcal{I}_1, \mathcal{F}, \mathcal{X}) = Con\,(Kos\,(\rho_0), \delta_{e_1}).$$

But by Słodkowski's Lemma (Theorem 3 from § 9) there exists $\lambda_1 \in \mathbb{C}$ such that $Con\,(Kos(\rho_0), \delta_{e_1} - \lambda_1)$ is not exact. Particularly (by Corollary 1 (b′) from § 8) at least one row of the bicomplex whose totalization is $Con\,(Kos\,(\rho_0), \delta_{e_1} - \lambda_1)$ (see diagram (3) from § 8) is not exact. Hence $\lambda_1 \in \sigma(\rho(e_1))$ (see Remark 8 and Proposition 5, both from § 10). If $m = \dim[\mathcal{E}, \mathcal{E}] \geq 1$, then $e_1 \in [\mathcal{E}, \mathcal{E}]$, so $\rho(e_1) \in \rho([\mathcal{E}, \mathcal{E}]) = [\rho(\mathcal{E}), \rho(\mathcal{E})]$. Hence $\rho(e_1)$ is quasinilpotent by Theorem 4 from § 25. But $\lambda_1 \in \sigma(\rho(e_1))$. Consequently $\lambda_1 = 0$ if $m \geq 1$. Moreover, if we define the character $\widetilde{\lambda}_1 : \mathcal{I}_1 \to \mathbb{C}$ by $\widetilde{\lambda}_1(e_1) = \lambda_1$, then by Proposition 6 from § 10 we have

$$Tot_{\widetilde{\lambda}_1}(\mathcal{I}_1, \mathcal{F}, \mathcal{X}) = Con\,(Tot(\{0\}, \mathcal{F}, \mathcal{X}), \delta_{e_1} - \widetilde{\lambda}_1(e_1)),$$

so $Tot_{\widetilde{\lambda}_1}(\mathcal{I}_1, \mathcal{F}, \mathcal{X})$ is not exact.

Now assume that we obtained $\lambda_1, \ldots, \lambda_k \in \mathbb{C}$ such that $\lambda_i = 0$ whenever $i \leq m$ and moreover $Tot_{\widetilde{\lambda}_k}(\mathcal{I}_k, \mathcal{F}, \mathcal{X})$ is not exact, where $\widetilde{\lambda}_k : \mathcal{I}_k \to \mathbb{C}$ is the character defined by $\widetilde{\lambda}_k(e_i) := \lambda_i$ for $i = 1, \ldots, k$. Then by Słodkowski's Lemma (Theorem 3 from § 9) applied for the morphism of complexes (cf. Proposition 6 from § 8)

$$\delta_{e_{k+1}} : Tot_{\widetilde{\lambda}_k}(\mathcal{I}_k, \mathcal{F}, \mathcal{X}) \to Tot_{\widetilde{\lambda}_k}(\mathcal{I}_k, \mathcal{F}, \mathcal{X})$$

we obtain $\lambda_{k+1} \in \mathbb{C}$ such that the complex

$$Con\,(Tot_{\widetilde{\lambda}_k}(\mathcal{I}_k, \mathcal{F}, \mathcal{X}), \delta_{e_{k+1}} - \lambda_{k+1})$$

is not exact. As above we deduce $\lambda_{k+1} \in \sigma(\rho(e_{k+1}))$ and $\lambda_{k+1} = 0$ if $k + 1 \leq m$. Hence if we define a linear functional $\widetilde{\lambda}_{k+1} : \mathcal{I}_{k+1} \to \mathbb{C}$ by $\widetilde{\lambda}_{k+1}(e_{k+1}) := \lambda_{k+1}$ and $\widetilde{\lambda}_{k+1}|_{\mathcal{I}_k} := \widetilde{\lambda}_k$, then $\widetilde{\lambda}_{k+1}$ is a character of $\mathcal{I}_{k+1}$. Moreover, by the choice of λ_{k+1} and by Proposition 6 from § 10 we deduce that $Tot_{\widetilde{\lambda}_{k+1}}(\mathcal{I}_{k+1}, \mathcal{F}, \mathcal{X})$ is not exact.

Finally we obtain a character $\widetilde{\lambda}_n =: \widetilde{\lambda} \in \widehat{\mathcal{E}}$ such that $Tot_{\widetilde{\lambda}}(\mathcal{E}, \mathcal{F}, \mathcal{X})$ is not exact. Hence by Corollary 1 (a) from § 8 at least one of the complexes

$$\mathbf{H}_q(\mathcal{B}_{\widetilde{\lambda}}(\mathcal{E}, \mathcal{F}, \mathcal{X}), \alpha..)$$

is not exact. Then by Remark 5 from § 10 we obtain that $Kos\,(\rho - \widetilde{\lambda})$ is not exact and $p(\lambda_1, \ldots, \lambda_n) = 0$, where $\lambda_i := \widetilde{\lambda}(e_i)$ for $i = 1, \ldots, n$. Hence we obtained $\lambda := (\lambda_1, \ldots, \lambda_n) \in \sigma(T)$ (by Theorem 1 above) and $p(\lambda) = 0$, as desired.

2° At this stage we show how one can reduce the general case to the situation studied at the first stage. Let $\{f_1, \ldots, f_N\}$ be a basis in $\mathcal{E}$ with properties similar to (6). Then one can find linear polynomials $\varphi_1, \ldots, \varphi_n \in \mathbb{C}[X_1, \ldots, X_N]$ such that $e_i = \varphi_i(f_1, \ldots, f_N)$ for $i = 1, \ldots, n$. Denote

$$\varphi := (\varphi_1, \ldots, \varphi_n) \in (\mathbb{C}[X_1, \ldots, X_N])^n,$$

$S_j := \rho(f_j)$ for $j = 1, \ldots, n$ and $S := (S_1, \ldots, S_N) \in \mathcal{B}(\mathcal{X})^N$. Then $T = \varphi(S)$. Hence by Corollary 1 we deduce

$$\sigma(T) = \varphi(\sigma(S)).$$

Consequently

$$p(\sigma(T)) = (p \circ \varphi)(\sigma(S)). \tag{7}$$

But by stage 1° applied to the tuple S and the tuple of polynomials $p \circ \varphi$ we have

$$(p \circ \varphi)(\sigma(S)) = \sigma((p \circ \varphi)(S)). \tag{8}$$

But $(p \circ \varphi)(S) = p(\varphi(S)) = p(T)$, hence by (7) and (8) we deduce $p(\sigma(T)) = \sigma(p(T))$. □

As an immediate consequence of either Theorem 4 or Corollary 1 we have the following fact, known as the *projection property* of the Taylor spectrum.

Corollary 2. *Let $T \in \mathcal{B}(\mathcal{X})^n$ be an n-tuple generating a nilpotent Lie algebra. If $1 \leq k \leq n$, $1 \leq i_1 < \cdots < i_k \leq n$ and $\pi : \mathbb{C}^n \to \mathbb{C}^k$ is the projection defined by $(z_1, \ldots, z_n) \mapsto (z_{i_1}, \ldots, z_{i_k})$ then*

$$\sigma(T_{i_1}, \ldots, T_{i_k}) = \pi(\sigma(T)).$$

Now we can proceed with:

Proof of Theorem 1 from § 23. By Rosenblum's Theorem (Theorem 1 from § 13) and by Engel's Theorem it follows that $\mathcal{G}$ is a nilpotent Lie algebra. Now let $A \in \mathcal{A}_0(\mathcal{G})$. Then there exist $G_1, \ldots, G_n \in \mathcal{G}$ and a polynomial $p \in \mathbb{C}\langle X_1, \ldots, X_n\rangle$ in n noncommuting variables such that $p(0) = 0$ and $p(G_1, \ldots, G_n) = A$. We have $\sigma(G_i) = \{0\}$ by hypothesis, hence by the projection property (Corollary 2 above) it easily follows $\sigma(G_1, \ldots, G_n) = \{0\}$. Now we deduce $\sigma(A) = \{0\}$ by the spectral mapping theorem (Theorem 4).

Now assume that $\mathcal{G}$ consists only of nilpotent operators. Choose a basis $T_1, \ldots, T_m$ of $\mathcal{G}$ and let $N \geq 1$ be such that $T_i^N = 0$ for $i = 1, \ldots, m$. Then, by the Poincaré-Birkhoff-Witt Theorem, we get

$$\mathcal{A}_0(\mathcal{G}) = \{T_1^{k_1} \cdots T_m^{k_m} \mid k_1, \ldots, k_m \in \{0, \ldots, N-1\}\},$$

hence $\dim \mathcal{A}_0(\mathcal{G}) < \infty$. But every element of a finite-dimensional associative algebra is algebraic. So every element of $\mathcal{A}_0(\mathcal{G})$ is algebraic and quasinilpotent, i.e. nilpotent. □

In the following, as an application of the projection property we shall describe the structure of the Taylor spectrum of a tuple of compact operators generating a nilpotent Lie algebra (see Theorem 5 below). We begin with some simple facts concerning arbitrary operators.

Remark 2. If $C, P \in \mathcal{B}(\mathcal{X})$ have the properties $P^2 = P$ and $[[C,P],P] = 0$, then $[C,P] = 0$. (Indeed, we need only to write the relation $[[C,P],P] = 0$ under the form $CP + PC = 2PCP$ and to multiply the last relation at left, respectively at right by P. Then one gets $PC = PCP$, respectively $CP = PCP$; hence $PC = CP$.)

Proposition 1. *Let $A \in \mathcal{B}(\mathcal{X})$ and $B = (B_1, \dots, B_q) \in \mathcal{B}(\mathcal{X})^q$ be a commuting tuple. Assume that there exists $m \geq 1$ such that*

$$(ad B_{i_1}) \cdots (ad B_{i_N})A = 0 \text{ for every } i_1, \dots, i_m \in \{1, \dots, q\}.$$

Then for every function $f \in \mathcal{O}(\sigma(B))$ taking only the values 0 and 1 we have $[A, f(B)] = 0$.

Proof. We prove by descending induction that for $k = m, m-1, \dots, 0$ the property $P(k)$ holds, where

$$P(k) : \forall i_1, \dots, i_k \in \{1, \dots, q\}, \; [(ad B_{i_k}) \cdots (ad B_{i_1})A, f(B)] = 0.$$

(Here P(0) denotes just the conclusion $[A, f(B)] = 0$.) The property $P(m)$ follows immediately from the hypothesis. Now assume that $P(k)$ holds for a certain $k \in \{m, \dots, 1\}$. To prove $P(k-1)$, fix for the moment the indices $i_1, \dots, i_{k-1} \in \{1, \dots, q\}$ and denote $C := (ad B_{i_{k-1}}) \cdots (ad B_{i_1})A$. (Of course, if $k = 1$ we don't fix anything and denote C:=A.) From $P(k)$ we deduce that for every $i \in \{1, \dots, q\}$ we have $[[B_i, C], f(B)] = 0$, which implies $[[C, f(B)], B_i] = 0$. Then it follows that $[[C, f(B)], f(B)] = 0$ because $f(B)$ belongs to the bicommutant of the tuple B. But $(f(B))^2 = f(B)$ in view of the hypothesis on f. Hence by Remark 2 we can deduce $[C, f(B)] = 0$. Recalling the definition of C we obtain $P(k-1)$. □

The following result is a weak "non-commutative variant" of the well-known theorem concerning commuting tuples of operators with non-connected Taylor spectra.

Corollary 3. *Let $C = (A_1, \dots, A_p, B_1, \dots, B_q) \in \mathcal{B}(\mathcal{X})^{p+q}$ be a tuple of operators generating a nilpotent Lie algebra. We assume that $B := (B_1, \dots, B_q)$ is a commuting tuple with $\sigma(B) = F_1 \cup F_2$, where F_1 and F_2 are closed non-empty and disjoint subsets of $\mathbb{C}^q$. We denote by V_1 an arbitrary open neighbourhood of F_1, with the closure disjoint of F_2, and by f the characteristic function of V_1. Then $f \in \mathcal{O}(\sigma(B))$. If we denote $\mathcal{X}_1 := f(B)\mathcal{X}$, $\mathcal{X}_2 := (1 - f(B))\mathcal{X}$ and $\pi : \sigma(C) \to \mathbb{C}^q$ is the projection on the last q coordinates, then we have:*

i) *$\mathcal{X}_1, \mathcal{X}_2$ are joint invariant subspaces for the tuple C and $\mathcal{X} = \mathcal{X}_1 \oplus \mathcal{X}_2$;*

ii) *$\sigma(C, \mathcal{X}_j) = \pi^{-1}(F_j)$ for $j = 1, 2$.*

Proof. Since the tuple C generates a nilpotent Lie algebra, we can use Proposition 1 to deduce $[A_i, f(B)] = 0$ for any $i = 1, \ldots, p$ and the assertions from $i)$ follow immediately.

Next by $i)$ we deduce

$$Kos(C - \lambda, \mathcal{X}) = \begin{array}{c} Kos(C - \lambda, \mathcal{X}_1) \\ \oplus \\ Kos(C - \lambda, \mathcal{X}_2) \end{array} \qquad \text{for every } \lambda \text{ in } \mathbb{C}^{p+q}.$$

This fact easily implies that

$$\sigma(C) = \sigma(C, \mathcal{X}_1) \cup \sigma(C, \mathcal{X}_2). \tag{9}$$

(See Remark 2 c) from § 8.)

On the other hand, in view of the projection property (Corollary 2) we obtain that $\pi(\sigma(C, \mathcal{X}_j)) = \sigma(B, \mathcal{X}_j) = F_j$, so

$$\sigma(C, \mathcal{X}_j) \subseteq \pi^{-1}(F_j) \quad (j = 1, 2). \tag{10}$$

A new application of the projection property (Corollary 2) implies $\pi(\sigma(C)) = \sigma(B) = F_1 \cup F_2$, hence

$$\sigma(C) = \pi^{-1}(F_1) \cup \pi^{-1}(F_2). \tag{11}$$

Now by (9), (10) and (11) we deduce $\sigma(C, \mathcal{X}_j) = \pi^{-1}(F_j)$ $(j = 1, 2)$. □

We shall need also the following simple result which settles the situation of tuples of operators generating nilpotent Lie algebras which act on finite-dimensional spaces.

Lemma 2. *Let $T = (T_1, \ldots, T_n)$ be a tuple of operators generating a nilpotent Lie algebra of operators on the finite-dimensional space $\mathcal{Y}$. For every $\lambda = (\lambda_1, \ldots, \lambda_n) \in \sigma(T)$ there exists $y \in \mathcal{Y} \setminus \{0\}$ such that $T_j y = \lambda_j y$ for $1 \leq j \leq n$. (In other words λ is a joint eigenvalue of the tuple T.)*

Proof. Since the complex $Kos(T - \lambda, \mathcal{X})(= Kos(id_{\mathcal{E}(T-\lambda)})$ is not exact, from the equivalence $(iii) \Leftrightarrow (ii')$ of Corollary 2 from § 11 we deduce $\cap_{j=1}^{n} Ker(T_j - \lambda_j) \neq \{0\}$. Hence we can choose a vector y with the desired properties. □

Now we prove a variant of the above fact which holds on infinite-dimensional spaces.

Corollary 4. *Let $T = (T_1, \ldots, T_n)$ be a tuple of compact operators which generates a nilpotent Lie algebra. Then every $\lambda = (\lambda_1, \ldots, \lambda_n) \in \sigma(T) \setminus \{0\}$ is an isolated point of $\sigma(T)$ and a joint eigenvalue of the tuple T, i.e. there exists $x \in \mathcal{X} \setminus \{0\}$ such that $T_j x = \lambda_j x$ for $1 \leq j \leq n$.*

Proof. If $\dim \mathcal{X} < \infty$, then $\sigma(T_j)$ is finite for $1 \leq j \leq n$. But $\sigma(T) \subseteq \sigma(T_1) \times \cdots \times \sigma(T_n)$ by the projection property (Corollary 2) hence $\sigma(T)$ is also finite. Consequently each point of $\sigma(T)$ is isolated. Next one applies Lemma 2.

Now suppose that $\mathcal{X}$ is an infinite-dimensional space. Since $\lambda \neq 0$ we may assume $\lambda_n \neq 0$. Hence if we denote $F_1 := \{\lambda_n\}$ and $F_2 := \sigma(T_n)\backslash\{\lambda_n\}$, then we get two closed, non-empty and disjoint sets with $\sigma(T_n) = F_1 \cup F_2$. Let's remark that we can apply Corollary 3 for $C = (T_1, \dots, T_n)$ and $B = T_n$. With the notation of Corollary 3, remark that $\mathcal{X}_1 = \mathcal{X}_{T_n}(\{\lambda_n\})$ is finite-dimensional since T_n is a compact operator and $\lambda_n \neq 0$. Since $\sigma(T) = \pi^{-1}(F_1) \cup \pi^{-1}(F_2)$ (where $\pi : \mathbb{C}^n \to \mathbb{C}$, $(z_1, \dots, z_n) \mapsto z_n$) it is easily seen that

$$\sigma(T) \cap (\mathbb{C}^{n-1} \times V_1) = \pi^{-1}(F_1) = \sigma(T, \mathcal{X}_1).$$

But the last set is finite because $\dim \mathcal{X}_1 < \infty$ (see the beginning of the present proof). Since $\mathbb{C}^{n-1} \times V_1$ is a neighbourhood of λ, it then follows that λ is an isolated point of $\sigma(T)$. It follows also that $\lambda \in \sigma(T, \mathcal{X}_1)$ hence λ is a joint eigenvalue of T by Lemma 2. □

A last auxiliary result which we shall need is the following.

Lemma 3. *Let $T = (T_1, \dots, T_n) \in \mathcal{B}(\mathcal{X})^n$ be a tuple of compact operators which generates a nilpotent Lie algebra. If $0 \notin \sigma(T)$, then $\mathcal{X}$ is finite-dimensional.*

Proof. We proceed by induction on n. The assertion is well known for $n = 1$. Now suppose that it holds for $(n-1)$-tuples. The set $\sigma(T)$ is compact (cf. Theorem 2) and every point of $\sigma(T)$ is isolated (by Corollary 4 since $0 \notin \sigma(T)$). Hence the set $\sigma(T)$ is finite.

On the other hand we may assume $0 \in \sigma(T_n)$. (If $0 \notin \sigma(T_n)$, then the conclusion $\dim \mathcal{X} < \infty$ follows from the case $n = 1$.) The set $\sigma(T_n)$ is finite by the projection property (Corollary 2) because $\sigma(T)$ is finite. Hence we can apply Corollary 3 for $C = (T_1, \dots, T_n)$, $B = T_n$, $F_1 = \{0\}$, $F_2 = \sigma(T_n)\backslash\{0\}$ and it follows that

$$\mathcal{X} = \mathcal{X}_1 \oplus \mathcal{X}_2; \ \sigma(T, \mathcal{X}_j) = \pi^{-1}(F_j) \qquad (j = 1, 2).$$

(If $\sigma(T_n) = \{0\}$, then we take $F_2 = \emptyset$ and $\mathcal{X}_2 = \{0\}$.) Particularly, in view of the definition of $\mathcal{X}_2$ we have $0 \notin \sigma(T_n, \mathcal{X}_2)$, so $\dim \mathcal{X}_2 < \infty$.

On the other hand, since $\sigma(T_n, \mathcal{X}_1) = F_1 = \{0\}$, by the projection property we deduce

$$\sigma(T, \mathcal{X}_1) = \sigma((T_1, \dots, T_{n-1}), \mathcal{X}_1) \times \{0\}.$$

But $0 \notin \sigma(T, \mathcal{X})$ and $\sigma(T, \mathcal{X}_1) \subseteq \sigma(T, \mathcal{X})$ (cf. Corollary 3 *ii*)), hence the above equality implies $0 \notin \sigma((T_1, \dots, T_{n-1}), \mathcal{X}_1)$. Then we deduce $\dim \mathcal{X}_1 < \infty$ by the induction hypothesis. Since we proved that $\mathcal{X}_2$ is also finite-dimensional, it follows that so is $\mathcal{X}(= \mathcal{X}_1 \oplus \mathcal{X}_2)$. □

Now we can establish the main result concerning the Taylor spectrum of a tuple of compact operators generating a nilpotent Lie algebra.

Theorem 5. *Let $T \in \mathcal{B}(\mathcal{X})^n$ be a tuple of compact operators generating a nilpotent Lie algebra. Then $\sigma(T)\backslash\{0\}$ is an at most countable set of joint eigenvalues of T, which accumulates at most toward 0. Moreover, if $\mathcal{X}$ is infinite-dimensional, then $0 \in \sigma(T)$.*

Proof. Let $T = (T_1, \ldots, T_n)$. By the projection property (Corollary 2) it follows that $\sigma(T) \subseteq \sigma(T_1) \times \cdots \times \sigma(T_n)$. Hence $\sigma(T)$ is at most countable in view of the well-known spectral properties of the compact operators. The other assertions follow from Corollary 4 and Lemma 3. □

For the sake of completeness we establish the following fact, which allows us particularly to identify the spectrum of a finite-dimensional nilpotent Lie algebra with the set of classical weights of that representation (see § 20).

Corollary 5. *Let $\rho : \mathcal{E} \to \mathcal{B}(\mathcal{X})$ be a representation of the nilpotent Lie algebra $\mathcal{E}$ by compact operators and let $\widetilde{\lambda} \in \widehat{\mathcal{E}}$ be a character of $\mathcal{E}$. We assume that either $\widetilde{\lambda} \neq 0$ or $\mathcal{X}$ is finite-dimensional. Then $\widetilde{\lambda} \in \sigma(\rho)$ iff there exists a vector $x \in \mathcal{X} \backslash \{0\}$ such that $\rho(e)x = \widetilde{\lambda}(e)x$ for every $e \in \mathcal{E}$.*

Proof. If there exists a vector $x \in \mathcal{X} \backslash \{0\}$ such that $\rho(e)x = \widetilde{\lambda}(e)x$ for every $e \in \mathcal{E}$, then we apply Remark 2 from § 25 to deduce $\widetilde{\lambda} \in \sigma(\rho)$. Conversely, let's assume $\widetilde{\lambda} \in \sigma(\rho)$. Choose a basis $\{e_1, \ldots, e_n\}$ in $\mathcal{E}$ and denote $\widetilde{\lambda}(e_i) = \lambda_i$, $\rho(e_i) = T_i$ for $i = 1, \ldots, n$ and $\lambda = (\lambda_1, \ldots, \lambda_n) \in \mathbb{C}^n$, $T = (T_1, \ldots, T_n) \in \mathcal{B}(\mathcal{X})^n$. Then $\lambda \in \sigma(T)$ by Theorem 1; hence either by Corollary 4 (if $\widetilde{\lambda} \neq 0$) or by Lemma 2 (if $\dim \mathcal{X} < \infty$) we obtain $x \in \mathcal{X} \backslash \{0\}$ such that $T_i x = \lambda_i x$ for $1 \leq i \leq n$. Since $T_i = \rho(e_i)$, $\lambda_i = \widetilde{\lambda}(e_i)$ and $\{e_1, \ldots, e_n\}$ is a basis in $\mathcal{E}$, it then follows that $\rho(e)x = \widetilde{\lambda}(e)x$ for every $e \in \mathcal{E}$. □

Our next aim is to prove a multi-dimensional non-commutative variant of the well-known Ringrose Theorem (see Theorem 6 below). It allows us to compute the Taylor spectrum of a tuple of compact operators generating a nilpotent Lie algebra, by means of a nest of joint invariant subspaces. In order to obtain such a result we need a few facts about nests of invariant subspaces. More precisely, let $T = (T_1, \ldots, T_n) \in \mathcal{B}(\mathcal{X})^n$. We denote by *LatT* the lattice of all closed subspaces of $\mathcal{X}$ which are invariant for $T_1, \ldots, T_n$. Next let $\mathcal{N}$ be a *nest* in *LatT*, i.e. a subset of *LatT* which is totally ordered by inclusion. For $\mathcal{Y} \in \mathcal{N}$ we denote by $\mathcal{Y}^-$ the closed subspace spanned by those elements of $\mathcal{N}$ which are strictly contained in $\mathcal{Y}$. Then $\mathcal{Y}^- \in LatT$ and, if $\mathcal{Y}^-$ is distinct from $\mathcal{Y}$, we say that $\mathcal{Y}$ *determines the atom* $\mathcal{Y}/\mathcal{Y}^-$. If moreover $\dim(\mathcal{Y}/\mathcal{Y}^-) = 1$, then the operator induced by T_i on $\mathcal{Y}/\mathcal{Y}^-$ can be identified with a scalar $\lambda_i \in \mathbb{C}$ such that

$$(T_i - \lambda_i)\mathcal{Y} \subseteq \mathcal{Y}^- \qquad (1 \leq i \leq n). \tag{12}$$

In this case the point $\lambda = (\lambda_1, \ldots, \lambda_n) \in \mathbb{C}^n$ will be called the *point determined by the atom* $\mathcal{Y}/\mathcal{Y}^-$.

We shall need also the following fact which is used in the proof of the classical Ringrose Theorem.

Lemma 4. *Let $S \in \mathcal{B}(\mathcal{X})$ be a compact operator and $\mathcal{N}$ be a maximal nest in $LatS$. Let $\mu \in \sigma(S) \backslash \{0\}$ and $x \in \mathcal{X} \backslash \{0\}$ be such that $Sx = \mu x$. If $\mathcal{Y}$ denotes the intersection of those $\mathcal{Z} \in \mathcal{N}$ having the property $x \in \mathcal{Z}$, then we have $\mathcal{Y} \in \mathcal{N}$, $\mathcal{Y} = \mathbb{C}x \oplus \mathcal{Y}^-$ and $(S - \mu)\mathcal{Y} \subseteq \mathcal{Y}^-$.*

Now we can prove the following result.

Lemma 5. *Let $T = (T_1, \dots, T_n) \in \mathcal{B}(\mathcal{X})^n$ be a tuple of compact operators which generates a nilpotent Lie algebra and let $\mathcal{N}$ be a maximal nest in $Lat T$. Then every atom of $\mathcal{N}$ has dimension 1. Moreover if we denote by $\{\lambda_\alpha \mid \alpha \in A\}$ the set of points determined by the atoms of $\mathcal{N}$, then we have $\sigma(T)\backslash\{0\} = \{\lambda_\alpha \mid \alpha \in A\}\backslash\{0\}$.*

Proof. The fact that every atom of $\mathcal{N}$ has dimension 1 follows by Corollary 3 from § 21. Now, for $\alpha \in A$ denote by $\mathcal{Y}_\alpha$ the element of $\mathcal{N}$ which determines the point λ_α and set $\lambda_\alpha =: (\lambda_{\alpha 1}, \dots, \lambda_{\alpha n}) \in \mathbb{C}^n$. Then by (12) we obtain for every $\alpha \in A$

$$(T_i - \lambda_{\alpha i})\mathcal{Y}_\alpha \subseteq \mathcal{Y}_\alpha^- \qquad (1 \le i \le n)$$

hence

$$(T_1 - \lambda_{\alpha 1})\mathcal{Y}_\alpha + \cdots + (T_n - \lambda_{\alpha n})\mathcal{Y}_\alpha \subseteq \mathcal{Y}_\alpha^- \neq \mathcal{Y}_\alpha.$$

This implies (see Example 2 from § 10) that the complex $Kos\,(T|_{\mathcal{Y}_\alpha} - \lambda_\alpha)$ is not exact at the first term, where $T|_{\mathcal{Y}_\alpha} := (T_1|_{\mathcal{Y}_\alpha}, \dots, T_n|_{\mathcal{Y}_\alpha})$. Consequently $\lambda_\alpha \in \sigma(T, \mathcal{Y}_\alpha)$. Hence, if $\lambda_\alpha \neq 0$, then λ_α is a joint eigenvalue of $T|_{\mathcal{Y}_\alpha}$ by Theorem 5. Particularly λ_α is a joint eigenvalue of T and this implies $\lambda_\alpha \in \sigma(T)$ (by Corollary 5 and Theorem 1). Consequently $\{\lambda_\alpha \mid \alpha \in A\}\backslash\{0\} \subseteq \sigma(T)$.

Conversely, let $\lambda = (\lambda_1, \dots, \lambda_n) \in \sigma(T)\backslash\{0\}$. Then by Theorem 5 there exists $x \in \mathcal{X}\backslash\{0\}$ such that

$$T_i x = \lambda_i x \qquad (1 \le i \le n). \tag{13}$$

Let $\mathcal{Y}$ be the intersection of those $\mathcal{Z} \in \mathcal{N}$ having the property $x \in \mathcal{Z}$. Now let $i \in \{1, \dots, n\}$ be such that $\lambda_i \neq 0$. Since every atom of $\mathcal{N}$ has dimension 1, it follows that the nest $\mathcal{N}$ is maximal also in $Lat T_i$. Hence we can apply Lemma 4 to deduce

$$\mathcal{Y} \in \mathcal{N} \text{ and } \mathcal{Y} = \mathbb{C}x \oplus \mathcal{Y}^- \tag{14}$$

and moreover

$$(T_i - \lambda_i)\mathcal{Y} \subseteq \mathcal{Y}^-. \tag{15}$$

Now if $i \in \{1, \dots, n\}$ is such that $\lambda_i = 0$, then (15) also holds (by (13) and (14)). Consequently (15) holds for $i = 1, \dots, n$ and then (14) shows that $\mathcal{Y} \in \mathcal{N}$ determines an atom and λ is just the point determined by that atom. Particularly $\lambda = \lambda_\alpha$ for some $\alpha \in A$. Hence $\sigma(T)\backslash\{0\} \subseteq \{\lambda_\alpha \mid \alpha \in A\}$. □

Theorem 6. *In the situation from Lemma 5 the following assertions hold.*
a) *If $\mathcal{X}$ is finite-dimensional, then $\sigma(T) = \{\lambda_\alpha \mid \alpha \in A\}$.*
b) *If $\mathcal{X}$ is infinite-dimensional, then $\sigma(T) = \{\lambda_\alpha \mid \alpha \in A\} \cup \{0\}$.*

Proof. The assertion *b*) follows from Lemma 5 and from the last sentence of Theorem 5.

Now assume that $\mathcal{X}$ is finite-dimensional. By Lemma 5 it suffices to prove that $0 \in \sigma(T)$ iff there exists $\alpha \in A$ such that $\lambda_\alpha = 0$. If such an α exists, then, as in the first part of the proof of Lemma 5 (with that notation) we obtain

$0=\lambda_\alpha\in\sigma(T,\mathcal{Y}_\alpha)$. Hence by Lemma 2 there exists $y\in\mathcal{Y}_\alpha\backslash\{0\}$ such that $T_iy=0$ for $1\le i\le n$. Then $0\in\sigma(T)$ by Corollary 5 and Theorem 1.

Conversely, suppose $0\in\sigma(T)$. First remark that $A=\{1,\ldots,N\}$, where $N:=\dim\mathcal{X}$. Let $\alpha\in A$ be the lowest index with the property $0\in\sigma(T,\mathcal{Y}_\alpha)$. Then $0\notin\sigma(T,\mathcal{Y}_{\alpha-1})$. Particularly $Kos\,(T|_{\mathcal{Y}_{\alpha-1}})$ is exact at the first term hence $\mathcal{Y}_{\alpha-1}=T_1(\mathcal{Y}_{\alpha-1})+\cdots+T_n(\mathcal{Y}_{\alpha-1})$ (see Example 2 from § 10). On the other hand $0\in\sigma(T,\mathcal{Y}_\alpha)$ implies that $Kos\,(T|_{\mathcal{Y}_\alpha})$ is not exact, hence by Corollary 2 from § 11 (the equivalence $(i')\Leftrightarrow(iii)$) we deduce $T_1(\mathcal{Y}_\alpha)+\cdots+T_n(\mathcal{Y}_\alpha)\neq\mathcal{Y}_\alpha$. Hence

$$\mathcal{Y}_{\alpha-1}=\sum_{i=1}^n T_i(\mathcal{Y}_{\alpha-1})\subseteq\sum_{i=1}^n T_i(\mathcal{Y}_\alpha)\subsetneq\mathcal{Y}_\alpha.$$

But $\dim(\mathcal{Y}_\alpha/\mathcal{Y}_\alpha^-)=1$, so $T_i(\mathcal{Y}_\alpha)\subseteq\mathcal{Y}_{\alpha-1}$ for $1\le i\le n$. This implies $\lambda_\alpha=0$ (see (12)) for our α. □

We finish with a result concerning the Taylor spectrum of a tensor product of tuples acting on Hilbert spaces.

Theorem 7. *Let $\mathcal{H}_1$ and $\mathcal{H}_2$ be complex Hilbert spaces. Suppose that the tuples $T=(T_1,\ldots,T_n)\in\mathcal{B}(\mathcal{H}_1)^n$, respectively $S=(S_1,\ldots,S_m)\in\mathcal{B}(\mathcal{H}_2)^m$, generate nilpotent Lie algebras. If we denote*

$$R:=(T_1\overline{\otimes}I,\ldots,T_n\overline{\otimes}I,I\overline{\otimes}S_1,\ldots,I\overline{\otimes}S_m)\in\mathcal{B}(\mathcal{H}_1\overline{\otimes}\mathcal{H}_2)^{n+m},$$

then R generates a nilpotent Lie algebra and $\sigma(R)=\sigma(T)\times\sigma(S)$.

Proof. Let $\mathcal{E}_1$ be a nilpotent Lie algebra generated by $e_1,\ldots,e_n\in\mathcal{E}_1$ and $\rho_1:\mathcal{E}_1\to\mathcal{B}(\mathcal{H}_1)$ be a representation such that $\rho_1(e_i)=T_i$ for $i=1,\ldots,n$. (For example $\mathcal{E}_1:=\mathcal{E}(T)$ and $\rho_1=id_{\mathcal{E}(T)}:\mathcal{E}(T)\to\mathcal{B}(\mathcal{H}_1)$.) Similarly, let $\mathcal{E}_2$ be a nilpotent Lie algebra generated by $f_1,\ldots,f_m\in\mathcal{E}_2$ and $\rho_2:\mathcal{E}_2\to\mathcal{B}(\mathcal{H}_2)$ be a representation such that $\rho_2(f_j)=S_j$ for $j=1,\ldots,m$. If we identify canonically $\mathcal{E}_1\cong\mathcal{E}_1\times\{0\}\subset\mathcal{E}_1\times\mathcal{E}_2$ and $\mathcal{E}_2=\{0\}\times\mathcal{E}_2\subset\mathcal{E}_1\times\mathcal{E}_2$, then we can write $e_i,f_j\in\mathcal{E}_1\times\mathcal{E}_2$ and

$$(\rho_1\overline{\otimes}\rho_2)(e_i)=T_i\overline{\otimes}I_{\mathcal{H}_2},\ (\rho_1\overline{\otimes}\rho_2)(f_j)=I_{\mathcal{H}_1}\overline{\otimes}S_j\ (1\le i\le n,1\le j\le m).$$

Moreover $\mathcal{E}_1\times\mathcal{E}_2$ is generated by $e_1,\ldots,e_n,f_1,\ldots,f_m$ and $[e_i,f_j]=0$ for all i,j. This implies that $\mathcal{E}_1\times\mathcal{E}_2$ is nilpotent, hence R generates a nilpotent Lie algebra. Moreover by Theorem 1 we deduce

$$\sigma(R)=\{(\lambda(e_1),\ldots,\lambda(e_n),\lambda(f_1),\ldots,\lambda(f_m))\mid\lambda\in\sigma(\rho_1\overline{\otimes}\rho_2)\},$$

$$\sigma(T)=\{(\lambda_1(e_1),\ldots,\lambda_1(e_n))\mid\lambda_1\in\sigma(\rho_1)\}$$

and

$$\sigma(S)=\{(\lambda_2(f_1),\ldots,\lambda_2(f_m))\mid\lambda_2\in\sigma(\rho_2)\}.$$

Hence by Theorem 9 from § 25 we deduce

$$\begin{aligned}\sigma(R) &= \{((\lambda_1 \otimes \lambda_2)(e_1), \ldots, (\lambda_1 \otimes \lambda_2)(e_n), (\lambda_1 \otimes \lambda_2)(f_1), \ldots, (\lambda_1 \otimes \lambda_2)(f_m)) \mid \\ &\qquad \lambda_1 \in \sigma(\rho_1), \lambda_2 \in \sigma(\rho_2)\} \\ &= \{(\lambda_1(e_1), \ldots, \lambda_1(e_n), \lambda_2(f_1), \ldots, \lambda_2(f_m)) \mid \lambda_1 \in \sigma(\rho_1), \lambda_2 \in \sigma(\rho_2)\} \\ &= \sigma(T) \times \sigma(S),\end{aligned}$$

as desired. □

§ 27 The Cartan-Taylor spectrum of a locally solvable Lie algebra of operators

The main aim of the present paragraph is to introduce a concept of spectrum for a finite-dimensional solvable Lie algebra of operators. In the case of nilpotent Lie algebras, this concept agres with the Taylor spectrum from § 26 (see the remarks after Definition 1 below). But this is no longer true for non-nilpotent Lie algebras. Actually, for the spectrum $\Sigma(\cdot)$ that is introduced below, the projection property on each Lie subalgebra holds (cf. Theorem 1 below). Hence for non-nilpotent Lie algebras the spectrum defined by means of a Koszul complex does not agree with $\Sigma(\cdot)$ (compare Theorem 5 from § 25 and Theorem 1 below).

The main idea of the spectrum $\Sigma(\cdot)$ (cf. Definition 1 below) is to use Cartan subalgebras for reducing to nilpotent Lie algebras, a situation settled in § 25 and § 26. The main technical tool is the conjugation theorem for Cartan subalgebras (Theorem 1 from § 5). We shall use it by means of a few remarks collected in the following lemma.

Lemma 1. *Let $\mathcal{G}$ be a finite-dimensional Lie subalgebra of $\mathcal{B}(\mathcal{X})$.*

a) *For every $G \in \mathcal{G}$ and $\varphi \in Aut_e(\mathcal{G})$ the operators G and $\varphi(G)$ are similar, hence they have equal spectra.*

b) *If $\mathcal{G}$ is solvable, then for every quasinilpotent operator $Q \in \mathcal{G}$ we have $Q + r(\mathcal{G}) = r(\mathcal{G})$.*

Proof. *a*) The automorphism φ is the composition of a finite number of automorphisms of the form $\exp(ad\,A)$ with $A \in \mathcal{G}$. But by Remark 1 from § 15 we have

$$(\exp(ad\,A))B = (\exp A)B(\exp(-A)) \text{ for } A, B \in \mathcal{B}(\mathcal{X})$$

hence there exists an invertible operator $U \in \mathcal{B}(\mathcal{X})$ such that $\varphi(G) = UGU^{-1}$ for $G \in \mathcal{G}$.

b) By Rosenblum's Theorem (Theorem 1 from § 13) the operator $ad\,Q : \mathcal{G} \to \mathcal{G}$ is nilpotent, hence the conclusion follows by Lemma 2 (*a*) from § 5. □

We need also the following:

Notation 1. If $\mathcal{G}$ is a subset of $\mathcal{B}(\mathcal{X})$, then $\mathcal{Q}_{\mathcal{G}}$ will denote the set of all quasinilpotent elements of $\mathcal{G}$.

Lemma 2. *If $\mathcal{G}$ is a finite-dimensional solvable Lie subalgebra of $\mathcal{B}(\mathcal{X})$, then $\mathcal{Q}_{\mathcal{G}}$ is an ideal of $\mathcal{G}$ containing $[\mathcal{G}, \mathcal{G}]$.*

Proof. Since $\mathcal{G}$ is solvable we have $[\mathcal{G}, \mathcal{G}] \subseteq \mathcal{Q}_{\mathcal{G}}$ by Theorem 4 from § 25, hence it remains to prove that $\mathcal{Q}_{\mathcal{G}}$ is a vector space. Let $Q_1, Q_2 \in \mathcal{Q}_{\mathcal{G}}$ and denote $\mathcal{L} := \mathbb{C}Q_1 + \mathbb{C}Q_2 + [\mathcal{G}, \mathcal{G}]$. Then $\mathcal{L}$ is an ideal of $\mathcal{G}$. Let us choose, by Lie's Theorem, a basis in $\mathcal{G}$ such that any operator $adG : \mathcal{G} \to \mathcal{G}$ $(G \in \mathcal{L})$ can be represented by an upper triangular matrix. Then by Rosenblum's Theorem (Theorem 1 from § 13) any operator adG $(G \in \mathcal{G})$ will be represented by the sum of three upper triangular matrices whose diagonals consist only of zeros. So $adG : \mathcal{G} \to \mathcal{G}$ is a nilpotent operator for any G in $\mathcal{L}$. Since $(adG)(\mathcal{L}) \subseteq \mathcal{L}$ for any G in $\mathcal{L}$, one can apply the Engel Theorem to deduce that $\mathcal{L}$ is a nilpotent Lie algebra. Hence $\sigma(Q_1 + Q_2) = \{0\}$ by either Corollary 1 or Theorem 4 from § 26. □

Now we establish a fact which will ensure the correctness of the definition of the spectrum $\Sigma(\cdot)$.

Lemma 3. *Let $\mathcal{G}$ be a finite-dimensional solvable Lie subalgebra of $\mathcal{B}(\mathcal{X})$. If $\mathcal{H}$ is a Cartan subalgebra of $\mathcal{G}$ and $\mathcal{G} = \mathcal{H} \oplus C_{\mathcal{H}}$ is the associated Cartan decomposition, then we define the set*

$$\Sigma_{\mathcal{H}} := \{f : \mathcal{G} \to \mathbb{C} \text{ linear} \mid f|_{\mathcal{H}} \in \sigma(id_{\mathcal{H}}) \text{ and } f|_{C_{\mathcal{H}}} = 0\}$$

Then for any Cartan subalgebras $\mathcal{H}, \mathcal{K}$ of $\mathcal{G}$ we have $\Sigma_{\mathcal{H}} = \Sigma_{\mathcal{K}} \subset \widehat{\mathcal{G}}$.

Proof. First we prove that $\Sigma_{\mathcal{H}} \subseteq \widehat{\mathcal{G}}$. Let $\mathcal{G} = \mathcal{H} \oplus C_{\mathcal{H}}$ be the Cartan decomposition of $\mathcal{G}$ with respect to $\mathcal{H}$, where $C_{\mathcal{H}} = \oplus_{\alpha \neq 0} \mathcal{G}^{\alpha}$ (cf. the notations of Proposition 4 from § 5). Then

$$[\mathcal{G}, \mathcal{G}] = [\mathcal{H}, \mathcal{H}] + \sum_{\alpha} [\mathcal{H}, \mathcal{G}^{\alpha}] + \sum_{\alpha, \beta} [\mathcal{G}^{\alpha}, \mathcal{G}^{\beta}].$$

But for $\alpha, \beta \in R$ we have $[\mathcal{G}^{\alpha}, \mathcal{G}^{\beta}] \subseteq \mathcal{G}^{\alpha+\beta}$ and $\mathcal{G}^0 = \mathcal{H}$. Hence

$$[\mathcal{G}, \mathcal{G}] \subseteq [\mathcal{H}, \mathcal{H}] + \sum_{\alpha} [\mathcal{G}^{\alpha}, \mathcal{G}^{-\alpha}] + \sum_{\alpha \neq 0} \mathcal{G}^{\alpha} \subseteq ([\mathcal{G}, \mathcal{G}] \cap \mathcal{H}) + C_{\mathcal{H}}.$$

Finally by Theorem 4 from § 25 we get $[\mathcal{G}, \mathcal{G}] \subseteq \mathcal{Q}_{\mathcal{H}} + C_{\mathcal{H}}$.

Now let $f \in \Sigma_{\mathcal{H}}$. Then $f|_{C_{\mathcal{H}}} = 0$ by the definition of the set $\Sigma_{\mathcal{H}}$. Moreover, $f|_{\mathcal{Q}_{\mathcal{H}}} = 0$ since $f|_{\mathcal{H}} \in \sigma(id_{\mathcal{H}})$ so $f(Q) = 0$ for every quasinilpotent operator $Q \in \mathcal{H}$ (see Corollary 2 from § 25). Consequently f vanishes on $[\mathcal{G}, \mathcal{G}]$, i.e. $f \in \widehat{\mathcal{G}}$.

Now we show that $\Sigma_{\mathcal{H}} = \Sigma_{\mathcal{K}}$. By the conjugation theorem for the Cartan subalgebras of $\mathcal{G}$ it follows that there exists an automorphism $\varphi \in Aut_e(\mathcal{G})$ such that $\varphi(\mathcal{H}) = \mathcal{K}$. By the proof of Lemma 1 *a*), this implies that the representations $id_{\mathcal{K}} \circ \varphi$ and $id_{\mathcal{H}}$ of $\mathcal{H}$ are intertwined by an invertible operator. Hence by the formula (3) from § 25 and by Proposition 3 from § 10 we obtain $\sigma(id_{\mathcal{H}}) = \{f \circ \varphi \mid f \in \sigma(id_{\mathcal{K}})\}$. Since $\varphi(C_{\mathcal{H}}) = C_{\mathcal{K}}$ (Proposition 4 (*b*) from § 5), we deduce

$$\Sigma_{\mathcal{H}} = \{f \circ \varphi \mid f \in \Sigma_{\mathcal{K}}\}.$$

Moreover for any $G \in \mathcal{G}$ we have $G - \varphi(G) \in [\mathcal{G}, \mathcal{G}]$ (see Lemma 1 from § 5). But we proved above that $\Sigma_{\mathcal{K}} \subseteq \widehat{\mathcal{G}}$, so for $f \in \Sigma_{\mathcal{K}}$ we have $f(G - \varphi(G)) = 0$. In other words $f \circ \varphi = f$ for any f in $\Sigma_{\mathcal{K}}$, so $\Sigma_{\mathcal{H}} = \Sigma_{\mathcal{K}}$. □

Now we can introduce the main concept.

Definition 1. Let $\mathcal{G}$ be a finite-dimensional solvable Lie subalgebra of $\mathcal{B}(\mathcal{X})$. Then the *Cartan-Taylor spectrum* of $\mathcal{G}$ is defined by

$$\Sigma(\mathcal{G}) := \Sigma_{\mathcal{H}},$$

where $\mathcal{H}$ is a Cartan subalgebra of $\mathcal{G}$ (see Lemma 3).

Concerning this definition let's remark that $\Sigma(\mathcal{G}) \subseteq \widehat{\mathcal{G}}$ by Lemma 3. Moreover, if $\mathcal{G}$ is nilpotent, then $\Sigma(\mathcal{G}) = \sigma(id_{\mathcal{G}})$, hence the above definition agrees with Definition 1 from § 25.

Now we begin to prove the properties of the spectrum $\Sigma(\cdot)$. To this end we need two auxiliary results.

Lemma 4. *If $\mathcal{G}$ is a finite-dimensional solvable Lie subalgebra of $\mathcal{B}(\mathcal{X})$, $Q \in \mathcal{Q}_{\mathcal{G}}$ and $f \in \Sigma(\mathcal{G})$, then $f(Q) = 0$.*

Proof. Let $\mathcal{H}$ be a Cartan subalgebra of $\mathcal{G}$ and let $\mathcal{G} = \mathcal{H} \oplus \mathcal{C}_{\mathcal{H}}$ be the corresponding Cartan decomposition. Hence there exist $Q' \in \mathcal{H}$ and $Q'' \in \mathcal{C}_{\mathcal{H}}$ such that $Q' + Q'' = Q$. By Theorem 4 from § 25 and Proposition 4 (*c*) from § 5 we get $Q'' \in \mathcal{Q}_{\mathcal{G}}$. Consequently by Lemma 2 we obtain $Q' = Q - Q'' \in \mathcal{Q}_{\mathcal{G}}$. Hence Q' is a quasinilpotent element of $\mathcal{H}$. Since $f|_{\mathcal{H}} \in \sigma(id_{\mathcal{H}})$, it then follows that $f(Q') = 0$ (see Corollary 2 from § 25). On the other hand $f(Q'') = 0$ by the definition of $\Sigma(\mathcal{G})$, since $Q'' \in \mathcal{C}_{\mathcal{H}}$. Consequently $f(Q) = f(Q') + f(Q'') = 0$. □

Lemma 5. *If $\mathcal{G}$ is a finite-dimensional solvable Lie subalgebra of $\mathcal{B}(\mathcal{X})$ and $\mathcal{I}$ is a Lie ideal of $\mathcal{G}$, then $\Sigma(\mathcal{G})|_{\mathcal{I}} = \Sigma(\mathcal{I})$.*

Proof. Since the Lie algebra $\mathcal{G}$ is solvable we may suppose without loss of generality that $\dim(\mathcal{G}/\mathcal{I}) = 1$. Let $\mathcal{H}$ be a Cartan subalgebra of $\mathcal{G}$. By Proposition 5 from § 5 it then follows that $\mathcal{H} \cap \mathcal{I}$ is contained in some Cartan subalgebra $\mathcal{K}$ of $\mathcal{I}$. Since $\mathcal{H} \cap \mathcal{I} \subseteq \mathcal{K} \subseteq (\mathcal{H} \cap \mathcal{I}) \oplus \mathcal{C}_{\mathcal{H}}$, it is easily seen that

$$\mathcal{K} = (\mathcal{H} \cap \mathcal{I}) \oplus (\mathcal{K} \cap \mathcal{C}_{\mathcal{H}}). \tag{1}$$

Now let $\mathcal{I} = \mathcal{K} \oplus \mathcal{C}_{\mathcal{K}}$ be the Cartan decomposition of $\mathcal{I}$ with respect to $\mathcal{K}$ and let us denote $\Sigma = \Sigma(\mathcal{G})|_{\mathcal{I}}$. To prove that $\Sigma = \Sigma(\mathcal{I})$ we must check that

$$\Sigma|_{\mathcal{C}_{\mathcal{K}}} = \{0\} \text{ and } \Sigma|_{\mathcal{K}} = \sigma(id_{\mathcal{K}}). \tag{2}$$

Since $\mathcal{C}_{\mathcal{K}}$ is a set of quasinilpotent elements of $\mathcal{K}$ (cf. Proposition 4 (*c*) from § 5 and Theorem 4 from § 25), the first of the above equalities is a consequence of Lemma 4. For proving the second equality from (2), we shall use the decomposition (1) of $\mathcal{K}$. First, using the definition of $\Sigma(\mathcal{G}) = \Sigma_{\mathcal{H}}$ (see Lemma 3) we get

$$\Sigma|_{\mathcal{H}\cap\mathcal{I}} = \sigma(id_{\mathcal{H}})|_{\mathcal{K}\cap\mathcal{I}}.$$

But $\mathcal{H}\cap\mathcal{I}$ is a subalgebra of the nilpotent Lie algebras $\mathcal{H}$ and $\mathcal{K}$, hence by Corollary 1 from § 25 we obtain

$$\sigma(id_{\mathcal{H}})|_{\mathcal{H}\cap\mathcal{I}} = \sigma(id_{\mathcal{H}\cap\mathcal{I}}) = \sigma(id_{\mathcal{K}})|_{\mathcal{H}\cap\mathcal{I}}.$$

Hence $\Sigma|_{\mathcal{H}\cap\mathcal{I}} = \sigma(id_{\mathcal{K}})|_{\mathcal{H}\cap\mathcal{I}}$. Moreover, the functionals from Σ vanish on $\mathcal{K}\cap\mathcal{C}_{\mathcal{H}}$ (by Lemma 4) as well as those from $\sigma(id_{\mathcal{K}})$ (by Corollary 2 from § 25) since $\mathcal{C}_{\mathcal{H}}$ is a set of quasinilpotent operators. Hence $\Sigma|_{\mathcal{K}} = \sigma(id_{\mathcal{K}})$ by (1), so (2) is completely proved. □

Now we can prove the projection property of the spectrum $\Sigma(\cdot)$ on any Lie subalgebra. This property can be viewed as a variant of Corollary 1 from § 25, holding for solvable Lie algebras.

Theorem 1. *Let $\mathcal{G}$ be a finite-dimensional solvable Lie subalgebra of $\mathcal{B}(\mathcal{X})$. For every Lie subalgebra $\mathcal{L}$ of $\mathcal{G}$ we have $\Sigma(\mathcal{G})|_{\mathcal{L}} = \Sigma(\mathcal{L})$.*

Proof. We proceed by induction on the dimension of $\mathcal{G}$. The conclusion is obvious if $\dim\mathcal{G} = 1$. Now suppose that the assertion holds for solvable Lie algebras of dimension strictly less than $\dim\mathcal{G}$. Let $\mathcal{H}_0$ be a Cartan subalgebra of $\mathcal{L}$ and $\mathcal{L} = \mathcal{H}_0 \oplus \mathcal{C}_{\mathcal{H}_0}$ be the corresponding Cartan decomposition. Since $\mathcal{C}_{\mathcal{H}_0}$ is a set of quasinilpotent operators (by Proposition 4 (*c*) from § 5 and Theorem 4 from § 25), we have $\Sigma(\mathcal{G})|_{\mathcal{C}_{\mathcal{H}_0}} = \{0\}$ by Lemma 4. Hence for proving $\Sigma(\mathcal{G})|_{\mathcal{L}} = \Sigma_{\mathcal{H}_0}(= \Sigma(\mathcal{L})$ cf. Definition 1) it remains only to check that $\Sigma(\mathcal{G})|_{\mathcal{H}_0} = \sigma(id_{\mathcal{H}_0})$. By Proposition 3 from § 5 only the following situations can occur.

i)The subalgebra $\mathcal{H}_0$ is contained in a certain Cartan subalgebra $\mathcal{H}$ of $\mathcal{G}$. In this case we have $\Sigma(\mathcal{G}) = \Sigma_{\mathcal{H}}$ (cf. Definition 1) hence

$$\Sigma(\mathcal{G})|_{\mathcal{H}_0} = (\Sigma(\mathcal{G})|_{\mathcal{H}})|_{\mathcal{H}_0} = \sigma(id_{\mathcal{H}})|_{\mathcal{H}_0} = \sigma(id_{\mathcal{H}_0})$$

where the last equality follows by Corollary 1 from § 25.

ii)The subalgebra $\mathcal{H}_0$ is contained in a certain ideal $\mathcal{I}$ of $\mathcal{G}$ such that $\mathcal{I} \neq \mathcal{G}$. Then $\dim\mathcal{I} < \dim\mathcal{G}$, so we get $\Sigma(\mathcal{I})|_{\mathcal{H}_0} = \Sigma(\mathcal{H}_0) = \sigma(id_{\mathcal{H}_0})$ by the induction hypothesis and the remarks following Definition 1. Now an application of Lemma 5 shows that $\Sigma(\mathcal{G})|_{\mathcal{H}_0} = \sigma(id_{\mathcal{H}_0})$. □

Now we notice a property of the spectrum $\Sigma(\cdot)$, property which will allow us to define the spectrum of a locally solvable Lie algebra of operators.

Corollary 1. *If $\mathcal{G}$ is a finite-dimensional solvable Lie subalgebra of $\mathcal{B}(\mathcal{X})$, then $\Sigma(\mathcal{G})$ is a compact subset of $\widehat{\mathcal{G}}$ (and $\Sigma(\mathcal{G}) \neq \emptyset$ if $\mathcal{X} \neq \{0\}$).*

Proof. Let $\mathcal{H}$ be a Cartan subalgebra of $\mathcal{G}$. Then $\Sigma(\mathcal{G}) = \Sigma_{\mathcal{H}}$ (cf. Definition 1). But by definition (see Lemma 3) it is easily seen that the map

$$\Sigma_{\mathcal{H}} \to \sigma(id_{\mathcal{H}}),\ f \mapsto f|_{\mathcal{H}}$$

is a homeomorphism. Now the desired conclusion follows by means of Theorem 3 from § 25. □

Now we can prove the following fact.

Theorem 2. *If $\mathcal{X} \neq \{0\}$, then there exists a unique map $\Sigma(\cdot)$ which associates to any locally solvable Lie subalgebra $\mathcal{G}$ of $\mathcal{B}(\mathcal{X})$ a compact non-empty subset $\Sigma(\mathcal{G})$ of $\widehat{\mathcal{G}}$ such that the following conditions are fulfilled.*

a) *If $\mathcal{G}$ is a finite-dimensional nilpotent Lie subalgebra of $\mathcal{B}(\mathcal{X})$, then $\Sigma(\mathcal{G}) = \sigma(id_{\mathcal{G}})$.*

b) *If $\mathcal{L}$ and $\mathcal{G}$ are locally solvable Lie subalgebras of $\mathcal{B}(\mathcal{X})$ such that $\mathcal{L} \subseteq \mathcal{G}$, then $\Sigma(\mathcal{G})|_{\mathcal{L}} = \Sigma(\mathcal{L})$.*

Proof. Existence. For finite-dimensional solvable Lie subalgebras we define $\Sigma(\cdot)$ by the Definition 1. Now let $\mathcal{G}$ be a locally solvable Lie subalgebra of $\mathcal{B}(\mathcal{X})$ and let $\{\mathcal{G}_i\}_{i\in I}$ be the local system of all the finite-dimensional solvable Lie subalgebras of $\mathcal{G}$, partially ordered by inclusion (see Definition 5 a) from § 2). Then by Theorem 1 we can define for $i \leq j$ the restriction map $p_{ij} : \Sigma(\mathcal{G}_j) \to \Sigma(\mathcal{G}_i)$, which will be an onto map. Then $\{\Sigma(\mathcal{G}_i)\}_{i\in I}$ becomes a projective system of compact spaces (cf. Corollary 1) whose maps are onto. In these conditions it is well known that the corresponding projective limit

$$\mathop{proj\,lim}_{i\in I} \Sigma(\mathcal{G}_i)$$

is a compact topological space and the natural projections

$$p_j : \mathop{proj\,lim}_{i\in I} \Sigma(\mathcal{G}_i) \to \Sigma(\mathcal{G}_j)$$

are onto. Obviously we can identify the projective limit with a subset of $\widehat{\mathcal{G}}$, which we define to be $\Sigma(\mathcal{G})$. That is

$$\Sigma(\mathcal{G}) = \mathop{proj\,lim}_{i\in I} \Sigma(\mathcal{G}_i) = \{f \in \widehat{\mathcal{G}} \mid \forall i \in I,\ f|_{\mathcal{G}_i} \in \Sigma(\mathcal{G}_i)\}. \tag{3}$$

(Note that $\mathcal{G}$ is the inductive limit of the inductive system of the finite-dimensional vector spaces $\{\mathcal{G}_i\}_{i\in I}$ endowed with the inclusion maps. Then $\widehat{\mathcal{G}}$ is the projective limit of the finite-dimensional vector spaces $\{\widehat{\mathcal{G}}_i\}_{i\in I}$ endowed with the restriction maps. We consider $\widehat{\mathcal{G}}$ endowed with this projective limit topology and then $\Sigma(\mathcal{G})$ is a topological subspace of $\widehat{\mathcal{G}}$. Obviously the projective limit topology on $\widehat{\mathcal{G}}$ is the locally convex topology defined by the family of seminorms $\{\|\cdot\|_i\}_{i\in I}$, where $\|f\|_i := \|f|_{\mathcal{G}_i}\|$ for $f \in \widehat{\mathcal{G}}$ and $i \in I$.)

Now we have to verify the properties $a)$ and $b)$. The property $a)$ follows by the remarks after Definition 1. For proving the property $b)$, let $\mathcal{L}$ be an arbitrary Lie subalgebra of $\mathcal{G}$. Notice that $\{\mathcal{L} \cap \mathcal{G}_i\}_{i\in I}$ is the local system of all finite-dimensional solvable Lie subalgebras of $\mathcal{L}$. Consequently, by the above definition of $\Sigma(\cdot)$ for locally solvable Lie algebras we have

$$\Sigma(\mathcal{L}) = \mathop{proj\,lim}_{i\in I} \Sigma(\mathcal{L} \cap \mathcal{G}_i) = \{f \in \widehat{\mathcal{L}} \mid \forall i \in I,\ f|_{\mathcal{L}\cap\mathcal{G}_i} \in \Sigma(\mathcal{L} \cap \mathcal{G}_i)\}. \tag{4}$$

On the other hand, $\mathcal{L} \cap \mathcal{G}_i$ is a subalgebra of the finite-dimensional solvable Lie algebra $\mathcal{G}_i$ for every $i \in I$, hence the restriction maps

$$\Sigma(\mathcal{G}_i) \to \Sigma(\mathcal{L} \cap \mathcal{G}_i),\ f \mapsto f|_{\mathcal{L} \cap \mathcal{G}_i} \qquad (i \in I) \tag{5}$$

are onto by Theorem 1. Since the topological spaces $\Sigma(\mathcal{L} \cap \mathcal{G}_i)$ and $\Sigma(\mathcal{G}_i)$ are compact Hausdorff it then follows that the projective limit of the maps (5) (which is the restriction map $\Sigma(\mathcal{G}) \to \Sigma(\mathcal{L})$, $f \mapsto f|_{\mathcal{L}}$) is onto, hence the desired property *b*) follows.

Uniqueness. Let $\Sigma'(\cdot)$ be another map with the properties *a*) and *b*). First we check that for a finite-dimensional solvable Lie subalgebra $\mathcal{G}$ of $\mathcal{B}(\mathcal{X})$ we have $\Sigma(\mathcal{G}) = \Sigma'(\mathcal{G})$. To this end let $\mathcal{H}$ be a Cartan subalgebra of $\mathcal{G}$ and let $\mathcal{G} = \mathcal{H} \oplus \mathcal{C}_{\mathcal{H}}$ be the associated Cartan decomposition. By the properties *a*) and *b*) of $\Sigma'(\cdot)$ we have

$$\Sigma'(\mathcal{G})|_{\mathcal{H}} = \Sigma'(\mathcal{H}) = \sigma(id_{\mathcal{H}}). \tag{6}$$

On the other hand if we apply the properties *a*) and *b*) to the one-dimensional subalgebra $\mathcal{S} := \mathbb{C}Q$ spanned by a quasinilpotent element Q, then we deduce

$$\Sigma'(\mathcal{G})|_{\mathcal{S}} = \Sigma'(\mathcal{S}) = \sigma(id_{\mathcal{S}}) \cong \sigma(Q) = \{0\}$$

(see Remark 1 from § 25). Particularly, since $\mathcal{C}_{\mathcal{H}}$ is a set of quasinilpotent operators (by Proposition 4 (*c*) from § 5 and Theorem 4 from § 25), we obtain $\Sigma'(\mathcal{G})|_{\mathcal{C}_{\mathcal{H}}} = \{0\}$. Then by (6) we get $\Sigma'(\mathcal{G}) = \Sigma_{\mathcal{H}} = \Sigma(\mathcal{G})$ (see Lemma 3 and Definition 1).

Now let $\mathcal{G}$ be a locally solvable Lie subalgebra of $\mathcal{B}(\mathcal{X})$ and let $\{\mathcal{G}_i\}_{i\in I}$ be as at the beginning of the proof. Then $\Sigma(\mathcal{G}_i) = \Sigma'(\mathcal{G}_i)$ $(i \in I)$ by what we have already proved. Since $\Sigma'(\cdot)$ has the property *b*) it then follows that $\Sigma'(\mathcal{G})|_{\mathcal{G}_i} = \Sigma(\mathcal{G}_i)$ for every $i \in I$, hence by (3) we get

$$\Sigma'(\mathcal{G}) \subseteq \Sigma(\mathcal{G}). \tag{7}$$

Assume that the above inclusion is strict. Since the respective sets are compact Hausdorff spaces it follows that there exists $f_0 \in \Sigma(\mathcal{G})$ and a neighbourhood V of f_0 such that V and $\Sigma'(\mathcal{G})$ are disjoint. Consequently there exist $\varepsilon > 0$ and $i_0 \in I$ such that the following implication holds:

$$f \in \Sigma(\mathcal{G}) \text{ and } \|f|_{\mathcal{G}_{i_0}} - f_0|_{\mathcal{G}_{i_0}}\| < \varepsilon \Longrightarrow f \in \Sigma'(\mathcal{G}). \tag{8}$$

(see the above remarks concerning the projective limit topology of $\Sigma(\mathcal{G})$).

On the other hand, since both $\Sigma(\cdot)$ and $\Sigma'(\cdot)$ have the property *b*), we have

$$f_0|_{\mathcal{G}_{i_0}} \in \Sigma(\mathcal{G})|_{\mathcal{G}_{i_0}} = \Sigma(\mathcal{G}_{i_0}) = \Sigma'(\mathcal{G}_{i_0}) = \Sigma'(\mathcal{G})|_{\mathcal{G}_{i_0}}. \tag{9}$$

(Recall that we have already proved that $\Sigma(\cdot)$ and $\Sigma'(\cdot)$ coincide on finite-dimensional solvable Lie algebras.) By (9) we can find $f \in \Sigma'(\mathcal{G})$ such that $f|_{\mathcal{G}_{i_0}} = f_0|_{\mathcal{G}_{i_0}}$, which contradicts the implication (8). Consequently we must have equality in (7), and the proof ends. □

In view of the preceding theorem we can introduce the concept of spectrum of a locally solvable Lie algebra of operators. This concept contains as a special case the concept given by Definition 1.

Definition 2. Let $\mathcal{X} \neq \{0\}$ and $\Sigma(\cdot)$ be the map referred to in Theorem 2. For every locally solvable Lie subalgebra $\mathcal{G}$ of $\mathcal{B}(\mathcal{X})$ we call $\Sigma(\mathcal{G})$ the *Cartan-Taylor spectrum* of $\mathcal{G}$.

Corollary 2. *Let $\mathcal{G}$ be a locally solvable Lie subalgebra of $\mathcal{B}(\mathcal{X})$. Assume that $f : \mathcal{G} \to \mathbb{C}$ and $(x_\alpha)_{\alpha \in A}$ is a net consisting of unit vectors from $\mathcal{X}$ such that for every $G \in \mathcal{G}$ we have*

$$\lim_\alpha (Gx_\alpha - f(G)x_\alpha) = 0.$$

Then $f \in \Sigma(\mathcal{G})$.

Proof. By (3) it suffices to prove that $f|_\mathcal{L} \in \Sigma(\mathcal{L})$ for an arbitrary finite-dimensional solvable subalgebra $\mathcal{L}$ of $\mathcal{G}$. To this end let $\mathcal{H}$ be a Cartan subalgebra of $\mathcal{L}$ and let $\mathcal{L} = \mathcal{H} \oplus \mathcal{C}_\mathcal{H}$ be the corresponding Cartan decomposition. We have $f(G) \in \sigma(G)$ for every $G \in \mathcal{G}$, hence $f|_{\mathcal{C}_\mathcal{H}} = 0$ because $\mathcal{C}_\mathcal{H}$ is a set of quasinilpotent operators (by Proposition 4 (c) from § 5 and Theorem 4 from § 25). On the other hand it is easily seen that f is linear; it vanishes on $[\mathcal{G}, \mathcal{G}]$ by Theorem 4 from § 25. Hence f is a character and then $f|_\mathcal{H} \in \sigma(id_\mathcal{H})$ by Remark 2 from § 25. Consequently, $f|_\mathcal{L} \in \Sigma_\mathcal{H} = \Sigma(\mathcal{L})$ (see Lemma 3 and Definition 1). □

Next we introduce a concept of joint spectrum for families of operators generating locally solvable Lie algebras.

Definition 3. If $T = (T_j)_{j \in J} \in \mathcal{B}(\mathcal{X})^J$ is a family of operators generating a locally solvable Lie subalgebra $\mathcal{L}$ of $\mathcal{B}(\mathcal{X})$, then we define the *Cartan-Taylor joint spectrum* of T by

$$\sigma(T) := \{(f(T_j))_{j \in J} \mid f \in \Sigma(\mathcal{L})\} \,(\subset \mathbb{C}^J).$$

The following theorem contains some basic properties of the above introduced spectrum.

Theorem 3. *Let $T = (T_j)_{j \in J} \in \mathcal{B}(\mathcal{X})^J$ be a family of operators generating a locally solvable Lie subalgebra $\mathcal{L}$ of $\mathcal{B}(\mathcal{X})$.*

1° *If $\mathcal{X} \neq \{0\}$, then $\sigma(T)$ is a compact nonempty subset of $\mathbb{C}^J$.*

2° *If $\mathcal{G}$ is an arbitrary locally solvable Lie subalgebra of $\mathcal{B}(\mathcal{X})$ with $\{T_j \mid j \in J\} \subseteq \mathcal{G}$, then $\sigma(T) := \{(f(T_j))_{j \in J} \mid f \in \Sigma(\mathcal{G})\}$.*

3° *(Projection property.) Let $J_0 \subseteq J$, $T_{J_0} := (T_j)_{j \in J_0} \in \mathcal{B}(\mathcal{X})^{J_0}$ and $\pi_{J_0} : \mathbb{C}^J \to \mathbb{C}^{J_0}$ be the natural projection. Then $\sigma(T_{J_0}) = \pi_{J_0}(\sigma(T))$.*

4° *If J is a finite set and the Lie algebra $\mathcal{L}$ is nilpotent (particularly, if T is a finite, commuting tuple), then $\sigma(T)$ reduces to the (Taylor) joint spectrum of T defined by Definition 1 from § 26.*

Proof. 1° We endow $\mathbb{C}^J$ with the usual product topology and $\Sigma(\mathcal{L})$ with the projective limit topology (see the proof of Theorem 2). Then the map

$$\Sigma(\mathcal{L}) \to \mathbb{C}^J,\ f \mapsto (f(T_j))_{j\in J},$$

is obviously continuous and the desired conclusion follows since $\Sigma(\mathcal{L})$ is compact and nonempty by Theorem 2.

2° We have $\mathcal{L} \subseteq \mathcal{G}$ and the conclusion follows by property *b*) in Theorem 2.

3° Let $\mathcal{L}_0$ be the Lie algebra generated by the system T_{J_0}. Then $\mathcal{L}_0 \subseteq \mathcal{L}$ hence by 2° we can write

$$\sigma(T_{J_0}) := \{(f(T_j))_{j\in J_0} \mid f \in \Sigma(\mathcal{L})\}.$$

This implies the desired conclusion in view of the definition of $\sigma(T)$.

4° If $\mathcal{L}$ is finite-dimensional nilpotent, then $\Sigma(\mathcal{L}) = \sigma(id_{\mathcal{L}})$ (see the remarks following Definition 1). Hence we have

$$\sigma(T) := \{(f(T_j))_{j\in J} \mid f \in \sigma(id_{\mathcal{L}})\}$$

and the assertion follows by Theorem 1 from § 26. □

For the sake of completeness we explicitly state the following immediate consequence of the assertions 2° and 4° from Theorem 3.

Corollary 3. *If $\mathcal{G}$ is a locally solvable Lie subalgebra of $\mathcal{B}(\mathcal{X})$ (particularly, if $\mathcal{G}$ is finite-dimensional solvable), then for any $G \in \mathcal{G}$ we have*

$$\sigma(G) = \{f(G) \mid f \in \Sigma(\mathcal{G})\}.$$

Now we establish two immediate consequences of the above fact.

Corollary 4. *Let $\mathcal{G}$ be a locally solvable Lie subalgebra of $\mathcal{B}(\mathcal{X})$ and endow $\mathcal{G}$ with the norm inherited from $\mathcal{B}(\mathcal{X})$. Then every $f \in \Sigma(\mathcal{G})$ is a bounded linear functional $f : \mathcal{G} \to \mathbb{C}$ with $\|f\| \leq 1$.*

Proof. By Corollary 3 we get $|f(G)| \leq \|G\|$ for every $f \in \Sigma(\mathcal{G})$ and $G \in \mathcal{G}$. □

Corollary 5. *Let $T, Q \in \mathcal{B}(\mathcal{X})$ be operators generating a finite-dimensional solvable Lie subalgebra of $\mathcal{B}(\mathcal{X})$. If Q is quasinilpotent then $\sigma(T+Q) = \sigma(T)$.*

Proof. Let $\mathcal{G}$ be the Lie subalgebra of $\mathcal{B}(\mathcal{X})$ generated by T and Q. By Corollary 3 we have

$$\begin{aligned}\sigma(T+Q) &= \{f(T+Q) \mid f \in \Sigma(\mathcal{G})\} = \{f(T) + f(Q) \mid f \in \Sigma(\mathcal{G})\}\\ &= \{f(T) \mid f \in \Sigma(\mathcal{G})\} = \sigma(T).\end{aligned}$$

and we are done. □

Concerning the above fact let's consider a simple example showing that, in the hypotheses of Corollary 5, the operators T and $T+Q$ need not be quasinilpotent equivalent.

Example 1. Let $\mathcal{X} = \mathbb{C}^2$ and

$$T = \begin{pmatrix} b+\gamma & 0 \\ 0 & b \end{pmatrix}, \quad Q = \begin{pmatrix} 0 & 1 \\ 0 & 0 \end{pmatrix}$$

where $b, \gamma \in \mathbb{C}$. Then $[T, Q] = \gamma Q$ and the hypotheses of Corollary 5 are satisfied. But for $\gamma \neq 0$ we have

$$\mathcal{X}_T(\{b\}) = \mathbb{C} \cdot \begin{pmatrix} 0 \\ 1 \end{pmatrix} \neq \mathbb{C} \cdot \begin{pmatrix} 1 \\ -\gamma \end{pmatrix} = \mathcal{X}_{T+Q}(\{b\}),$$

hence the operators T and $T+Q$ cannot be quasinilpotent equivalent (see Corollary 2.3.5 from the book of I. Colojoară and C. Foiaş [1]).

Finally we consider joint spectral properties specific to families and Lie algebras of compact operators. We shall prove variants of Theorems 5 and 6 from § 26, which hold for solvable Lie algebras (or even for quasisolvable Lie algebras). To this end we begin with an auxiliary fact.

Lemma 6. *Let $\mathcal{G}$ be a quasisolvable Lie subalgebra of $\mathcal{B}(\mathcal{X})$ consisting only of compact operators. Let $\mathcal{N}$ be a maximal nest in Lat$\mathcal{G}$. Then every atom associated to $\mathcal{N}$ has dimension 1. If $\{\mathcal{Y}_\alpha\}_{\alpha \in A}$ is the set of all elements of $\mathcal{N}$ generating atoms, then for every $\alpha \in A$ there exists a character $\varphi_\alpha \in \widehat{\mathcal{G}}$ such that*

$$(G - \varphi_\alpha(G))\mathcal{Y}_\alpha \subsetneq \mathcal{Y}_\alpha$$

for every $G \in \mathcal{G}$.

Proof. The first assertion is an immediate consequence of Corollary 5 from § 22. So, if we denote by $\mathcal{Y}_\alpha^-$ the closure of the union of all $\mathcal{Z} \in \mathcal{N}$ with $\mathcal{Z} \subset \mathcal{Y}_\alpha$ and $\mathcal{Z} \neq \mathcal{Y}_\alpha$, then we have a natural representation

$$\varphi_\alpha : \mathcal{G} \to \mathcal{B}(\mathcal{Y}_\alpha / \mathcal{Y}_\alpha^-)$$

defined by restriction and factorization of operators. Since we have already observed that the atom $\mathcal{Y}_\alpha / \mathcal{Y}_\alpha^-$ has dimension 1, the representation φ_α is actually a character of $\mathcal{G}$; moreover

$$(G - \varphi_\alpha(G))\mathcal{Y}_\alpha \subsetneq \mathcal{Y}_\alpha$$

for every $G \in \mathcal{G}$. □

Now we can state a variant of Theorem 6 from § 26.

Lemma 7. *Let $\mathcal{G}$ be a finite-dimensional nilpotent Lie subalgebra of $\mathcal{B}(\mathcal{X})$ consisting only of compact operators. Let $\mathcal{N}$ be a maximal nest in Lat$\mathcal{G}$ and $\{\varphi_\alpha\}_{\alpha \in A}$ be as in Lemma 6. Then the following assertions hold.*

1° *If $\dim \mathcal{X} < \infty$, then $\sigma(id_{\mathcal{G}}) = \{\varphi_\alpha \mid \alpha \in A\}$.*
2° *If $\dim \mathcal{X} = \infty$, then $\sigma(id_{\mathcal{G}}) = \{\varphi_\alpha \mid \alpha \in A\} \cup \{0\}$.*

Proof. First we choose a finite system of generators of $\mathcal{G}$. Then we apply Theorems 1 and 6 from § 26. □

Theorem 4. *Let $\mathcal{G}$ be a quasisolvable Lie subalgebra of $\mathcal{B}(\mathcal{X})$ consisting only of compact operators. Let $\mathcal{N}$ be a maximal nest in Lat$\mathcal{G}$ and $\{\varphi_\alpha\}_{\alpha\in A}$ be as in Lemma 6. Then the following assertions hold.*

1° *If* $\dim \mathcal{X} < \infty$, *then* $\Sigma(\mathcal{G}) = \{\varphi_\alpha \mid \alpha \in A\}$.

2° *If* $\dim \mathcal{X} = \infty$, *then* $\Sigma(\mathcal{G}) = \{\varphi_\alpha \mid \alpha \in A\} \cup \{0\}$.

Proof. By (3) (see the proof of Theorem 2) it suffices to prove that for each finite-dimensional solvable Lie subalgebra $\mathcal{L}$ of $\mathcal{G}$ we have $\varphi_\alpha|_{\mathcal{L}} \in \Sigma(\mathcal{L})$ (for $\alpha \in A$) and, if $\dim \mathcal{X} = \infty$ then $0 \in \Sigma(\mathcal{L})$. To this end consider a Cartan subalgebra $\mathcal{H}$ of $\mathcal{L}$ and let $\mathcal{L} = \mathcal{H} \oplus \mathcal{C}_{\mathcal{H}}$ be the corresponding Cartan decomposition. Then $\mathcal{C}_{\mathcal{H}}$ is a set of quasinilpotent operators (by Proposition 4 (c) from § 5 and Theorem 4 from § 25) hence $\varphi_\alpha|_{\mathcal{C}_{\mathcal{H}}} = 0$. (Indeed, for every $G \in \mathcal{G}$ we have $\varphi(G) \in \sigma(G)$ by the classical Ringrose Theorem.) On the other hand $\varphi_\alpha|_{\mathcal{H}} \in \sigma(id_{\mathcal{H}})$ by Lemma 7. Hence $\varphi_\alpha|_{\mathcal{L}} \in \Sigma_{\mathcal{H}} = \Sigma(\mathcal{L})$ (see Lemma 3 and Definition 1). Finally, if $\dim \mathcal{X} = \infty$ then $0 \in \sigma(id_{\mathcal{H}})$ by Lemma 7 hence $0 \in \Sigma_{\mathcal{H}} = \Sigma(\mathcal{L})$. □

Now we state a variant of Theorem 5 from § 26, which holds for Lie algebras instead of systems of operators.

Theorem 5. *Let $\mathcal{G}$ be a finite-dimensional solvable Lie subalgebra of $\mathcal{B}(\mathcal{X})$ consisting only of compact operators. Then $\Sigma(\mathcal{G})$ is a compact, at most countable subset of $\widehat{\mathcal{G}}$, accumulating at most towards 0. If* $\dim \mathcal{X} = \infty$ *then* $0 \in \Sigma(\mathcal{G})$.

Proof. If $\mathcal{G}$ is nilpotent then the conclusion follows by choosing a finite system of generators of $\mathcal{G}$ and then applying Theorems 5 and 1 from § 26.

In the general case consider a Cartan subalgebra $\mathcal{H}$ of $\mathcal{G}$. Let $\mathcal{L} = \mathcal{H} \oplus \mathcal{C}_{\mathcal{H}}$ be the corresponding Cartan decomposition. Then $\Sigma(\mathcal{G}) = \Sigma_{\mathcal{H}}$ (cf. Definition 1) and the map

$$\Sigma_{\mathcal{H}} \to \sigma(id_{\mathcal{H}}),\ f \mapsto f|_{\mathcal{H}},$$

is a homeomorphism which takes 0 in 0 (see the definition of $\Sigma_{\mathcal{H}}$ in Lemma 3). But the set $\sigma(id_{\mathcal{H}})(= \Sigma(\mathcal{H}))$ has all the desired properties, since $\mathcal{H}$ is nilpotent (see the observation from the beginning of the present proof). Hence the conclusion follows. □

Finally we note the following variant of Theorem 5 from § 26, which holds for tuples of operators generating finite-dimensional solvable Lie algebras.

Theorem 6. *Let $T = (T_1, \ldots, T_n) \in \mathcal{B}(\mathcal{X})^n$ be a system of compact operators generating a finite-dimensional solvable Lie subalgebra $\mathcal{G}$ of $\mathcal{B}(\mathcal{X})$. Let $\mathcal{N}$ be a maximal nest in Lat$T$(= Lat$\mathcal{G}$) and $\{\lambda_\alpha\}_{\alpha\in A}$ be the family of points of $\mathbb{C}^n$ determined by the atoms of $\mathcal{N}$. Then the following assertions hold.*

1° *If* $\dim \mathcal{X} < \infty$, *then* $\sigma(T) = \{\lambda_\alpha \mid \alpha \in A\}$.

2° *If* $\dim \mathcal{X} = \infty$, *then* $\sigma(T) = \{\lambda_\alpha \mid \alpha \in A\} \cup \{0\}$.

3° *The spectrum $\sigma(T)$ is a compact, at most countable subset of $\mathbb{C}^n$, accumulating at most towards 0. If* $\dim \mathcal{X} = \infty$, *then* $0 \in \sigma(T)$.

Proof. The assertions follow from Theorems 5 and 4, in view of the obvious fact that

$$\Psi : \Sigma(\mathcal{G}) \to \sigma(T),\ \varphi \mapsto (\varphi(T_1), \dots, \varphi(T_n)),$$

is a homeomorphism which takes 0 in 0 and $\Psi(\varphi_\alpha) = \lambda_\alpha$ for $\alpha \in A$, where $\{\varphi_\alpha\}_{\alpha \in A}$ is the family from Lemma 6. □

Corollary 6. *Let $\mathcal{X} = \mathbb{C}^N$ and $T_i \in \mathcal{B}(\mathcal{X})$ be given by a triangular matrix*

$$T_i = \begin{pmatrix} \lambda_i & * & * \\ 0 & \ddots & * \\ 0 & 0 & \omega_i \end{pmatrix} \qquad (1 \le i \le n).$$

Then the n-tuple $T = (T_1, \dots, T_n)$ generates a solvable Lie subalgebra of $\mathcal{B}(\mathcal{X})$ and we have

$$\sigma(T) = \{(\lambda_1, \dots, \lambda_n), \dots, (\omega_1, \dots, \omega_n)\}\ (\subset \mathbb{C}^n).$$

§ 28 Lie ideals of generalized spectral operators

The aim of this paragraph is to show that, in a finite-dimensional solvable Lie algebra of operators, the elements of certain classes of generalized spectral operators (e.g. the quasinilpotents, the nilpotents, the $\mathcal{C}^0(\mathbb{C})$-spectral ones etc.) constitute Lie ideals. See Corollary 3 below.

Throughout the present paragraph, for a subset $\mathcal{G}$ of $\mathcal{B}(\mathcal{X})$ we use the following notation:

$$\begin{aligned} \mathcal{Q}_\mathcal{G} &:= \{Q \in \mathcal{G} \mid Q \text{ is quasinilpotent}\}, \\ \mathcal{N}_\mathcal{G} &:= \{N \in \mathcal{G} \mid N \text{ is nilpotent}\}, \\ \mathcal{J}_\mathcal{G} &:= \{T \in \mathcal{G} \mid T \text{ is Jordan operator}\}, \\ \mathcal{S}^0_\mathcal{G} &:= \{T \in \mathcal{G} \mid T \text{ is } \mathcal{C}^0(\mathbb{C})\text{-spectral operator}\}. \end{aligned}$$

Also we shall use the notation $\mathcal{E}(\mathcal{A})$, respectively $\mathcal{E}(T)$ or $\mathcal{E}(T_1, \dots, T_n)$, for the Lie subalgebra of $\mathcal{B}(\mathcal{X})$ generated by a set $\mathcal{A} \subseteq \mathcal{B}(\mathcal{X})$, respectively by an n-tuple $T = (T_1, \dots, T_n) \in \mathcal{B}(\mathcal{X})^n$.

Theorem 1. *Let $\mathcal{G}$ be a finite-dimensional Lie subalgebra of $\mathcal{B}(\mathcal{X})$. Then $\mathcal{G}$ is solvable iff $\mathcal{Q}_\mathcal{G}$ is an ideal of $\mathcal{G}$ containing $[\mathcal{G}, \mathcal{G}]$.*

Proof. The necessity follows by Lemma 2 from § 27. Now we prove the sufficiency. Since $\mathcal{Q}_\mathcal{G}$ is a finite-dimensional Lie algebra containing only quasinilpotent operators, it is nilpotent by Rosenblum's Theorem (Theorem 1 from § 13) and by Engel's Theorem. But $[\mathcal{G}, \mathcal{G}] \subseteq \mathcal{Q}_\mathcal{G}$, so $[\mathcal{G}, \mathcal{G}]$ is also a nilpotent Lie algebra. This last fact implies that $\mathcal{G}$ is solvable. □

Theorem 2. *If $\mathcal{G}$ is a finite-dimensional Lie subalgebra of $\mathcal{B}(\mathcal{X})$, then $\mathcal{N}_\mathcal{G}$ is a nilpotent ideal of $\mathcal{G}$. Moreover if $\mathcal{H}$ is a Cartan subalgebra of $\mathcal{G}$ and $\mathcal{G} = \mathcal{H} \oplus \mathcal{C}_\mathcal{H}$ is the corresponding Cartan decomposition, then $\mathcal{C}_\mathcal{H} \subseteq \mathcal{N}_\mathcal{G}$.*

Proof. The proof will consist of two steps.

1° First we assume that $\mathcal{G}$ is a nilpotent Lie algebra. Moreover, we assume that $\mathcal{N}_{\mathcal{G}} \neq \{0\}$. We proceed by induction on $\dim \mathcal{G} =: d$. For $d = 1$ the conclusion is obvious. Next assume that the conclusion holds for nilpotent Lie algebras of dimension strictly less than d. Theorem 1 from § 19 implies that there exists $N \in \mathcal{N}_{\mathcal{G}} \setminus \{0\}$ such that $[N, \mathcal{G}] = \{0\}$. Choose $m \geq 1$ such that $N^{m-1} \neq 0 = N^m$ and denote $\mathcal{X}_i := Ker N^i$ for $i = 0, \dots, m$. Then

$$\{0\} = \mathcal{X}_0 \subset \mathcal{X}_1 \subset \cdots \subset \mathcal{X}_m = \mathcal{X}$$

is a nest of invariant subspaces for $\mathcal{G}$ because $[N^i, \mathcal{G}] = \{0\}$ for every $i \geq 0$. Now for $1 \leq i \leq m$ consider the representations

$$\varphi_i : \mathcal{G} \to \mathcal{B}(\mathcal{X}_i / \mathcal{X}_{i-1})$$

defined by restriction and factorization of operators. It is easily seen that for every $G \in \mathcal{G}$ the following equivalence holds:

$$G \text{ is nilpotent} \Leftrightarrow \forall i \in \{1, \dots, m\}, \varphi_i(G) \text{ is nilpotent.} \tag{1}$$

Moreover, remark that $N \in Ker \varphi_i$, hence $\dim(\varphi_i(\mathcal{G})) < \dim \mathcal{G}$ for $i = 1, \dots, m$. Particularly, by the induction hypothesis we deduce

$$\mathcal{N}_{\varphi_i(\mathcal{G})} \text{ is an ideal of } \varphi_i(\mathcal{G}) \qquad (1 \leq i \leq m). \tag{2}$$

Now, let $G \in \mathcal{G}$, $N_1, N_2 \in \mathcal{N}_{\mathcal{G}}$ and $\alpha_1, \alpha_2 \in \mathbb{C}$. Then by (1) we deduce that $\varphi_i(N_1), \varphi_i(N_2) \in \mathcal{N}_{\varphi_i(\mathcal{G})}$ for every $i \in \{1, \dots, m\}$. Hence by (2) we have

$$\varphi_i(\alpha_1 N_1 + \alpha_2 N_2) = \alpha_1 \varphi_i(N_1) + \alpha_2 \varphi_i(N_2) \in \mathcal{N}_{\varphi_i(\mathcal{G})}$$

and

$$\varphi_i([G, N_1]) = [\varphi_i(G), \varphi_i(N_1)] \in \mathcal{N}_{\varphi_i(\mathcal{G})}.$$

Since these facts hold for every $i \in \{1, \dots, m\}$, by (1) we deduce that $\alpha_1 N_1 + \alpha_2 N_2$ and $[G, N_1]$ belong to $\mathcal{N}_{\mathcal{G}}$. Hence $\mathcal{N}_{\mathcal{G}}$ is an ideal of $\mathcal{G}$.

2° Now we treat the general case. Assume that $\mathcal{G}$ is solvable and consider a Cartan subalgebra $\mathcal{H}$ of $\mathcal{G}$. Let $\mathcal{G} = \mathcal{H} \oplus \mathcal{C}_{\mathcal{H}}$ be the corresponding Cartan decomposition, where $\mathcal{C}_{\mathcal{H}} = \oplus_{\alpha \neq 0} \mathcal{G}^{\alpha}$ (see Proposition 4 from § 5). Since $\mathcal{N}_{\mathcal{G}} \subseteq \mathcal{Q}_{\mathcal{G}}$, we have trivially $\mathcal{N}_{\mathcal{G}} = \mathcal{N}_{\mathcal{Q}_{\mathcal{G}}}$. But $\mathcal{Q}_{\mathcal{G}}$ is a nilpotent Lie algebra (by Theorem 1 and its proof), hence $\mathcal{N}_{\mathcal{G}}$ is an ideal of $\mathcal{Q}_{\mathcal{G}}$ by the preceding step of the proof.

On the other hand let $H \in \mathcal{H} \cap r(\mathcal{G})$. Then for every root $\alpha \neq 0$ it is well known that $\alpha(H) \neq 0$ and

$$\mathcal{G}^{\alpha} = Ker((ad H - \alpha(H))^{\dim \mathcal{G}})$$

(see also the proof of Proposition 4 from § 5). Consequently for every $\alpha \neq 0$ we have $\mathcal{G}^{\alpha} \subseteq \mathcal{N}_{\mathcal{G}}$ by Theorem 4 from § 17. But we have already proved that $\mathcal{N}_{\mathcal{G}}$ is a

Lie algebra, so $\oplus_{\alpha\neq 0}\mathcal{G}^{\alpha} = \mathcal{C}_{\mathcal{H}} \subseteq \mathcal{N}_{\mathcal{G}}$. Since $\mathcal{G} = \mathcal{H} \oplus \mathcal{C}_{\mathcal{H}}$ and $\mathcal{N}_{\mathcal{G}}$ is a vector space, the last inclusion implies $\mathcal{N}_{\mathcal{G}} = (\mathcal{N}_{\mathcal{G}} \cap \mathcal{H}) \oplus \mathcal{C}_{\mathcal{H}}$, i.e.

$$\mathcal{N}_{\mathcal{G}} = \mathcal{N}_{\mathcal{H}} \oplus \mathcal{C}_{\mathcal{H}} \,(\subset \mathcal{Q}_{\mathcal{G}}). \tag{3}$$

Consequently

$$[\mathcal{G},\mathcal{N}_{\mathcal{G}}] = [\mathcal{H}\oplus\mathcal{C}_{\mathcal{H}},\mathcal{N}_{\mathcal{G}}] = [\mathcal{H},\mathcal{N}_{\mathcal{G}}] + [\mathcal{C}_{\mathcal{H}},\mathcal{N}_{\mathcal{G}}] = [\mathcal{H},\mathcal{N}_{\mathcal{H}}] + [\mathcal{H},\mathcal{C}_{\mathcal{H}}] + [\mathcal{C}_{\mathcal{H}},\mathcal{N}_{\mathcal{G}}]. \tag{4}$$

But $[\mathcal{H},\mathcal{N}_{\mathcal{H}}] \subseteq \mathcal{N}_{\mathcal{H}}$ (by step 1° of the proof since $\mathcal{H}$ is nilpotent), so $[\mathcal{H},\mathcal{N}_{\mathcal{H}}] \subseteq \mathcal{N}_{\mathcal{G}}$. Moreover $[\mathcal{H},\mathcal{C}_{\mathcal{H}}] \subseteq \mathcal{C}_{\mathcal{H}} \subseteq \mathcal{N}_{\mathcal{G}}$ (see Proposition 4 (*c*) from § 5 and (3)) and $[\mathcal{C}_{\mathcal{H}},\mathcal{N}_{\mathcal{G}}] \subseteq \mathcal{N}_{\mathcal{G}}$ (also by (3)). Hence by (4) we deduce $[\mathcal{G},\mathcal{N}_{\mathcal{G}}] \subseteq \mathcal{N}_{\mathcal{G}} + \mathcal{N}_{\mathcal{G}} + \mathcal{N}_{\mathcal{G}} = \mathcal{N}_{\mathcal{G}}$, (because we have already proved that $\mathcal{N}_{\mathcal{G}}$ is a vector space). This implies that $\mathcal{N}_{\mathcal{G}}$ is an ideal of $\mathcal{G}$. □

Next we study the set of Jordan operators from a finite-dimensional Lie subalgebra of $\mathcal{B}(\mathcal{X})$. For proving that this set is an ideal (see Theorem 2 below), first we establish a sequence of auxiliary results.

Lemma 1. *Let $\mathcal{G}$ be a finite-dimensional solvable Lie subalgebra of $\mathcal{B}(\mathcal{X})$. Let $T \in \mathcal{J}_{\mathcal{G}}$ with the Jordan decomposition $T = S + Q$. Then $\mathcal{E}(\mathcal{G} \cup \{S,Q\})$ is a finite-dimensional solvable Lie algebra.*

Proof. For $\lambda \in \mathbb{C}$ denote

$$\mathcal{G}^{\lambda} := \{G \in \mathcal{G} \mid \exists n \geq 1 : \,(adT - \lambda)^n G = 0\}.$$

If we denote $\sigma(adT|_{\mathcal{G}}) = \{\lambda_0,\ldots,\lambda_m\}$, then it follows that $\mathcal{G} = \oplus_{j=0}^{m}\mathcal{G}^{\lambda_j}$. Now, the operators $adT, adS : \mathcal{B}(\mathcal{X}) \to \mathcal{B}(\mathcal{X})$ are quasinilpotent equivalent ($[adT, adS] = 0$ and $adT - adS = adQ$ is quasinilpotent by Theorem 1 from § 13 since Q is quasinilpotent). Consequently these operators have the same spectral maximal subspaces. (Note that adT, adS have the single-valued extension property by Proposition 1 from § 13 because both S and Q are decomposable by Corollary 5 from § 14.) Particularly

$$\mathcal{B}(\mathcal{X})_{adT}(\{\lambda\}) = \mathcal{B}(\mathcal{X})_{adS}(\{\lambda\}) \text{ for all } \lambda \in \mathbb{C}. \tag{5}$$

On the other hand $adS : \mathcal{B}(\mathcal{X}) \to \mathcal{B}(\mathcal{X})$ is a normal-equivalent operator (Proposition 4 from § 14) hence by Corollary 4 from § 14 we get

$$\mathcal{B}(\mathcal{X})_{adS}(\{\lambda\}) = Ker(adS - \lambda) \text{ for all } \lambda \in \mathbb{C}. \tag{6}$$

But $\mathcal{G}^{\lambda} \subseteq \mathcal{B}(\mathcal{X})_{adT}(\{\lambda\})$ hence (5) and (6) imply

$$[S,G] = \lambda G \text{ for all } G \in \mathcal{G}^{\lambda}, \lambda \in \mathbb{C}. \tag{7}$$

Particularly we have $(adS)\mathcal{G}^{\lambda} \subseteq \mathcal{G}^{\lambda}$. But obviously $(adT)\mathcal{G}^{\lambda} \subseteq \mathcal{G}^{\lambda}$, hence also $(adQ)\mathcal{G}^{\lambda} \subseteq \mathcal{G}^{\lambda}$. Consequently

$$\mathcal{E}(\mathcal{G} \cup \{S,Q\}) = \mathbb{C} \cdot S + \mathcal{G}^{\lambda_0} + \cdots + \mathcal{G}^{\lambda_m} = \mathbb{C} \cdot S + \mathcal{G}. \tag{8}$$

By (7) and (8) we deduce $[\mathcal{E}(\mathcal{G} \cup \{S, Q\}), \mathcal{E}(\mathcal{G} \cup \{S, Q\})] \subseteq \mathcal{G}$. But $\mathcal{G}$ is solvable, so $\mathcal{E}(\mathcal{G} \cup \{S, Q\})$ is also solvable. Moreover, this algebra is finite-dimensional by (8), and the proof ends. □

Lemma 2. *Let $\mathcal{G}$ be a finite-dimensional solvable Lie subalgebra of $\mathcal{B}(\mathcal{X})$ such that* $\dim(\mathcal{G}/\mathcal{Q}_{\mathcal{G}}) = 1$. *If $\mathcal{G}$ is not nilpotent, then the set of all regular elements of $\mathcal{G}$ has the following description*

$$r(\mathcal{G}) = \{\alpha T + Q \mid \alpha \in \mathbb{C}\backslash\{0\}, Q \in \mathcal{Q}_{\mathcal{G}}\}$$

for any $T \in \mathcal{G}\backslash\mathcal{Q}_{\mathcal{G}}$.

Proof. Choose by Lie's Theorem a basis in $\mathcal{G}$ such that every operator $ad\,G$ $(G \in \mathcal{G})$ be represented by an upper triangular matrix. Then by Rosenblum's Theorem (Theorem 1 from § 13) the matrix of $ad\,Q$ has only zeros on the diagonal for every $Q \in \mathcal{Q}_{\mathcal{G}}$. Consequently, for every $\alpha \in \mathbb{C}\backslash\{0\}$ and $Q \in \mathcal{Q}_{\mathcal{G}}$, the matrices of $ad\,T$ and $ad\,(\alpha T + Q)$ have the same number of zeros on the diagonal, and this number is strictly less than $\dim \mathcal{G}$ because $\mathcal{G}$ is not nilpotent. Particularly for every $\alpha \in \mathbb{C}\backslash\{0\}$ and $Q, Q' \in \mathcal{Q}_{\mathcal{G}}$ we have

$$\dim \mathit{Ker}((ad\,(\alpha T + Q))^N) = \dim \mathit{Ker}((ad\,T)^N) < N = \dim \mathit{Ker}((ad\,Q')^N)$$

where $N := \dim \mathcal{G}$. Since $\mathcal{G} = \mathbb{C}T + \mathcal{Q}_{\mathcal{G}}$, it then follows that the set of those $G \in \mathcal{G}$ with minimal $\dim \mathit{Ker}((ad\,G)^N)$ is $\{\alpha T + Q \mid \alpha \in \mathbb{C}\backslash\{0\}, Q \in \mathcal{Q}_{\mathcal{G}}\}$ and the proof ends. □

Lemma 3. *Let $S, Q \in \mathcal{B}(\mathcal{X})$ be operators generating a finite-dimensional solvable Lie subalgebra of $\mathcal{B}(\mathcal{X})$. If S is a normal-equivalent operator and Q is quasinilpotent, then $T := S + Q$ is a Jordan operator (but of course $T = S + Q$ is the Jordan decomposition of T iff $[S, Q] = 0$, which may not hold).*

Proof. First observe that if the Lie algebra generated by S and Q is even nilpotent then $(ad\,S)^n Q = 0$ for some $n \geq 1$. This implies $Q \in \mathcal{B}(\mathcal{X})_{ad\,S}(\{0\})$, hence $[S, Q] = 0$ by (6) (see the proof of Lemma 1). Hence in this case T is a Jordan operator with the Jordan decomposition $T = S + Q$.

Now let's come back to the proof of the general case. We have obviously

$$\mathcal{G} = \mathbb{C}S + \mathbb{C}Q + [\mathcal{G}, \mathcal{G}]. \tag{9}$$

By Theorem 1 it follows that $\mathbb{C}Q + [\mathcal{G}, \mathcal{G}] \subseteq \mathcal{Q}_{\mathcal{G}}$. We may suppose $S \notin \mathcal{Q}_{\mathcal{G}}$. (Otherwise $S = 0$ by Theorem 6 from § 14.) Then S is a regular element of $\mathcal{G}$ by Lemma 2. So there exists a Cartan subalgebra $\mathcal{H}$ of $\mathcal{G}$ such that $S \in \mathcal{H}$. Then $\mathbb{C}S \subseteq \mathcal{H} \subseteq \mathbb{C}S \oplus \mathcal{Q}_{\mathcal{G}} (= \mathcal{G})$, so $\mathcal{H} = \mathbb{C}S \oplus (\mathcal{H} \cap \mathcal{Q}_{\mathcal{G}})$. Since $\mathcal{H}$ is nilpotent, by the observation from the beginning of the proof it follows that $[\mathbb{C}S, \mathcal{H} \cap \mathcal{Q}_{\mathcal{G}}] = \{0\}$. Particularly every operator of the form $\alpha S + Q'$ ($\alpha \in \mathbb{C}\backslash\{0\}$, $Q' \in \mathcal{H} \cap \mathcal{Q}_{\mathcal{G}}$) is a Jordan operator by definition. Hence by Lemma 2 the set $r(\mathcal{G}) \cap \mathcal{H}$ consists only of Jordan operators.

On the other hand $S+Q \in r(\mathcal{G})$ by Lemma 2, hence there exists a Cartan subalgebra $\mathcal{K}$ of $\mathcal{G}$ such that $S+Q \in \mathcal{K}$. By Theorem 1 from § 5 and the proof of Lemma 1 $a)$ from § 27, there exists an invertible operator $A \in \mathcal{B}(\mathcal{X})$ such that the map $G \mapsto AGA^{-1}$ defines an automorphism φ of $\mathcal{G}$ such that $\varphi(\mathcal{K}) = \mathcal{H}$. Then $\varphi(r(\mathcal{G})) = r(\mathcal{G})$ (since φ is an automorphism) and $S+Q \in \mathcal{K} \cap r(\mathcal{G})$, hence $A(S+Q)A^{-1} = \varphi(S+Q) \in \mathcal{H} \cap r(\mathcal{G})$. Hence $A(S+Q)A^{-1}$ is a Jordan operator in view of what we have proved above. This easily implies that $S+Q$ is also a Jordan operator. □

Corollary 1. *Let $\mathcal{G}$ be a finite-dimensional solvable Lie subalgebra of $\mathcal{B}(\mathcal{X})$, $\mathcal{H}$ be a Cartan subalgebra of $\mathcal{G}$ and $\mathcal{G} = \mathcal{H} \oplus \mathcal{C}_{\mathcal{H}}$ be the corresponding Cartan decomposition. If $\pi : \mathcal{G} \to \mathcal{H}$ is the natural projection corresponding to this decomposition, then we have*

$$T \text{ is a Jordan operator} \Leftrightarrow \pi(T) \text{ is a Jordan operator}$$

for every $T \in \mathcal{G}$.

Proof. Assume that T is a Jordan operator and let $T = S+Q$ be its Jordan decomposition. The Lie algebra $\mathcal{E}(\mathcal{G} \cup \{S,Q\})$ is finite-dimensional solvable by Lemma 1. Then by Theorem 1 the operator $Q_1 := (\pi(T)-T)+Q$ is quasinilpotent, since $\pi(T) - T \in \mathcal{C}_{\mathcal{H}}$ is nilpotent by Theorem 2. But $T = S+Q$, hence $Q_1 + S = \pi(T)$. Now $\pi(T)$ is a Jordan operator by Lemma 3 because S and Q_1 are elements of the finite-dimensional solvable Lie algebra $\mathcal{E}(\mathcal{G} \cup \{S,Q\})$.

The converse implication can be proved similarly. □

Now we can prove:

Theorem 3. *Let $\mathcal{G}$ be a finite-dimensional solvable Lie subalgebra of $\mathcal{B}(\mathcal{X})$. Then $\mathcal{J}_{\mathcal{G}}$ is an ideal of $\mathcal{G}$ containing $[\mathcal{G},\mathcal{G}]$.*

Proof. Every quasinilpotent operator is a Jordan operator, hence $[\mathcal{G},\mathcal{G}] \subseteq \mathcal{J}_{\mathcal{G}}$ by Theorem 1. Moreover every scalar multiple of a Jordan operator is also a Jordan operator. Hence it remains to prove that if $T_1, T_2 \in \mathcal{B}(\mathcal{X})$ are Jordan operators generating a finite-dimensional solvable Lie algebra $\mathcal{G}$ then T_1+T_2 is also a Jordan operator. The proof of this fact will consist of two steps.

1° First assume that $\mathcal{G}$ is a nilpotent Lie algebra. Let $T_j = S_j + Q_j$ be the Jordan decomposition of T_j for $j = 1,2$. Two successive applications of Lemma 1 show that $\mathcal{E}(\mathcal{G} \cup \{S_1, Q_1, S_2, Q_2\})$ is a finite-dimensional solvable Lie algebra. Particularly $Q_1 + Q_2$ is a quasinilpotent operator by Theorem 1.

On the other hand, since $T_1, T_2 \in \mathcal{G}$ and $\mathcal{G}$ is nilpotent, it follows that $(adT_2)^n T_1 = 0$ for some $n \geq 1$. Consequently $T_1 \in \mathcal{B}(\mathcal{X})_{adT_2}(\{0\})$. This implies, as in the proof of Lemma 1 (see (5) and (6)), that $[T_1, S_2] = 0$. But S_1 belongs to the bicommutant of T_1 (see Theorem 7 from § 14) hence $[S_1, S_2] = 0$. This implies that $S_1 + S_2$ is a normal-equivalent operator (Corollary 6 from § 14). Since we have already proved that $Q_1 + Q_2$ is a quasinilpotent operator and $S_1 + S_2$, $Q_1 + Q_2$ belong to the finite-dimensional solvable Lie algebra $\mathcal{E}(\mathcal{G} \cup \{S_1, Q_1, S_2, Q_2\})$, by Lemma 3 we get that $S_1 + S_2 + Q_1 + Q_2$ is a Jordan operator. But this operator is just $T_1 + T_2$.

2° Now we treat the general case. Let $\mathcal{H}$ be a Cartan subalgebra of $\mathcal{G}$. By Corollary 1, the elements $\pi(T_1), \pi(T_2) \in \mathcal{H}$ are Jordan operators. Since $\mathcal{H}$ is a nilpotent Lie algebra, the operator $\pi(T_1) + \pi(T_2)$ is a Jordan operator by step 1° of the proof. But $\pi : \mathcal{G} \to \mathcal{H}$ is a linear map, hence actually $\pi(T_1 + T_2)$ is a Jordan operator. Now again by Corollary 1 we obtain that $T_1 + T_2$ is a Jordan operator. □

Now one can establish a variant of Theorem 3, which holds for $C^0(\mathbb{C})$-spectral operators instead of Jordan operators (see Theorem 4 below). This variant is based on corresponding variants of Lemma 3 and Corollary 1 which we state below. We omit the proofs because they are very similar to those given above.

Lemma 4. *Let $S, Q \in \mathcal{B}(\mathcal{X})$ be operators generating a finite-dimensional solvable Lie subalgebra of $\mathcal{B}(\mathcal{X})$. If S is a $C^0(\mathbb{C})$-scalar operator and Q is quasinilpotent then $T := S + Q$ is a $C^0(\mathbb{C})$-spectral operator.*

Corollary 2. *Let $\mathcal{G}$ be a finite-dimensional solvable Lie subalgebra of $\mathcal{B}(\mathcal{X})$. Let $\mathcal{H}$ and π be as in Corollary 1. Then we have*

$$T \text{ is a } C^0(\mathbb{C})\text{-spectral operator} \Leftrightarrow \pi(T) \text{ is a } C^0(\mathbb{C})\text{-spectral operator}$$

for every $T \in \mathcal{G}$.

Theorem 4. *Let $\mathcal{G}$ be a finite-dimensional solvable Lie subalgebra of $\mathcal{B}(\mathcal{X})$. Then $\mathcal{S}^0_{\mathcal{G}}$ is an ideal of $\mathcal{G}$ containing $[\mathcal{G}, \mathcal{G}]$.*

Now we can summarize the above results in the following way:

Corollary 3. *If $\mathcal{G}$ is a finite-dimensional solvable Lie subalgebra of $\mathcal{B}(\mathcal{X})$, then it has the following nest of ideals consisting of generalized spectral operators:*

$$\mathcal{N}_{\mathcal{G}} \subseteq \mathcal{Q}_{\mathcal{G}} \subseteq \mathcal{S}^0_{\mathcal{G}} \subseteq \mathcal{J}_{\mathcal{G}}.$$

Moreover $[\mathcal{G}, \mathcal{G}] \subseteq \mathcal{Q}_{\mathcal{G}}$ and if $\mathcal{N}_{\mathcal{G}} = \{0\}$ then $\mathcal{G}$ is nilpotent.

Proof. The respective sets are ideals of $\mathcal{G}$ by Theorems 1-4. The inclusion $[\mathcal{G}, \mathcal{G}] \subseteq \mathcal{Q}_{\mathcal{G}}$ follows by Theorem 1. The last assertion follows by Theorem 2 since for any Cartan decomposition $\mathcal{G} = \mathcal{H} \oplus \mathcal{C}_{\mathcal{H}}$ we have $\mathcal{C}_{\mathcal{H}} \subseteq \mathcal{N}_{\mathcal{G}} = \{0\}$, so $\mathcal{G} = \mathcal{H}$. But the Cartan subalgebra $\mathcal{H}$ is nilpotent, so $\mathcal{G}$ must be nilpotent. □

Finally we prove Proposition 1 from § 24. We need the following simple fact:

Lemma 5. *Let $\mathcal{A}$ be a complex unital Banach algebra and $\mathcal{N}$ a finite-dimensional Lie subalgebra of $\mathcal{A}$ containing only nilpotent elements. Then there exists $m \geq 1$ such that $b_1 \cdots b_m = 0$ for every $b_1, \ldots, b_m \in \mathcal{N}$.*

Proof. Let $\rho : \mathcal{A} \to \mathcal{B}(\mathcal{A})$ be the left regular representation of $\mathcal{A}$, hence $\rho(b)a = ba$ for $a, b \in \mathcal{A}$. Then ρ is a faithful representation, so $\rho(\mathcal{N})$ is a finite-dimensional Lie subalgebra of $\mathcal{B}(\mathcal{A})$ consisting only of nilpotent operators. Then by Theorem 1 from § 23 we deduce that the associative algebra generated by $\rho(\mathcal{N})$ is a finite-dimensional nil-algebra (i.e. containing only nilpotent elements). Since ρ is a faithful representation, the associative subalgebra generated by $\mathcal{N}$ in $\mathcal{A}$ will be also a finite-dimensional nil-algebra. Now it is straightforward to deduce the desired conclusion. □

Proof of Proposition 1 from § 24. As in the proof of Lemma 5 above, let $\rho : \mathcal{A} \to \mathcal{B}(\mathcal{A})$ be the left regular representation of $\mathcal{A}$. Since this is a faithful representation, we deduce that $\rho(\mathcal{L})$ is a Lie subalgebra of $\mathcal{B}(\mathcal{A})$ and $\rho(\mathcal{I})$ is a finite-dimensional solvable ideal of $\rho(\mathcal{L})$. Moreover we have $\rho(\mathcal{N}_{\mathcal{I}}) = \mathcal{N}_{\rho(\mathcal{I})}$. Hence the set $\rho(\mathcal{N}_{\mathcal{I}})$ is an ideal of $\rho(\mathcal{I})$ by Theorem 2 above. This implies that $\mathcal{N}_{\mathcal{I}}$ is an ideal of $\mathcal{I}$ since ρ is faithful. Hence it remains to prove that for an arbitrary $l \in \mathcal{L}$ we have $[l, \mathcal{N}_{\mathcal{I}}] \subseteq \mathcal{N}_{\mathcal{I}}$.

To this end remark that, since $\mathcal{I}$ is a finite-dimensional solvable ideal of $\mathcal{L}$, it follows that $\mathcal{I}_1 := \mathbb{C}l + \mathcal{I}$ is a finite-dimensional solvable Lie algebra. Hence, as above, $\mathcal{N}_{\mathcal{I}_1}$ is an ideal of $\mathcal{I}_1$. Particularly $[l, \mathcal{N}_{\mathcal{I}_1}] \subseteq \mathcal{N}_{\mathcal{I}_1}$, so the proof is finished if $\mathcal{N}_{\mathcal{I}_1} = \mathcal{N}_{\mathcal{I}}$. Next assume that $\mathcal{N}_{\mathcal{I}_1} \neq \mathcal{N}_{\mathcal{I}}$. We have obviously $\mathcal{N}_{\mathcal{I}_1} \cap \mathcal{I} = \mathcal{N}_{\mathcal{I}}$, hence our assumption implies $\mathcal{N}_{\mathcal{I}_1} \backslash \mathcal{I} \neq \emptyset$. Let's take $b \in \mathcal{N}_{\mathcal{I}_1} \backslash \mathcal{I}$. Since $\dim(\mathcal{I}_1/\mathcal{I}) \leq 1$, it then follows that $\mathcal{I}_1 := \mathbb{C}b + \mathcal{I} = \mathcal{N}_{\mathcal{I}_1} + \mathcal{I}$. So

$$1 \geq \dim((\mathcal{N}_{\mathcal{I}_1} + \mathcal{I})/\mathcal{I}) = \dim(\mathcal{N}_{\mathcal{I}_1}/(\mathcal{N}_{\mathcal{I}_1} \cap \mathcal{I})) = \dim(\mathcal{N}_{\mathcal{I}_1}/\mathcal{N}_{\mathcal{I}}).$$

But $b \in \mathcal{N}_{\mathcal{I}_1} \backslash \mathcal{I} \subseteq \mathcal{N}_{\mathcal{I}_1} \backslash \mathcal{N}_{\mathcal{I}}$, so

$$\mathcal{N}_{\mathcal{I}_1} = \mathbb{C}b + \mathcal{N}_{\mathcal{I}}. \tag{10}$$

But $\mathcal{N}_{\mathcal{I}_1}$ contains only nilpotent elements, hence it is a nilpotent Lie algebra (see e.g. Theorem 1 from § 23). Then by (10) and by Lemma 2 from § 25 we deduce that $\mathcal{N}_{\mathcal{I}}$ is an ideal of $\mathcal{N}_{\mathcal{I}_1}$. So

$$[l, \mathcal{N}_{\mathcal{I}}] \subseteq [\mathcal{I}_1, \mathcal{N}_{\mathcal{I}}] = [\mathcal{N}_{\mathcal{I}_1} + \mathcal{I}, \mathcal{N}_{\mathcal{I}}] = [\mathcal{N}_{\mathcal{I}_1}, \mathcal{N}_{\mathcal{I}}] + [\mathcal{I}, \mathcal{N}_{\mathcal{I}}] \subseteq \mathcal{N}_{\mathcal{I}} + \mathcal{N}_{\mathcal{I}} = \mathcal{N}_{\mathcal{I}}.$$

Hence $\mathcal{N}_{\mathcal{I}}$ is an ideal of $\mathcal{L}$. Since $\mathcal{N}_{\mathcal{I}}$ is a finite-dimensional Lie algebra containing only nilpotent elements, by Lemma 5 we get

$$\underbrace{\mathcal{N}_{\mathcal{I}} \cdots \mathcal{N}_{\mathcal{I}}}_{m \text{ times}} = \{0\}.$$

On the other hand, since $\mathcal{N}_{\mathcal{I}}$ is a Lie ideal we have $\mathcal{N}_{\mathcal{I}} \cdot \mathcal{L} \subseteq \mathcal{L} \cdot \mathcal{N}_{\mathcal{I}} + \mathcal{N}_{\mathcal{I}}$, hence $\mathcal{N}_{\mathcal{I}} \cdot \mathcal{A}(\mathcal{L}) = \mathcal{A}(\mathcal{L}) \cdot \mathcal{N}_{\mathcal{I}}$. Consequently

$$\mathcal{N}_{\mathcal{I}} \cdot \mathcal{A}(\mathcal{L}) \cdots \mathcal{N}_{\mathcal{I}} \cdot \mathcal{A}(\mathcal{L}) = \mathcal{N}_{\mathcal{I}} \cdots \mathcal{N}_{\mathcal{I}} \cdot \mathcal{A}(\mathcal{L}) \cdots \mathcal{A}(\mathcal{L}) = \{0\}.$$

Particularly $(ab)^m = 0$ for every $a \in \mathcal{N}_{\mathcal{I}}, b \in \mathcal{A}(\mathcal{L})$. Here m is independent of b, so we get $(ab)^m = 0$ also for $a \in \mathcal{N}_{\mathcal{I}}$, $b \in \overline{\mathcal{A}(\mathcal{L})}$. Hence $\mathcal{N}_{\mathcal{I}} \subseteq rad\, \overline{\mathcal{A}(\mathcal{L})}$. □

Notes

Concerning § 25: The notion of spectrum for a representation (Definiton 1 from § 25) was introduced by C. Ott [1], [2], [3] and D. Beltiţă [1]. It extends a concept introduced by E. Boasso and A. Larotonda [1]. Remark 2 is taken from A.S. Fainshtein [1]. Theorems 1-2 were proved by E. Boasso and A. Larotonda [1]. For another proof of Theorem 2 see C. Ott [1], [2], [3], where a gap in the original proof of Theorem 2 was pointed out. However our Remark 4 "fills" the gap and the original proof is "saved" in this way. For a proof of Lemma 2 see e.g. C. Ott [3]. Corollary 1 was essentially proved by A.S. Fainshtein [1]; see also C. Ott [3]. Theorems 3-4 were proved by E. Boasso and A. Larotonda [1]; see also C. Ott [1], [3]. It is worth noticing that Theorem 4 follows also by Corollary 1 from § 24. Proposition 1, Theorems 5-7 and Corollary 5 were proved by D. Beltiţă [4]. Lemma 3 *(b)* was proved by M. Şabac [4]. Results related to Corollary 3 and Example 1 were obtained by E. Boasso [2]; see also the examples given in E. Boasso and A. Larotonda [1], E. Boasso [3], A.S. Fainshtein [1], C. Ott [2], [3]. Theorem 8 was proved by E. Boasso [1]; see also C. Ott [3]. Corollary 6 was proved for the first time by A.S. Fainshtein [1]; see also E. Boasso [1] and C. Ott [3]. Theorem 9 was proved by E. Boasso [5]; see C. Ott [3] for some generalizations. Finally it is worth mentioning that E. Boasso [1], C. Ott [3] and A.A. Dosiev [2] defined the Słodkowski spectra for a representation of a Lie algebra; for commuting n-tuples of operators, these spectra were introduced by Z. Słodkowski [1]. We note also that A.A. Dosiev [2] introduced the concept of essential spectrum of a representation. For other properties of the spectrum of a representation see C. Ott [3].

The Taylor joint spectrum for commuting tuples of operators was introduced by J.L. Taylor [1] and it gave rise to a wide literature; see e.g. Şt. Frunză [1], F.-H. Vasilescu [5], J. Eschmeier [1]. All the facts from the first part of § 26, from Definition 1 to Corollary 2 inclusive, are due to A.S. Fainshtein [1]. Theorem 5 was proved by D. Beltiţă [3]; in the same paper was proved Theorem 6 under an equivalent form. The proof of Lemma 4, as well as the classical Ringrose's Theorem, can be found e.g. in the book of H.R. Dowson [1]. For Theorem 7 see C. Ott [3].

All the facts from § 27 are due to D. Beltiţă; particularly the Cartan-Taylor spectrum for quasisolvable Lie algebras of operators was introduced in D. Beltiţă [3]. Definition 3 and Theorem 3 appear here for the first time. The existence of Example 1 was stated in D. Beltiţă [3], where can be found all the other facts contained in § 27. We thank to A.A. Dosiev for pointing out an error in the proof of Proposition 2.1 from D. Beltiţă [3]; this error is overcome in the present proof of Lemma 3. The name "Cartan-Taylor" for the spectra introduced in Definitions 1 and 2 was proposed by A.A. Dosiev and Yu.V. Turovskiĭ.

Theorem 1 from § 28 was proved in D. Beltiţă [3]. All the other facts contained in § 28 are due to D. Beltiţă and appear here for the first time.

Chapter V

Semisimple Lie Algebras of Operators

§ 29 Lie subalgebra with involution consisting of bounded operators on a complex Banach space. Normal elements given by a space of self-adjoint operators

As we have seen, it is possible to define a structure of complex Lie algebra with involution depending on some real vector subspaces with special properties (see §7). In the special case of Lie subalgebras of bounded operators on a complex Banach space, we can obtain a Lie subalgebra with involution in the following way.

As usual, let $\mathcal{X}$ be a complex Banach space and $\mathcal{B}(\mathcal{X})$ the associative algebra of all bounded linear operators on $\mathcal{X}$ with the Lie algebra structure given by $[A, B] = AB - BA$. For $A \in \mathcal{B}(\mathcal{X})$, $\sigma(A)$ will be as usual the spectrum of A.

Lemma 1. *Let $\mathcal{A}$ be a real vector subspace of $\mathcal{B}(\mathcal{X})$ such that*

1. $[A, B] \in i\mathcal{A}$ for every $A, B \in \mathcal{A}$;

2. $\{0\} \neq \sigma(A) \subset \mathbb{R}$ if $0 \neq A \in \mathcal{A}$.

Then $\mathcal{A} \cap i\mathcal{A} = \{0\}$ and $\mathcal{L} := \mathcal{A} \oplus i\mathcal{A}$ is a complex Lie algebra with involution.

Proof. It is an easy exercise to verify the condition of Lemma 1 from §7. □

Definition 1. A *space of self-adjoint operators* is a real vector subspace $\mathcal{A} \subset \mathcal{B}(\mathcal{X})$ consisting of scalar generalized operators such that

1. $[A, B] \in i\mathcal{A}$ for every $A, B \in \mathcal{A}$
2. $\{0\} \neq \sigma(A) \subset \mathbb{R}$ if $0 \neq A \in \mathcal{A}$.

Example 1. If $\mathcal{X}$ is a complex Hilbert space, then the set of all bounded self-adjoint operators or the set of all self-adjoint operators from a certain ideal of compact operators provide examples of spaces of self-adjoint operators.

Example 2. If $\mathcal{X} = l^p(\mathbb{N})$ $(1 \le p \le \infty)$, then an example of a space of self-adjoint operators can be constructed as follows. Denote by $\mathcal{F}$ the set of all Hermitian matrices $(a_{ij})_{i,j\ge 0}$ with only a finite number of non-zero elements. Then $\mathcal{F}$ naturally acting on $l^p(\mathbb{N})$ can be considered as a space $\mathcal{A}_p$ of self-adjoint operators on $l^p(\mathbb{N})$. Actually any operator of $\mathcal{A}_p$ is a finite-rank, (D)-scalar operator (see Definition 3 from §14).

Example 3. For an arbitrary Banach space $\mathcal{X}$, let $\mathcal{D}$ be a real Lie algebra of quasi-skew-adjoint operators (see Corollary 2 and Remark 5 from §14). We recall that any non-zero operator from $i\mathcal{D}$ is a generalized scalar operator with non-zero real spectrum. Then $i\mathcal{D}$ is a space of self-adjoint operators.

Definition 2. If $\mathcal{A} \subset \mathcal{B}(\mathcal{X})$ is a space of self-adjoint operators, then the normal elements of the complex Lie *-algebra $\mathcal{L} := \mathcal{A} \oplus i\mathcal{A}$ will be called *$\mathcal{A}$-normal operators.*

The following proposition contains a characterization of the normal elements in a complex Lie *-algebra given by a space of self-adjoint operators.

Proposition 1. *Let $\mathcal{A} \subset \mathcal{B}(\mathcal{X})$ be a space of self-adjoint operators and $\mathcal{L} := \mathcal{A} \oplus i\mathcal{A}$ the corresponding complex Lie *-algebra. The $\mathcal{A}$-normal operators (i.e. normal elements in $\mathcal{L}$) are generalized scalar operators $G \in \mathcal{L}$ having a spectral distribution $\mathbf{U}^G$ with $\mathbf{U}^G_{Re\,\lambda}, \mathbf{U}^G_{Im\,\lambda} \in \mathcal{A}$.*

Proof. If G is a generalized scalar operator with a spectral distribution $\mathbf{U}^G$, we recall that $G = \mathbf{U}^G_\lambda$ (the value of the spectral distribution $\mathbf{U}^G$ on the identity function λ of $\mathbb{C} = \mathbb{R}^2$, see Definition 1 from §14). So we have $G = \mathbf{U}^G_{Re\,\lambda} + i\mathbf{U}^G_{Im\,\lambda}$ (where $\mathbf{U}^G_{Re\,\lambda}, \mathbf{U}^G_{Im\,\lambda}$ are the values of the spectral distribution on the functions $Re\,\lambda$, $Im\,\lambda$; these are the $\mathcal{C}^\infty$-functions defined on $\mathbb{C} = \mathbb{R}^2$ by $\lambda \mapsto Re\,\lambda$, $\lambda \mapsto Im\,\lambda$).

If $\mathbf{U}^G_{Re\,\lambda}, \mathbf{U}^G_{Im\,\lambda} \in \mathcal{A}$, then

$$G^* = \mathbf{U}^G_{Re\,\lambda} - i\mathbf{U}^G_{Im\,\lambda} = \mathbf{U}^G_{\bar\lambda},$$

where $\bar\lambda$ denotes the $\mathcal{C}^\infty$-function defined on $\mathbb{R}^2$ by $\lambda \mapsto \bar\lambda$. Obviously we have $[G, G^*] = 0$ because $\mathbf{U}^G_\varphi \mathbf{U}^G_\psi = \mathbf{U}^G_\psi \mathbf{U}^G_\varphi = \mathbf{U}^G_{\varphi\psi}$ for every $\varphi, \psi \in \mathcal{C}^\infty$, particularly for $\varphi(\lambda) = \lambda$ and $\psi(\lambda) = \bar\lambda$.

Conversely, if G is a normal element of $\mathcal{L}$, then we have $G = A + iB$ with $A, B \in \mathcal{A}$, A and B being generalized scalar operators with spectrum on the real line and $[G, G^*] = 0$ implies $[A, B] = 0$. Then G is a generalized scalar operator being the sum of the generalized scalar operators A and iB with thin spectra (see Theorems 4.1.11, 4.3.3 and Corollary 4.3.4 from the book of I. Colojoară and C. Foiaş [1]). A spectral distribution $\mathbf{U}^G$ of G can be obtained from the spectral distributions $\mathbf{U}^A, \mathbf{U}^B$ of A, B as follows:

$$\mathbf{U}^G_\varphi = (\mathbf{U}^A \otimes \mathbf{U}^B)_{\varphi\circ(\zeta\otimes 1 + 1\otimes i\mu)} \quad \text{for every } \varphi \in \mathcal{C}^\infty(\mathbb{R}^2),$$

where $\mathbb{R}^2$ is identified with $\mathbb{C}$ and both ζ and μ denote the identity function on $\mathbb{C}$. Particularly we have

$$\mathbf{U}^G_{Re\,\lambda} = \mathbf{U}^A_{Re\,\zeta} + \mathbf{U}^B_{-Im\,\mu}.$$

But $\mathbf{U}^A_\varphi = \mathbf{U}^A_{\varphi|_R}, \mathbf{U}^B_\varphi = \mathbf{U}^B_{\varphi|_R}$ for every $\varphi \in \mathcal{C}^\infty(\mathbb{C})$ because $A = \mathbf{U}^A_\lambda, B = \mathbf{U}^B_\lambda$ are generalized scalar operators having spectra on the real line.

Particularly we have

$$\mathbf{U}^A_{Re\,\zeta} = \mathbf{U}^A_{\zeta|_\mathbb{R}} = A, \qquad \mathbf{U}^B_{-Im\,\mu} = \mathbf{U}^B_{-Im\,\mu|_\mathbb{R}} = 0.$$

Hence $\mathbf{U}^G_{Re\,\lambda} \in \mathcal{A}$ and in an analogous manner we can deduce that $\mathbf{U}^G_{Im\,\lambda} \in \mathcal{A}$. □

Proposition 2. *If $\mathcal{A} \subset \mathcal{B}(\mathcal{X})$ is a space of self-adjoint operators and G is an $\mathcal{A}$-normal operator, then there exists a unique spectral distribution $\mathbf{U}^G$ with the properties $\mathbf{U}^G_\lambda = G$ and $\mathbf{U}^G_{Re\,\lambda}, \mathbf{U}^G_{Im\,\lambda} \in \mathcal{A}$.*

Proof. We prove uniqueness of the spectral distribution given by Proposition 1. Indeed, $\mathbf{U}^G_\lambda = G$ and $\mathbf{U}^G_{\bar\lambda} = G^*$ and the values of $\mathbf{U}^G$ are uniquely determined on polynomials in λ and $\bar\lambda$. On the other hand, the distribution $\mathbf{U}^G$ is uniquely determined by its values on polynomials in λ and $\bar\lambda$ since the set of these polynomials is dense in $\mathcal{C}^\infty(\mathbb{C})$. □

The general Theorem 1 from §7 has the following consequences.

Theorem 1. *Let $\mathcal{L} = \mathcal{A} \oplus i\mathcal{A}$ be a complex Lie *-algebra which is a Lie subalgebra of $\mathcal{B}(\mathcal{X})$. If $\mathcal{G}$ is a complex Lie subalgebra of $\mathcal{L}$ consisting of generalized scalar operators G having spectral distributions $\mathbf{U}^G$ such that $\mathbf{U}^G_{Re\,\lambda}, \mathbf{U}^G_{Im\,\lambda} \in \mathcal{A}$, then $[\mathbf{U}^{G_1}_\lambda, \mathbf{U}^{G_2}_{\bar\lambda}] = 0$ for every $G_1, G_2 \in \mathcal{G}$. If $[G_1, G_2] = 0$, then $[\mathbf{U}^{G_1}_\varphi, \mathbf{U}^{G_2}_\varphi] = 0$ for every $\varphi \in \mathcal{C}^\infty$. By Proposition 1 the same conclusion holds if $\mathcal{A}$ is a space of self-adjoint operators and $\mathcal{G}$ consists of $\mathcal{A}$-normal operators.*

Proof. We have $G = \mathbf{U}^G_\lambda = \mathbf{U}^G_{Re\,\lambda} + i\mathbf{U}^G_{Im\,\lambda}$ and $G^* = \mathbf{U}^G_{\bar\lambda}$ because $\mathbf{U}^G_{Re\,\lambda}$ and $\mathbf{U}^G_{Im\,\lambda}$ belong to $\mathcal{A}$. We deduce $[G, G^*] = 0$ as in the proof of Proposition 1 because $\mathbf{U}^G$ is multiplicative. By Theorem 1 from §7 we deduce that $[G_1, G_2^*]$, $[iG_1, G_2^*]$ are skew-adjoint because $\mathcal{G}$ is a Lie subalgebra of $\mathcal{L}$ consisting of normal elements. Hence we deduce

$$([G_1, G_2^*])^* = -[G_1, G_2^*]$$

and

$$(i[G_1, G_2^*])^* = -i([G_1, G_2^*])^* = i[G_1, G_2^*].$$

On the other hand,

$$(i[G_1, G_2^*])^* = ([iG_1, G_2^*])^* = -[iG_1, G_2^*] = -i[G_1, G_2^*].$$

Therefore

$$i[G_1, G_2^*] = -i[G_1, G_2^*],$$

so $[G_1, G_2^*] = 0$ for every $G_1, G_2 \in \mathcal{G}$.

The last part of the theorem is an easy consequence of the density in $\mathcal{C}^\infty(\mathbb{C})$ of the set of polynomials in λ and $\bar\lambda$. □

Theorem 2. *Let $\mathcal{G}$ be an ideally finite semisimple Lie subalgebra of a complex Lie *-algebra $\mathcal{L} = \mathcal{A} \oplus i\mathcal{A} \subseteq \mathcal{B}(\mathcal{X})$, where $\mathcal{A}$ is a space of self-adjoint operators. If*

$\mathcal{G}$ consists only of $\mathcal{A}$-normal operators (i.e. generalized scalar operators G having a spectral distribution $\mathbf{U}^G$ with $\mathbf{U}^G_{Re\,\lambda}, \mathbf{U}^G_{Im\lambda} \in \mathcal{A}$), then for every $G \in \mathcal{G}$ we may uniquely write $G = G_0 + G_1$; G_0 is $\mathcal{A}$-normal operator and $G_1 \in i\mathcal{A}$ (i.e. G_1 is skew-adjoint). Moreover for any $H \in \mathcal{G}$, if we similarly write $H = H_0 + H_1$, then $[H_0, G_1] = [H_1, G_0] = 0$ and $[G_0, H_0^*] = 0$ (i.e. $[\mathbf{U}^{G_0}_\lambda, \mathbf{U}^{H_0}_{\bar{\lambda}}] = 0$).

Proof. We can apply Theorem 1 from §7. □

In the following we will consider a Lie *-algebra $\mathcal{L} \subseteq \mathcal{B}(\mathcal{X})$ given by a space $\mathcal{A}$ of self-adjoint operators as above: $\mathcal{L} = \mathcal{A} \oplus i\mathcal{A}$. First we study sufficient conditions for a Lie subalgebra of $\mathcal{A}$-normal operators to be abelian. These conditions are related to the following well-known assertion in complex Hilbert spaces for usual normal operators:

> *If a set of normal operators on a complex Hilbert space is a complex vector space, then its elements commute pairwise.*

The proofs are based on the following lemma.

Lemma 2. *If G is an $\mathcal{A}$-normal operator and G is quasinilpotent, then $G = 0$.*

Proof. By the spectral mapping theorem for C^∞-functional calculus (see Theorem 3.2.1 from the book of I. Colojoară and C. Foiaş [1]), from $\sigma(G) = \{0\}$ it follows that $\sigma(\mathbf{U}^G_{Re\,\lambda}) = \sigma(\mathbf{U}^G_{Im\,\lambda}) = \{0\}$. By the definition of a space of self-adjoint operators it then follows that $\mathbf{U}^G_{Re\,\lambda} = \mathbf{U}^G_{Im\,\lambda} = 0$ so $G = \mathbf{U}^G_{Re\,\lambda} + i\mathbf{U}^G_{Im\,\lambda} = 0$. □

Corollary 1. *Let $\mathcal{G}$ be a real Lie subalgebra of $\mathcal{A}$-normal operators. If $\mathcal{G}$ is either nilpotent or $[\mathcal{G}, \mathcal{G}]$ is abelian (in any case $\mathcal{G}$ may be infinite-dimensional), then $\mathcal{G}$ is abelian.*

Proof. By Lemma 2 it suffices to show that $[\mathcal{G}, \mathcal{G}]$ contains only quasinilpotent operators. If $\mathcal{G}$ is nilpotent, this follows by Theorem 4 from §25. On the other hand, if $[\mathcal{G}, \mathcal{G}]$ is abelian, then any element of $[\mathcal{G}, \mathcal{G}]$ is quasinilpotent by Proposition 1.13 of Yu.V. Turovskiĭ [4]. □

Corollary 2. *Let $\mathcal{G}$ be a real Lie subalgebra of $\mathcal{A}$-normal operators. If $\mathcal{G}$ is solvable (it may be infinite-dimensional) then it is abelian.*

Proof. The conclusion follows from Corollary 1 by induction on the length of the derived series of $\mathcal{G}$. □

The last result has a variant holding for locally solvable Lie algebras.

Theorem 3. *Any locally solvable real Lie algebra of $\mathcal{A}$-normal operators is abelian. Particularly, any locally nilpotent, quasisolvable or quasinilpotent Lie algebra of $\mathcal{A}$-normal operators is abelian.*

Proof. Let $\mathcal{G}$ be a locally solvable real Lie algebra of $\mathcal{A}$-normal operators. Any pair of elements of $\mathcal{G}$ is contained in a certain solvable Lie subalgebra of $\mathcal{G}$ (see Definition 5 a) from §2). Hence it commutes by Corollary 2. □

We note that, even on a Hilbert space, there exist real Lie algebras of normal operators which are not abelian. For example the real Lie algebra $su(2)$ is not abelian. We will discuss now the case of the complex Lie algebras of $\mathcal{A}$-normal operators. We can obtain the same conclusion as in Theorem 3 for complex locally finite Lie algebras.

Proposition 3. *Any locally finite complex Lie algebra of $\mathcal{A}$-normal operators is abelian.*

Proof. Proceeding as in the proof of Theorem 3 it suffices to prove the statement for $\mathcal{G}$ of finite dimension.

In this case we suppose that $\mathcal{G}$ is not nilpotent. Then one can find $G \in \mathcal{G}$ such that the operator $adG : \mathcal{G} \to \mathcal{G}$ is not nilpotent. So it has a non-zero eigenvector $H \in \mathcal{G}$ with the corresponding eigenvalue $\lambda \neq 0$, i.e. $[G, H] = \lambda H$. By the nilpotence property given in Corollary 2 from §16 we deduce that H is nilpotent, so $H = 0$ by Lemma 2. This contradiction shows that $\mathcal{G}$ is a nilpotent Lie algebra. Hence it is abelian by Corollary 1. □

Proposition 4. *Let us consider $\mathcal{G}$ a complex Lie algebra of $\mathcal{A}$-normal operators. For $G \in \mathcal{G}$ we denote by $\mathbf{U}^G$ the spectral distribution of G and by $ord\mathbf{U}^G$ the order of this distribution. We suppose also that there exists a subset $\mathcal{S}$ of $\mathcal{G}$ with the following properties:*

1) *$\mathcal{S}$ generates $\mathcal{G}$ as a complex vector space;*
2) *$N := \sup\{ord\mathbf{U}^G | G \in \mathcal{S}\} < \infty$.*

Then we have

$$(adG)^{2N+1} = 0 \text{ on } \mathcal{G}$$

for any $G \in \mathcal{S}$.

Remark 1. If $\mathcal{G}$ is finite-dimensional then $\mathcal{S}$ may be any basis of $\mathcal{G}$.

Proof of Proposition 4. We have by Proposition 1 for any $G \in \mathcal{G}$

$$G^* = \mathbf{U}^G_{\bar{\lambda}}$$

hence G^* is a generalized spectral operator with the spectral distribution $\mathbf{U}^{G^*}$,

$$\varphi \mapsto \mathbf{U}^G_{\varphi \circ \bar{\lambda}}$$

having order at most $ord\mathbf{U}^G$. Moreover $\mathbf{U}^{G^*}_{\bar{\lambda}} = \mathbf{U}^G_{\lambda} = G$. Hence by Theorem 1 for any $G_1, G_2 \in \mathcal{S}$ we successively have

$$[\mathbf{U}^{G_1}_{\lambda}, \mathbf{U}^{G_2}_{\bar{\lambda}}] = 0 \Leftrightarrow [\mathbf{U}^{G_1}_{\lambda}, \mathbf{U}^{G_2^*}_{\lambda}] = 0 \Leftrightarrow (ad\mathbf{U}^{G_2^*}_{\lambda})\mathbf{U}^{G_1}_{\lambda} = 0.$$

By the commutativity Theorem 1 from §14 it then follows

$$(ad\mathbf{U}^{G_2^*}_{\bar{\lambda}})^{2N+1}\mathbf{U}^{G_1}_{\lambda} = 0 \text{ for any } G_1, G_2 \in \mathcal{S},$$

that is

$$(ad\,G_2)^{2N+1}G_1 = 0 \text{ for any } G_1, G_2 \in \mathcal{S}.$$

Since G_1 and G_2 are arbitrary in $\mathcal{S}$ the conclusion follows by the hypothesis 1).□

Corollary 3. *Let $\mathcal{G}$ be a complex Lie algebra of $\mathcal{A}$-normal operators. If there exists a subset $\mathcal{S}$ of $\mathcal{G}$ such that*

1) *$\mathcal{S}$ generates $\mathcal{G}$ as a complex vector space;*
2) *$ord\,\mathbf{U}^G = 0$ for any $G \in \mathcal{S}$*

then $\mathcal{G}$ is abelian.

Corollary 4. *If $\mathcal{A}$ is a space of self-adjoint operators consisting only of Dunford scalar operators, then any complex Lie algebra of $\mathcal{A}$-normal operators is abelian.*

The semisimple ideally finite Lie algebras of $\mathcal{A}$-normal operators have interesting special properties.

Proposition 5. *If $\mathcal{G}$ is a semisimple ideally finite real Lie algebra of $\mathcal{A}$-normal operators, then it has the following properties.*

1° *The Lie algebra $\mathcal{G}$ contains only skew-adjoint operators, i.e. it is a Lie subalgebra of $i\mathcal{A}$.*
2° *The Killing form of $\mathcal{G}$ is negatively defined.*
3° *The Lie algebra $\mathcal{G}$ has a faithful representation by finite-rank skew-adjoint operators (in the classical sense on a certain Hilbert space).*

Proof. 1° First of all we observe that an ideally finite real Lie subalgebra $\mathcal{F}$ of $\mathcal{A} \oplus i\mathcal{A}$ which satisfies $[\mathcal{F}, \mathcal{F}^*] = \{0\}$ is abelian (i.e. $[\mathcal{F}, \mathcal{F}] = \{0\}$). Indeed, we can suppose $\mathcal{F}$ finite-dimensional. If $\widetilde{\mathcal{F}}$ is the complex vector space spanned by $\mathcal{F}$, then $\widetilde{\mathcal{F}}$ is a complex finite-dimensional Lie algebra of $\mathcal{A}$-normal operators because $[\mathcal{F}, \mathcal{F}^*] = \{0\}$. Hence $\widetilde{\mathcal{F}}$ is abelian by Proposition 3.

On the other hand by Theorem 2 for any $G \in \mathcal{G}$ we may uniquely write $G = G_0 + G_1$, where G_0 is $\mathcal{A}$-normal and $G_1 \in i\mathcal{A}$. It results also that $\mathcal{F} := \{G_0 \mid G \in \mathcal{G}\}$ is an ideally finite Lie subalgebra because $\mathcal{F} = \pi(\mathcal{G})$, where

$$\pi : \mathcal{G} \to \mathcal{A} \oplus i\mathcal{A}, \quad G \mapsto G_0,$$

is obviously a Lie morphism. Theorem 2 shows also that $[\mathcal{F}, \mathcal{F}^*] = \{0\}$, hence $\mathcal{F}$ is abelian. This means that $\pi([\mathcal{G}, \mathcal{G}]) = \{0\}$. Hence $\pi(\mathcal{G}) = \{0\}$ because $\mathcal{G} = [\mathcal{G}, \mathcal{G}]$, $\mathcal{G}$ being a semisimple ideally finite Lie algebra. Therefore, $G = G_1 \in i\mathcal{A}$ for every $G \in \mathcal{A}$ and the property 1° is proved.

2° By Theorem 3 from §4 we have $\mathcal{G} = \bigoplus_{k \in \Lambda} \mathcal{I}_k$, where each $\mathcal{I}_k$ is a finite-dimensional simple ideal of $\mathcal{G}$ and $[\mathcal{I}_k, \mathcal{I}_l] = \{0\}$ for $k \neq l$. Then for every $G_1, G_2 \in \mathcal{G}$ there exists a finite subset Λ_{G_1, G_2} of Λ such that $(ad\,G_1)(ad\,G_2)$ vanishes on $\bigoplus_{k \notin \Lambda_{G_1, G_2}} \mathcal{I}_k$. So for proving the assertion 2° we can suppose that $\mathcal{G}$ is finite-dimensional.

Let G be arbitrary in $\mathcal{G}$, $\widetilde{\mathcal{G}}$ be the complexification of $\mathcal{G}$ and $\lambda_1, \ldots, \lambda_k$ the eigenvalues (repeated according to the multiplicities) of the operator $ad\,G : \widetilde{\mathcal{G}} \to \widetilde{\mathcal{G}}$.

Then $\widetilde{\mathcal{G}} = \mathcal{G} \oplus i\mathcal{G}$ is a finite-dimensional subspace of $\mathcal{B}(\mathcal{X})$ and $\widetilde{\mathcal{G}}$ is invariant for the operator $adG : \mathcal{B}(\mathcal{X}) \to \mathcal{B}(\mathcal{X})$. By Rosenblum's theorem (Theorem 1 from §13) the spectrum of the last operator equals $\sigma(G) - \sigma(G)$, hence it is contained in $i\mathbb{R}$ because $\mathcal{G} \subseteq i\mathcal{A}$ by 1°. Then the spectrum of adG restricted to any invariant subspace is also contained in $i\mathbb{R}$. Hence $\{\lambda_j \mid 1 \leq j \leq k\} \subset i\mathbb{R}$. Then the trace of $(adG)^2 : \mathcal{G} \to \mathcal{G}$ is $Tr((adG)^2) = (\lambda_1)^2 + \cdots + (\lambda_k)^2 \leq 0$. But the Killing form is non-degenerate because $\mathcal{G}$ is semisimple. Therefore we have

$$Tr((adG)^2) = (\lambda_1)^2 + \cdots + (\lambda_k)^2 < 0$$

and the property 2° is proved.

3° Let $\widetilde{\mathcal{G}} := \mathcal{G} \oplus i\mathcal{G}$. Obviously $\widetilde{\mathcal{G}}$ is a semisimple ideally finite complex Lie algebra. We denote by $\mathcal{B}$ the Killing form of $\mathcal{G}$ and define on $\widetilde{\mathcal{G}}$ a sesquilinear form by

$$\langle A + iB, A' + iB' \rangle := -\mathcal{B}(A, A') - \mathcal{B}(B, B') - i(-\mathcal{B}(A, B') + \mathcal{B}(B, A')).$$

By 2°, this is a scalar product. Let $\mathcal{H}$ be the Hilbert space obtained by completion of $\widetilde{\mathcal{G}}$ with respect to this scalar product. For $G \in \mathcal{G}$ let $\rho(G) \in \mathcal{B}(\mathcal{H})$ be the finite-rank operator induced by adG. Since $\mathcal{B}(\cdot, \cdot)$ is an invariant bilinear form, one easily checks that

$$\langle \rho(G)H, K \rangle + \langle H, \rho(G)K \rangle = 0$$

for any $H, K \in \widetilde{\mathcal{G}}$ and $G \in \mathcal{G}$.

Hence $\rho : \mathcal{G} \to \mathcal{B}(\mathcal{H})$ is a representation of $\mathcal{G}$ by finite-rank skew-adjoint operators. This is a faithful representation, since the center of $\mathcal{G}$ reduces to $\{0\}$ because $\mathcal{G}$ is semisimple. □

Now we can give a characterization of the ideally finite Lie algebras having faithful representations by finite-rank skew-adjoint operators on Hilbert spaces.

Corollary 5. *Let $\mathcal{U}$ be an ideally finite real Lie algebra. The following assertions are equivalent.*

1° *There exists a space of self-adjoint operators $\mathcal{A}$ on a certain complex Banach space such that $\mathcal{U}$ and $i\mathcal{A}$ are isomorphic Lie algebras.*

2° *The Lie algebra $\mathcal{U}$ is the direct sum of an abelian real Lie algebra and a semisimple ideally finite real Lie algebra with negatively defined Killing form.*

3° *The Lie algebra $\mathcal{U}$ has a faithful representation by finite-rank skew-adjoint operators on a certain Hilbert space.*

Proof. 1° ⇒2°. There exist a quasisolvable ideal $\mathcal{R}$ and a semisimple ideally finite subalgebra $\mathcal{G}$ of $i\mathcal{A}$ such that $i\mathcal{A} = \mathcal{R} + \mathcal{G}$ (see the extension of Levi's theorem to semisimple ideally finite Lie algebras, i.e. Theorem 2 from §4). Then $\mathcal{R}$ is an abelian Lie algebra by Theorem 3 above. Moreover $[\mathcal{R}, [\mathcal{R}, \mathcal{G}]] \subseteq [\mathcal{R}, \mathcal{R}] = \{0\}$. By the Kleinecke-Sirokov Theorem (see Remark 1 from §17) and Lemma 1 above it then follows that $[\mathcal{R}, \mathcal{G}] = \{0\}$. Hence $i\mathcal{A}$ is the direct sum of the abelian Lie algebra $\mathcal{R}$ and the semisimple ideally finite Lie algebra $\mathcal{G}$ whose Killing form is negative (cf. Proposition 5) defined because $\mathcal{G}$ consists of $\mathcal{A}$-normal operators.

$2^\circ \Rightarrow 3^\circ$. One uses the same argument as in the proof of Proposition 5(3°) and the obvious fact that any abelian real Lie algebra can be faithfully represented by finite-rank skew-adjoint operators on a sufficiently large Hilbert space.

$3^\circ \Rightarrow 1^\circ$. This is obvious. □

Remark 2. In the conditions of Corollary 5, if $\mathcal{U}$ is finite-dimensional then we can specify:

– The Hilbert space from the assertion 3° can be chosen finite-dimensional.

– The Lie algebra $\mathcal{U}$ is isomorphic to a space $i\mathcal{A}$ of skew-adjoint operators on a certain Banach space if and only if $\mathcal{U}$ is the Lie algebra of a compact Lie group. (Indeed there exists a positive definite form f such that $f(x,[y,z]) = f([x,y],z)$ for any $x, y, z \in \mathcal{U}$ and we can apply the criterion from Remark 2 from §6.)

§ 30 Individual spectral properties in ideally finite semisimple Lie algebras of operators

In this paragraph we prove some results concerning an arbitrary ideally finite semisimple Lie algebra of bounded operators on a complex Banach space $\mathcal{X}$. These facts show that the situation is quite similar to what happens on finite-dimensional spaces: the operators of such an algebra are Dunford spectral of finite type and their spectra are finite (see Theorem 1 below). Moreover if our Lie algebra, say $\mathcal{G}$, is finite-dimensional, then the associative algebra generated by $\mathcal{G}$ is finite-dimensional (cf. Theorem 2 below) and the finite-dimensional invariant subspaces of $\mathcal{G}$ span the whole space $\mathcal{X}$ (Corollary 5 below).

In the present paragraph, whenever it is not otherwise stated, we denote by $\mathcal{G}$ a complex semisimple finite-dimensional Lie subalgebra of $\mathcal{B}(\mathcal{X})$ and by $\mathcal{G}_u$ a compact real form of $\mathcal{G}$ (see §6). We need the following auxiliary result.

Lemma 1. *Denote* $\Gamma_u := \{e^{T_1} \cdots e^{T_n} \mid T_1, \ldots, T_n \in \mathcal{G}_u,\ n \geq 1\}$. *Then* Γ_u *is a bounded subset of* $\mathcal{B}(\mathcal{X})$.

Proof. It is well known that Γ_u endowed with the norm operator topology has a structure of connected real Lie group whose Lie algebra is isomorphic to $\mathcal{G}_u$ (see e.g. Corollary 7.5 from the book of H. Upmeier [1]). Then Γ_u is compact by Remark 1 from §6. Consequently Γ_u is a bounded subset of $\mathcal{B}(\mathcal{X})$. □

Proposition 1. *Every element of* $\mathcal{G}_u$ *is a quasi-skew-adjoint operator. Moreover if* $\mathcal{H}_u$ *is a Cartan subalgebra of* $\mathcal{G}_u$, *then* $\mathcal{H} := \mathcal{H}_u + i\mathcal{H}_u$ *is a Cartan subalgebra of* $\mathcal{G}$ *and every element of* $\mathcal{H}$ *is a normal-equivalent operator.*

Proof. Let Γ_u be as in Lemma 1. Since Γ_u is a bounded subset of $\mathcal{B}(\mathcal{X})$, for every $T \in \mathcal{G}_u$ we have

$$\sup_{t\in\mathbb{R}} \|e^{tT}\| \leq \sup_{S\in\Gamma_u} \|S\| < \infty,$$

hence T is quasi-skew-adjoint. Furthermore, since Γ_u is a semisimple Lie algebra it follows that $[\mathcal{H}_u, \mathcal{H}_u] = \{0\}$. In view of what we have just proved, it then follows that every $H \in \mathcal{H} = \mathcal{H}_u + i\mathcal{H}_u$ is of the form $H = A + iB$ with A and B quasi-skew-adjoint operators. Consequently H is a normal-equivalent operator. Finally, it is well known that $\mathcal{H}$ is a Cartan subalgebra of $\mathcal{G}$. □

Particularly, Proposition 1 implies that every ideally finite semisimple Lie subalgebra of $\mathcal{B}(\mathcal{X})$ gives rise to a space of self-adjoint operators of the type mentioned in Example 3 from §29. More precisely we can state:

Corollary 1. *Let $\mathcal{S}$ be an ideally finite semisimple complex Lie subalgebra of $\mathcal{B}(\mathcal{X})$ and let $\{\mathcal{S}_j\}_{j\in J}$ be the family of the (finite-dimensional) simple ideals of $\mathcal{S}$. For every $j \in J$ let's denote by $\mathcal{D}_j$ a compact real form of $\mathcal{S}_j$. Then $\mathcal{A} := \bigoplus_{j\in J} i\mathcal{D}_j$ is a space of self-adjoint operators and $\mathcal{S} = \mathcal{A} \oplus i\mathcal{A}$.*

Proof. Let $T \in \bigoplus_{j\in J} \mathcal{D}_j$ be arbitrary. Then there exists a finite subset $\{j_1, \dots, j_m\}$ of J such that $T \in \mathcal{D}_{j_1} \oplus \cdots \oplus \mathcal{D}_{j_m} =: \mathcal{L}_u$. But $\mathcal{L}_u$ is obviously a compact real form of the semisimple Lie algebra $\mathcal{S}_{j_1} \oplus \cdots \oplus \mathcal{S}_{j_m}$, hence T is quasi-skew-adjoint by Proposition 1. Consequently $\bigoplus_{j\in J} \mathcal{D}_j$ is a Lie algebra of quasi-skew-adjoint operators. Hence $\mathcal{A}$ is a space of self-adjoint operators in view of Example 3 from §29. Furthermore the equality $\mathcal{S} = \mathcal{A} \oplus i\mathcal{A}$ follows from $\mathcal{S}_j = i\mathcal{D}_j \oplus \mathcal{D}_j$ for $j \in J$ by Theorem 3 from §4. □

Notation 1. We consider the space of self-adjoint operators $i\mathcal{G}_u$ (see the notations preceding Lemma 1) and denote by * the corresponding involution of $\mathcal{G} = i\mathcal{G}_u \oplus \mathcal{G}_u$. Hence for $T \in \mathcal{G}$ we have

$$T^* = T \Leftrightarrow T \in i\mathcal{G}_u$$

and

$$T^* = -T \Leftrightarrow T \in \mathcal{G}_u.$$

Proposition 2. *For every ($i\mathcal{G}_u$)-normal element $T \in \mathcal{G}$ (i.e. $[T, T^*] = 0$) there exists a Cartan subalgebra $\mathcal{H}$ of $\mathcal{G}$ such that $\mathcal{H}$ is a Lie *-subalgebra and $T \in \mathcal{H}$.*

Proof. Let's denote by $\mathcal{B}(\cdot,\cdot)$ the Killing form of $\mathcal{G}_u$. Let's endow $\mathcal{G}$ with a scalar product $\langle\cdot,\cdot\rangle$ defined by

$$\langle A + iB, A' + iB' \rangle := -\mathcal{B}(A, A') - \mathcal{B}(B, B') - i(\mathcal{B}(B, A') - \mathcal{B}(A, B'))$$

(see the proof of Proposition 5(3°) from §29). Since the Killing form is an invariant bilinear form on $\mathcal{G}_u$, it follows that $ad : \mathcal{G} \to \mathcal{B}(\mathcal{G})$ is a *-representation of $\mathcal{G}$ by operators on the Hilbert space $(\mathcal{G}, \langle\cdot,\cdot\rangle)$.

On the other hand, since $[T, T^*] = 0$ we have $T = A + iB$ with $A = A^*$, $B = B^*$ and $[A, B] = 0$. Consequently $ad(iA), ad(iB) \in \mathcal{B}(\mathcal{G})$ are skew-adjoint operators on the Hilbert space $\mathcal{G}$. Since $iA, iB \in \mathcal{G}_u$, it then follows that $ad(iA)$, $ad(iB) : \mathcal{G}_u \to \mathcal{G}_u$ are semisimple linear maps (Recall that a linear map L on some vector space $\mathcal{V}$ is semisimple whenever for every linear subspace $\mathcal{V}_1$ of $\mathcal{V}$

with $L(\mathcal{V}_1) \subseteq \mathcal{V}_1$ there exists a linear subspace $\mathcal{V}_2$ of $\mathcal{V}$ such that $L(\mathcal{V}_2) \subseteq \mathcal{V}_2$ and $\mathcal{V}$ is the direct sum of $\mathcal{V}_1$ and $\mathcal{V}_2$.) Since $[iA, iB] = 0$ it then follows (by Proposition 1.10.6 (*iii*) from the book of J. Dixmier [1]) that $iA, iB \in \mathcal{H}_u$ for a certain Cartan subalgebra $\mathcal{H}_u$ of $\mathcal{G}_u$. Then $\mathcal{H} := \mathcal{H}_u + i\mathcal{H}_u$ is a Cartan subalgebra of $\mathcal{G}$ with the desired properties. □

Lemma 2. *Let $A \in \mathcal{G}\backslash\{0\}$ and λ be a non-zero complex number such that we have $[[A, A^*], A] = \lambda A$. Then the following assertions hold.*

i) *λ is a real number and $[[A^*, A], A^*] = \lambda A^*$.*

ii) *Both A and A^* are nilpotent operators.*

iii) *For every positive integer n the formulas*

$$[A^*, A^n] = nA^{n-1}[A^*, A] - \lambda \cdot \frac{n(n-1)}{2} A^{n-1}$$

and

$$[(A^*)^n, A] = n(A^*)^{n-1}[A^*, A] + \lambda \cdot \frac{n(n-1)}{2} (A^*)^{n-1}$$

hold.

iv) $[AA^*, A^*A] = 0$.

Proof. *i*) Let's denote $B := [A, A^*]$. Then $B^* = [(A^*)^*, A^*] = [A, A^*] = B$, hence $B \in i\mathcal{G}_u$. Since $i\mathcal{G}_u$ is a space of self-adjoint operators (see Corollary 1 above) we deduce $\sigma(B) \subseteq \mathbb{R}$. Consequently the spectrum of $ad\,B : \mathcal{B}(\mathcal{X}) \to \mathcal{B}(\mathcal{X})$ is contained in $\mathbb{R}$ by Rosenblum's theorem (see Theorem 1 from §13). But $(ad\,B)A = \lambda A$ by hypothesis, so $\lambda \in \sigma(ad\,B) \subset \mathbb{R}$. Now the desired formula in *i*) follows from the hypothesis by means of the involution $*$.

ii) With the above notation, from $[B, A] = \lambda A$ it follows that A is nilpotent (see Corollary 2 from §16). The fact that A^* is nilpotent follows similarly in view of the formula from *i*).

iii) We note that $(ad\,A)^2(A^*) = -\lambda A$ by hypothesis. Particularly we have $(ad\,A)^3(A^*) = 0$. Then we compute $(ad(A^n))(A^*)$ by means of the asymptotic formula for the commutators. More precisely, we apply the last formula of Corollary 3 from §15 for the function $f \in \mathcal{O}(\mathbb{C})$, $f(z) = z^n$. Since $f^{(k)}(z) = n(n-1)\cdots(n-k+1)z^{n-k+1}$, we obtain

$$(ad(A^n))(A^*) = \sum_{k=1}^{2} \frac{(-1)^{k+1}}{k!} n(n-1)\cdots(n-k+1)A^{n-k}(ad\,A)^k(A^*),$$

hence

$$[A^n, A^*] = \sum_{k=1}^{2} (-1)^{k+1} \binom{n}{k} A^{n-k}(ad\,A)^k(A^*),$$

which is obviously equivalent to the first desired formula in *iii*) since $(ad\,A)^2(A^*) = -\lambda A$. The other formula of *iii*) can be proved similarly using the formula from *i*).

iv) By the formulas contained in the hypothesis and in *i*) we deduce

$$\begin{aligned}[AA^*, A^*A] &= [AA^* - A^*A, A^*A] = [[A, A^*], A^*A] \\ &= [[A, A^*], A^*]A + A^*[[A, A^*], A] = (-\lambda A^*)A + A^*(\lambda A) = 0\end{aligned}$$

and the proof finishes. □

Lemma 3. *Let $A \in \mathcal{G}\backslash\{0\}$ and λ be a non-zero complex number such that we have $[[A, A^*], A] = \lambda A$. If we denote $B_n := A^n(A^*)^n$ and $D_n := (A^*)^n A^n$ for $n \in \mathbb{N}$, then the following assertions hold.*

a) *For every n the formulas*

$$B_n AA^* = \frac{1}{n+1}B_{n+1} + \frac{n}{n+1}B_n A^*A + n \cdot \frac{\lambda}{2}B_n$$

and

$$D_n A^*A = \frac{1}{n+1}D_{n+1} + \frac{n}{n+1}D_n AA^* + n \cdot \frac{\lambda}{2}D_n$$

hold.

b) *We have*

$$[B_n, B_m] = [B_m, D_n] = [D_n, D_m] = 0$$

for every $m, n \in \mathbb{N}$.

c) *The formula*

$$B_m D_n AA^* = \frac{n+1}{m+n+1}B_{m+1}D_n + \frac{m}{m+n+1}B_m D_{n+1} + \lambda \cdot \frac{m(n+1)}{2}B_m D_n$$

holds for every $m, n \in \mathbb{N}$.

d) *For every positive integer n, the operator $(AA^*)^n$ is a linear combination of the products $B_p D_q$ with $1 \le p \le n, 0 \le q \le n$.*

e) *If n is a positive integer such that $A^n = (A^*)^n = 0$ (see the Lemma 2), then thc equality*

$$(AA^*)^n \prod_{1 \le p \le n, 0 \le q \le n} \left(AA^* - \lambda \cdot \frac{p(q+1)}{2}\right) = 0$$

holds.

Proof. a) By Lemma 2 *iii*) we obtain
$B_n AA^* = A^n(A^*)^n AA^* = A^{n+1}(A^*)^{n+1} + A^n[(A^*)^n, A]A^*$

$$= B_{n+1} + A^n\left(n(A^*)^{n-1}[A^*, A] + \lambda \cdot \frac{n(n-1)}{2}(A^*)^{n-1}\right)A^*$$

$$= B_{n+1} + \lambda \cdot \frac{n(n-1)}{2}B_n + nA^n(A^*)^{n-1}[A^*, A]A^*$$

$$= B_{n+1} + \lambda \cdot \frac{n(n-1)}{2} B_n + nA^n(A^*)^n[A^*, A] + nA^n(A^*)^{n-1}[[A^*, A], A^*]$$

$$= B_{n+1} + \lambda \cdot \frac{n(n-1)}{2} B_n + nB_nA^*A - nB_nAA^* + \lambda n B_n$$

$$= B_{n+1} + \lambda \cdot \frac{n(n+1)}{2} B_n + nB_nA^*A - nB_nAA^*.$$

Consequently

$$(n+1)B_nAA^* = B_{n+1} + \lambda \cdot \frac{n(n+1)}{2} B_n + nB_nA^*A$$

and the first desired formula in a) follows. The second formula can be proved similarly using the fact that A and A^* play symmetric roles.

b) If either $n = 0$ or $m = 0$ the result is obvious. Furthermore $[B_1, D_1] = 0$ by Lemma 2 iv). On the other hand, using the formulas from a) it is easy to prove by induction that for every positive integer n there exist two polynomials $p_n, q_n \in \mathbb{C}[X, Y]$ such that $B_n = p_n(B_1, D_1)$ and $D_n = q_n(B_1, D_1)$. Hence by $[B_1, D_1] = 0$ we deduce the desired relations.

c) For $n = 0$ the desired formula reduces to the first formula from a). If $m = 0$ the desired formula reduces to $D_nAA^* = B_1D_n$, that is $D_nB_1 = B_1D_n$, which follows from b). Next assume $m, n \geq 1$. Then from the first formula in a) we deduce, in view of b), that

$$B_mD_nAA^* = (B_mAA^*)D_n = \left(\frac{1}{m+1} B_{m+1} + \frac{m}{m+1} B_mA^*A + \lambda \cdot \frac{m}{2} B_m\right) D_n.$$

Hence

$$B_mD_nAA^* = \frac{1}{m+1} B_{m+1}D_n + \lambda \cdot \frac{m}{2} B_mD_n + \frac{m}{m+1} B_mD_nA^*A. \tag{1}$$

Moreover from the second formula in a) we deduce that

$$B_m(D_nAA^*) = B_m \left(\frac{n+1}{n} D_nA^*A - \frac{1}{n} D_{n+1} - \lambda \cdot \frac{n+1}{2} D_n\right).$$

Hence

$$B_mD_nAA^* = -\frac{1}{n} B_mD_{n+1} - \lambda \cdot \frac{n+1}{2} B_mD_n + \frac{n+1}{n} B_mD_nA^*A. \tag{2}$$

Now we eliminate $B_mD_nA^*A$ between (1) and (2) and obtain

$$\frac{m+1}{m} B_mD_nAA^* - \frac{1}{m} B_{m+1}D_n - \lambda \cdot \frac{m+1}{2} B_mD_n$$

$$= \frac{n}{n+1} B_mD_nAA^* + \frac{1}{n+1} B_mD_{n+1} + \lambda \cdot \frac{n}{2} B_mD_n,$$

or

$$\left(\frac{m+1}{m} - \frac{n}{n+1}\right) B_m D_n AA^* = \frac{1}{m} B_{m+1} D_n + \lambda \cdot \frac{m+1}{2} B_m D_n$$

$$+ \frac{1}{n+1} B_m D_{n+1} + \lambda \cdot \frac{n}{2} B_m D_n.$$

Hence

$$\frac{m+n+1}{m(n+1)} B_m D_n AA^* = \frac{1}{m} B_{m+1} D_n + \lambda \cdot \frac{m+n+1}{2} B_m D_n + \frac{1}{n+1} B_m D_{n+1}.$$

The last formula is obviously equivalent to that desired.

d) The assertion folows by induction on n using *b*) and *c*).

e) In view of *d*) it suffices to prove that for every pair $(p, q) \in \{1, \cdots, n\} \times \{0, \cdots, n\}$ we have

$$B_p D_q \prod_{1 \le k \le n, 0 \le m \le n} \left(AA^* - \lambda \cdot \frac{k(m+1)}{2}\right) = 0.$$

Now let's define the *degree* of $B_p D_q$ as being $p + q$. Then

$$B_p D_q \left(AA^* - \lambda \cdot \frac{p(q+1)}{2}\right) = \frac{q+1}{p+q+1} B_{p+1} D_q + \frac{p}{p+q+1} B_p D_{q+1}$$

(see *c*)) is a linear combination of terms having degrees greater than $p + q$. Hence the relation (3) can be obtained by descending induction, using the fact that $B_n = D_n = 0$ by the hypothesis $A^n = (A^*)^n = 0$. □

Corollary 2. *If $A \in \mathcal{G}$ and there exists a non-zero complex number λ such that*

$$[[A, A^*], A] = \lambda A,$$

then the spectrum of the operator $[A, A^*]$ is a finite set.

Proof. By Lemma 2 *ii*) there exists a positive integer n such that $A^n = (A^*)^n = 0$. Hence by the Lemma 3 *e*) there exists a polynomial $p \in \mathbb{C}[X]$ such that $p(AA^*) = 0$. Consequently, the set $\sigma(AA^*)$ is finite. On the other hand, since A and A^* play symmetric roles (see Lemma 2 *i*)) we deduce also that $p(A^*A) = 0$. Hence the set $\sigma(AA^*)$ is also finite. Consequently, in view of the projection property of the Taylor joint spectrum (see e.g. Corollary 2 from §26) of the commuting system (A^*A, AA^*) (cf. Lemma 2 *iv*)) we deduce that this joint spectrum $\sigma(A^*A, AA^*)$ is finite. Now let's consider the polynomial $f \in \mathbb{C}[X, Y]$, $f(X, Y) = X - Y$. Then $[A^*, A] = f(A^*A, AA^*)$ hence by the spectral mapping theorem (see e.g. Corollary 1 or Theorem 4 from §26) we get

$$\sigma([A^*, A]) = f(\sigma(A^*A, AA^*)).$$

But we have already seen that $\sigma(A^*A, AA^*)$ is a finite set, hence $\sigma([A^*, A])$ is also finite. □

Proposition 3. *Let $\mathcal{H}$ be a Cartan subalgebra of $\mathcal{G}$ such that $\mathcal{H}$ is also a *-subalgebra of $\mathcal{G}$. Then the spectrum of any element of $\mathcal{H}$ is a finite set.*

Proof. Let R be the set of the non-zero roots of $\mathcal{G}$ with respect to $\mathcal{H}$. Let $\{H_\alpha, E_\alpha \mid \alpha \in R\}$ be a system of generators for $\mathcal{G}$ as in Theorem 2 from §5. Particularly we have

$$[H, E_\alpha] = \alpha(H)E_\alpha, \quad \alpha(H_\alpha) \neq 0, \tag{3}$$

$$[E_\alpha, E_{-\alpha}] = -H_\alpha, \tag{4}$$

$$(E_\alpha)^* = -E_{-\alpha}, \tag{5}$$

$$[H_\alpha, H_\beta] = 0 \tag{6}$$

for every $\alpha, \beta \in R$ and $H \in \mathcal{H}$. By (4) and (5) we deduce $[E_\alpha, (E_\alpha)^*] = H_\alpha$. Hence by (3) we obtain $[[E_\alpha, (E_\alpha)^*], E_\alpha] = \alpha(H_\alpha)E_\alpha$ with $\alpha(H_\alpha) \neq 0$. Consequently the spectrum of the operator $H_\alpha = [E_\alpha, (E_\alpha)^*]$ is finite in view of Corollary 2. Hence $(H_\alpha)_{\alpha\in R}$ is a commuting system (see (6)) of operators with finite spectra. Consequently its Taylor joint spectrum is finite by the projection property (see e.g. Corollary 9 from §26). By the spectral mapping theorem it then follows that the spectrum of every linear combination of the operators H_α ($\alpha \in R$) is finite (compare with the end of the proof of Corollary 2). But $(H_\alpha)_{\alpha\in R}$ is a basis of $\mathcal{H}$, hence the desired conclusion follows. □

Corollary 3. *If $T \in \mathcal{G}$ is $(i\mathcal{G}_u)$-normal (i.e. $[T, T^*] = 0$), then T is a Dunford scalar operator with finite spectrum.*

Proof. By Proposition 2 there exists a Cartan subalgebra $\mathcal{H}$ of $\mathcal{G}$ such that $T \in \mathcal{H}$ and $\mathcal{H}$ is a *-subalgebra of $\mathcal{G}$. Hence $\sigma(T)$ is a finite set by Proposition 3. On the other hand, T is a normal-equivalent operator by Proposition 1, hence the conclusion easily follows by Corollary 4 from §14. □

Corollary 4. *If $T \in \mathcal{G}$ and $adT : \mathcal{G} \to \mathcal{G}$ is a semisimple map, then T is a Dunford scalar operator with finite spectrum.*

Proof. Since the Lie algebra $\mathcal{G}$ is semisimple, the hypothesis implies the existence of a Cartan subalgebra $\mathcal{K}$ of $\mathcal{G}$ such that $T \in \mathcal{K}$. On the other hand, let $\mathcal{H}$ be a Cartan subalgebra of $\mathcal{G}$ such that $\mathcal{H}$ is a *-subalgebra (see Proposition 1). Then every element of $\mathcal{H}$ is $(i\mathcal{G}_u)$-normal (since $\mathcal{H}$ is abelian), hence $\mathcal{H}$ is a set of Dunford scalar operators with finite spectra, by Corollary 3.

But by Theorem 1 from §5 and the proof of Lemma 1 *a*) from §27 there exists an invertible operator $S \in \mathcal{B}(\mathcal{X})$ such that $S \cdot \mathcal{H} \cdot S^{-1} = \mathcal{K}$. Since $T \in \mathcal{K}$ and every element of $\mathcal{H}$ is a Dunford scalar operator with finite spectrum (as we have just seen), the desired conclusion follows immediately. □

The following result concerns another type of elements of $\mathcal{G}$.

Proposition 4. *If $T \in \mathcal{G}$ and $adT : \mathcal{G} \to \mathcal{G}$ is a nilpotent map, then the operator T is nilpotent.*

Proof. Let $\mathcal{H}$ be a Cartan subalgebra of $\mathcal{G}$ and R be the set of roots of $\mathcal{G}$ with respect to $\mathcal{H}$. Furthermore let $\{H_\alpha, E_\alpha \mid \alpha \in R\}$ be a system of generators for $\mathcal{G}$ as in Theorem 2 from §5 and endow R with a total ordering corresponding to T as in Theorem 3 from §5. Denote $R_+ = \{\alpha \in R \mid \alpha > 0\}$ and $R_- = \{\alpha \in R \mid \alpha < 0\}$. Consequently (see the assertion 3 in Theorem 3 from §5) we have

$$T \in \bigoplus_{\alpha \in R_+} \mathcal{G}^\alpha.$$

Moreover (by Theorem 3(2°) from §5) let $H \in \mathcal{H}$ be such that

$$\alpha(H) > 0 \text{ for every } \alpha \in R_+ .$$

We have (by Theorem 2 from §5)

$$\prod_{\alpha \in R_+} (ad\, H - \alpha(H)) = 0 \text{ on } \bigoplus_{\alpha \in R_+} \mathcal{G}^\alpha$$

hence if we denote $R_+ = \{\alpha_1, \ldots, \alpha_m\}$ and $a_i := \alpha_i(H) > 0\,(1 \leq i \leq m)$, we deduce

$$(ad\, H - a_1) \cdots (ad\, H - a_m)T = 0. \tag{7}$$

Since $H \in \mathcal{H}$, the map $ad\, H : \mathcal{G} \to \mathcal{G}$ is semisimple. Hence by Corollary 4 above and by Proposition 1 from §13, the operator $ad\, H : \mathcal{B}(\mathcal{X}) \to \mathcal{B}(\mathcal{X})$ has the single-valued extension property. Then by (7) and by the spectral mapping theorem for the local spectrum (see Theorem 1.1.6 from the book of I. Colojoară and C. Foiaş [1]) it is easy to deduce

$$\sigma_{ad\, H}(T) \subseteq \{a_1, \ldots, a_m\} \subseteq [a, \infty),$$

where

$$a := \min_{1 \leq i \leq m} a_i > 0.$$

This relation together with Corollary 1 from §13 implies by induction that

$$T^p \mathcal{X} = T^p \left(\mathcal{X}_H(\sigma(H))\right) \subseteq \mathcal{X}_H(\sigma(H) + [pa, \infty))$$

for every positive integer p. Hence for $p > \frac{1}{a} diam(\sigma(H))$ we have $T^p = 0$. □

Now we can establish one of the main facts of the present paragraph.

Theorem 1. *Let $\mathcal{G}$ be an ideally finite semisimple complex Lie subalgebra of $\mathcal{B}(\mathcal{X})$. Then every element of $\mathcal{G}$ is a Dunford scalar operator of finite type (hence generalized scalar operator) with finite spectrum. Moreover for $T \in \mathcal{G}$ the following assertions hold.*

i) *If* $ad T : \mathcal{G} \to \mathcal{G}$ *is a semisimple map, then there exist a finite family* $\mathcal{X}_1, \ldots, \mathcal{X}_m$ *of invariant subspaces of* T *and a finite family* $\lambda_1, \ldots, \lambda_m$ *of complex numbers such that* $\mathcal{X} = \mathcal{X}_1 \oplus \cdots \oplus \mathcal{X}_m$ *and* $T = \lambda_1 I_{\mathcal{X}_1} \oplus \cdots \oplus \lambda_m I_{\mathcal{X}_m}$.

ii) *If* $ad T : \mathcal{G} \to \mathcal{G}$ *is a nilpotent map, then* T *is a nilpotent operator.*

Proof. Let $T \in \mathcal{G}$ be arbitrary. Since $\mathcal{G}$ equals the direct sum of its simple finite-dimensional ideals (by Theorem 3 from §4), T will belong to the sum of a finite family of such ideals. Consequently we may assume without loss of generality that $\mathcal{G}$ itself is finite-dimensional.

Now, since $\mathcal{G}$ is semisimple, it is well known (see e.g. the definition 1.6.8 from the book of J. Dixmier [1]) that for $T \in \mathcal{G}$ we can find two elements $S, N \in \mathcal{G}$ such that $T = S + N$, $[S, N] = 0$, $ad S : \mathcal{G} \to \mathcal{G}$ is semisimple and $ad N : \mathcal{G} \to \mathcal{G}$ is nilpotent. The operator S is a Dunford scalar operator with finite spectrum by Corollary 4. (Particularly we have

$$\mathcal{X} = \bigoplus_{\lambda \in \sigma(S)} \mathcal{X}_S(\{\lambda\}) \text{ and } S = \bigoplus_{\lambda \in \sigma(S)} \lambda I_{\mathcal{X}_S(\{\lambda\})},$$

so the assertion *i*) holds.) Moreover, the operator N is nilpotent by Proposition 4 (which particularly implies the assertion *ii*)). Consequently, since $T = S + N$ and $[S, N] = 0$ we deduce that T is a Dunford scalar operator of finite type (since N is nilpotent); furthermore every Dunford scalar operator of finite type is a generalized scalar operator (see Theorem 4.3.6 from the book of I. Colojoară and C. Foiaş [1]). Also $\sigma(T) = \sigma(S)$ is a finite set. □

Concerning the associative hull of an ideally finite semisimple Lie algebra of operators we can prove:

Theorem 2. *Let* $\mathcal{A}$ *be the unital associative subalgebra of* $\mathcal{B}(\mathcal{X})$ *generated by the ideally finite semisimple Lie subalgebra* $\mathcal{G}$ *of* $\mathcal{B}(\mathcal{X})$. *Then every element of* $\mathcal{A}$ *is an algebraic operator, hence has finite spectrum. If moreover* $\mathcal{G}$ *is finite-dimensional, then* $\mathcal{A}$ *is also finite-dimensional.*

Proof. As in the proof of Theorem 1 we can assume that $\mathcal{G}$ is finite-dimensional. Then let $\mathcal{H}$ be a Cartan subalgebra of $\mathcal{G}$ and let

$$\mathcal{G} = \mathcal{N}_- \oplus \mathcal{H} \oplus \mathcal{N}_+$$

be one of the corresponding "triangular decompositions" of $\mathcal{G}$ (see Theorem 3(5°) from §5). Denote by $\{E_i^\pm \mid 1 \le i \le k\}$ a basis of $\mathcal{N}_\pm$ and by $\{H_j \mid 1 \le j \le r\}$ a basis of $\mathcal{H}$. Now the map $ad E_i^\pm : \mathcal{G} \to \mathcal{G}$ is nilpotent by the last assertion of Theorem 3 from §5, so, by Proposition 4 above, we deduce that there exists a positive integer m such that

$$(E_i^\pm)^m = 0 \quad \text{for} \quad 1 \le i \le k.$$

Consequently by the Poincaré-Birkhoff-Witt Theorem it follows that the unital associative algebra $\mathcal{A}$ (generated by $\mathcal{G}$) is spanned as a complex vector space by the following set

$$\{(E_1^-)^{\alpha_1} \cdots (E_k^-)^{\alpha_k} H_1^{\beta_1} \dots H_r^{\beta_r} (E_1^+)^{\gamma_1} \dots (E_k^+)^{\gamma_k} \mid$$

$$\alpha_1, \dots, \alpha_k, \gamma_1, \dots, \gamma_k \in \{0, \dots, m\} \text{ and } \beta_1, \dots, \beta_r \in \mathbb{N}\}.$$

Consequently, for proving that $\dim \mathcal{A} < \infty$ it will suffice to show that the unital associative algebra generated by $\mathcal{H}$ is finite-dimensional.

To this end, consider the commuting tuple $H = (H_1, \dots, H_r) \in \mathcal{B}(\mathcal{X})^r$. Since $H_1, \dots, H_r$ belong to the Cartan subalgebra $\mathcal{H}$, we can apply Theorem 1 *i*) to each of them. Then by the projection property we deduce that the Taylor joint spectrum $\sigma(H)$ is finite, let's say $\sigma(H) = \{\lambda_1, \dots, \lambda_q\}$ with $\lambda_j = (\lambda_{j1}, \dots, \lambda_{jr}) \in \mathbb{C}^r \quad (1 \le j \le q)$. Since $[H_k, H_l] = 0$ we have $H_k \mathcal{X}_{H_l}(\{\lambda_{jl}\}) \subseteq \mathcal{X}_{H_l}(\{\lambda_{jl}\})$ for $1 \le k, l \le r$ and $1 \le j \le q$. This easily implies

$$\mathcal{X} = \mathcal{X}_1 \oplus \cdots \oplus \mathcal{X}_q \tag{8}$$

where

$$\mathcal{X}_j := \bigcap_{l=1}^{r} Ker(H_l - \lambda_{jl}) \quad \text{for } 1 \le j \le q.$$

Hence

$$H_l = \lambda_{1l} I_{\mathcal{X}_1} \oplus \cdots \oplus \lambda_{ql} I_{\mathcal{X}_q} \quad (1 \le l \le r).$$

Then it is easily seen that the unital associative algebra generated by $H_1, \dots, H_r$ is contained in $\mathbb{C} I_{\mathcal{X}_1} \oplus \cdots \oplus \mathbb{C} I_{\mathcal{X}_q}$, hence it is finite-dimensional. As we have seen above, this finishes the proof of the fact that $\mathcal{A}$ is finite-dimensional. Particularly, since $\mathcal{A}$ is an associative finite-dimensional algebra, every element of it is algebraic. That is, for each $T \in \mathcal{G} \subset \mathcal{A}$ there exists a polynomial $p \in \mathbb{C}[X]$ such that $p(T) = 0$, and the proof finishes. □

Corollary 5. *Every finite-dimensional semisimple complex Lie subalgebra $\mathcal{G}$ of $\mathcal{B}(\mathcal{X})$ has finite-dimensional invariant subspaces. Actually $\mathcal{X}$ is the union of its finite-dimensional subspaces which are invariant for $\mathcal{G}$.*

Proof. Let $\mathcal{A}$ be the unital associative hull of $\mathcal{G}$. Then $\dim \mathcal{A} < \infty$ hence, for an arbitrary $x \in \mathcal{X}$, the set

$$\mathcal{A}x = \{Ax \mid A \in \mathcal{A}\}$$

is a finite-dimensional subspace containing x. Moreover, since $\mathcal{G} \cdot \mathcal{A}x \subseteq \mathcal{A} \cdot \mathcal{A}x \subseteq \mathcal{A}x$, it follows that $\mathcal{A}x$ is an invariant subspace for $\mathcal{G}$. □

§ 31 Semisimple Lie algebras of compact quasinilpotent operators

The present paragraph contains some results concerning norm closed Lie algebras of compact operators on a complex Banach space $\mathcal{X}$. We shall denote throughout by $\mathcal{K}(\mathcal{X})$ the set of all compact operators on $\mathcal{X}$. As usual, by *normed Lie algebra* we shall mean a Lie algebra endowed with a vector space norm such that its bracket (i.e. the Lie product) is a bounded bilinear map. By *representation of a normed Lie algebra* on a normed vector space we mean a bounded representation by bounded operators.

First of all let's notice a class of semisimple Lie algebras which are not necessarily ideally finite.

Proposition 1. *Let $\mathcal{G}$ be a normed Lie algebra. If $\mathcal{G}$ has no non-trivial closed ideals and* $\dim \mathcal{G} > 1$, *then $\mathcal{G}$ is semisimple (i.e. it has no non-trivial solvable ideal).*

Proof. First note that every non-zero ideal of $\mathcal{G}$ must be dense in $\mathcal{G}$ since the closure of an ideal is obviously a closed ideal. Now we assume that we can find a (not necessarily closed) solvable ideal $\mathcal{I}$ of $\mathcal{G}$, $\mathcal{I} \neq \{0\}$. Then there exists a positive integer n such that $\mathcal{I}^{(n-1)} \neq \{0\} = \mathcal{I}^{(n)}$. Since $\mathcal{I}^{(n)} = [\mathcal{I}^{(n-1)}, \mathcal{I}^{(n-1)}]$, it follows that $\mathcal{I}^{(n-1)}$ is an abelian non-zero ideal of $\mathcal{G}$. As we remarked above, the ideal $\mathcal{I}^{(n-1)}$ must be dense in $\mathcal{G}$. But this ideal is abelian, hence $\mathcal{G}$ is also abelian. But in this case every closed subspace of $\mathcal{G}$ is a closed ideal of $\mathcal{G}$, which contradicts the hypothesis. □

We shall need the following definition describing an extension of the concept of nilpotent Lie algebra.

Definition 1. Let $\mathcal{G}$ be a normed Lie algebra. Then $\mathcal{G}$ is called *ad-quasinilpotent Lie algebra* if $\lim_{n\to\infty} \|(adT)^n S\|^{\frac{1}{n}} = 0$ for every $T, S \in \mathcal{G}$.

Obviously every normed Lie algebra which is quasinilpotent (cf. Definition 5 from §2) is also ad-quasinilpotent.

As an easy consequence of the asymptotic formula for the commutators (see §15) we make the following observation.

Remark 1. Let $\mathcal{G}$ be an ad-quasinilpotent Lie subalgebra of $\mathcal{B}(\mathcal{X})$. (We consider $\mathcal{G}$ endowed with the operator norm.) If K is a spectral set of an operator $T \in \mathcal{G}$, then the spectral maximal subspace $\mathcal{X}_T(K)$ is invariant for the Lie algebra $\mathcal{G}$ by Theorem 9 from §21.

Now we can prove:

Theorem 1. *Let $\mathcal{G}$ be a closed Lie subalgebra of $\mathcal{K}(\mathcal{X})$ having no closed non-trivial ideals. If* $\dim \mathcal{G} > 1$ *and $\mathcal{G}$ is ad-quasinilpotent then every element of $\mathcal{G}$ is a quasinilpotent operator.*

Proof. Assume that for a certain $T \in \mathcal{G}$ we have $\sigma(T) \neq \{0\}$. Since T is a compact operator, it then follows that for a certain $\lambda \in \mathbb{C}\backslash\{0\}$ we have $0 < \dim \mathcal{X}_T(\{\lambda\}) <$

∞. But $\mathcal{X}_T(\{\lambda\})$ is an invariant subspace for $\mathcal{G}$ by the Remark 1. Consequently we can consider the following finite-dimensional representation of $\mathcal{G}$:

$$\rho : \mathcal{G} \to \mathcal{B}(\mathcal{X}_T(\{\lambda\})),\ S \mapsto S|_{\mathcal{X}_T(\{\lambda\})}.$$

Since $\rho(T) \neq 0$, it follows that $Ker\rho \neq \mathcal{G}$. Now, if $Ker\rho \neq \{0\}$ then $Ker\rho$ is a closed non-trivial ideal of $\mathcal{G}$, a contradiction. But if $Ker\rho = \{0\}$ then $\mathcal{G}$ is finite-dimensional (having a faithful finite-dimensional representation). Since $\mathcal{G}$ is ad-quasinilpotent, by Engel's Theorem it then follows that $\mathcal{G}$ is a nilpotent Lie algebra. But in this case $\mathcal{G}$ also has closed non-trivial ideals. □

Next we need the following auxiliary result.

Lemma 1. *Let $A, B \in \mathcal{B}(\mathcal{X})$. If A is a compact operator and there exists $\mu \in \mathbb{C}\backslash\{0\}$ such that $\lim_{n\to\infty} \|(ad\,A - \mu)^n B\|^{\frac{1}{n}} = 0$, then B is a finite-rank operator.*

Proof. Let $\sigma(A) \subset \cup_{i=0}^n D_i$ be a finite covering of $\sigma(A)$ with open discs of diameter strictly less than $|\mu|$; we may assume $0 \in D_0\backslash(\cup_{i=0}^n D_i)$. We write $F_i := D_i \cap \sigma(A)$ $(0 \le i \le n)$. Since A is a compact operator, each F_i is a closed set. Moreover, since A is particularly a decomposable operator we easily deduce

$$\mathcal{X} = \mathcal{X}_A(F_0) + \mathcal{X}_A(F_1) + \cdots + \mathcal{X}_A(F_m). \tag{1}$$

But $\dim \mathcal{X}_A(F_i) < \infty$ for $i = 1, \ldots, m$ because A is compact and $0 \notin F_1 \cup \cdots \cup F_m$. Particularly

$$\dim B(\mathcal{X}_A(F_i)) < \infty \quad (1 \le i \le m). \tag{2}$$

On the other hand the hypotheses imply $\sigma_{ad\,A}(B) = \{\mu\}$. Hence by Corollary 1 from §13 we deduce $B(\mathcal{X}_A(F_0)) \subseteq \mathcal{X}_A(F_0+\mu)$. But the diameter of F_0 is strictly less than $|\mu|$, so $(F_0 + \mu) \cap F_0 = \emptyset$. Since $0 \in F_0$ (we may assume $\dim \mathcal{X} = \infty$ without loss of generality) and A is a compact operator, it then follows that $\mathcal{X}_A(F_0 + \mu)$ is finite-dimensional and then by (1) and (2) we deduce $\dim B(\mathcal{X}) < \infty$. □

Now we can prove one of the main results of the present paragraph.

Theorem 2. *Let $\mathcal{G}$ be a norm closed Lie subalgebra of $\mathcal{K}(\mathcal{X})$, which has no closed non-trivial ideals. If $\dim \mathcal{G} > 1$ then $\mathcal{G}$ has at least one of the following properties:*

i) *Every element of $\mathcal{G}$ is a quasinilpotent operator.*
ii) *The finite-rank operators from $\mathcal{G}$ constitute a dense ideal which is locally finite.*

Proof. If $\mathcal{G}$ is ad-quasinilpotent then the property *i)* holds by Theorem 1. Next assume that we can find $A \in \mathcal{G}$ such that the operator $ad\,A|_{\mathcal{G}} : \mathcal{G} \to \mathcal{G}$ is not quasinilpotent. But the spectrum of the operator $ad\,A : \mathcal{B}(\mathcal{X}) \to \mathcal{B}(\mathcal{X})$ is at most countable by Rosenblum's theorem (see Theorem 1 from §13) because A is compact hence its spectrum is at most countable. On the other hand the topological boundary of the spectrum of the restriction of an operator to an invariant subspace consists of approximate eigenvalues, hence it is contained in the spectrum of the operator itself. Particularly the boundary of $\sigma(ad\,A|_{\mathcal{G}})$ is contained in the spectrum of $ad\,A$. Since this last set is at most countable it easily follows that $\sigma(ad\,A|_{\mathcal{G}})$ (coincides with its topological boundary and) is at most countable. Then we can take $\mu \in \sigma(ad\,A|_{\mathcal{G}})\backslash\{0\}$ such that the spectral maximal subspace $\mathcal{G}_{ad\,A|_{\mathcal{G}}}(\{\mu\})$ is non-zero. If we take a non-zero element B of this spectral maximal subspace then

$$\lim_{n\to\infty} \|(ad\,A - \mu)^n B\|^{\frac{1}{n}} = 0.$$

Hence $B \in \mathcal{G}$ is a finite-rank operator by Lemma 1. Particularly $\mathcal{I} \neq \{0\}$, where $\mathcal{I}$ denotes the set of the finite-rank operators from $\mathcal{G}$. It is well known that $\mathcal{I}$ is a locally finite ideal of $\mathcal{G}$. Next the norm closure of $\mathcal{I}$ is also an ideal of $\mathcal{G}$. But $\mathcal{G}$ has no closed non-trivial ideals by the hypothesis, hence $\mathcal{I}$ must be dense in $\mathcal{G}$. Consequently the property *ii*) holds. □

Our next aim is to prove that, in the framework of Theorem 2, if $\mathcal{G}$ has the property *i*) then it contains no non-zero finite-rank operator (see Theorem 3 below). Hence the property *i*) excludes the property *ii*). To this end we need some consequences of the well-known Lomonosov's Lemma which asserts that, if $\mathcal{A}$ is an associative subalgebra of $\mathcal{B}(\mathcal{X})$ without nontrivial invariant subspaces, then for each compact operator $B \in \mathcal{B}(\mathcal{X})$ with $B \neq 0$ there exist $x \in \mathcal{X}\backslash\{0\}$ and $A \in \mathcal{A}$ such that $ABx = x$.

Lemma 2. *Every associative subalgebra $\mathcal{A}$ of $\mathcal{B}(\mathcal{X})$ consisting only of compact quasinilpotent operators has a nontrivial invariant subspace if* $\dim \mathcal{X} > 1$.

Proof. If the conclusion does not hold then, by Lomonosov's Lemma, for $B \in \mathcal{A}$ we obtain an element $AB \in \mathcal{A}$ which is not quasinilpotent, thus contradicting the hypothesis. □

Corollary 1. *Every locally finite Lie subalgebra $\mathcal{G}$ of $\mathcal{B}(\mathcal{X})$ consisting only of compact quasinilpotent operators has a non-trivial invariant subspace if* $\dim \mathcal{X} > 1$. *Moreover every maximal nest in $Lat(\mathcal{G})$ is also maximal in the lattice $Lat\,\mathcal{X}$ of all closed subspaces of $\mathcal{X}$.*

Proof. The associative algebra generated by our Lie algebra $\mathcal{G}$ consists only of quasinilpotent operators by Theorem 2 from §23. Hence it has a non-trivial invariant subspace by Lemma 2. The last part of the statement is a straightforward consequence of the first one. □

Now we can prove:

Theorem 3. *Let $\mathcal{G}$ be a norm closed Lie subalgebra of $\mathcal{K}(\mathcal{X})$ which is separable and has no closed non-trivial ideal. If* $\dim \mathcal{G} > 1$ *and every element of $\mathcal{G}$ is a quasinilpotent operator, then $\mathcal{G}$ contains no non-zero finite-rank operator.*

Proof. Assume that $\mathcal{G}$ contains non-zero finite-rank operators. Then the ideal $\mathcal{I}$ of the finite-rank operators from $\mathcal{G}$ must be dense in $\mathcal{G}$ because its norm closure is a non-zero closed ideal of $\mathcal{G}$. But by Corollary 1 there exists a nest $\mathcal{N} \subseteq Lat(\mathcal{I})$ such that $\mathcal{N}$ is maximal in the lattice of all closed subspaces of $\mathcal{X}$. Since $\mathcal{I}$ is dense in $\mathcal{G}$ we have $Lat(\mathcal{I}) = Lat(\mathcal{G})$, so $\mathcal{N} \subseteq Lat(\mathcal{G})$.

First assume that the following condition is satisfied:

(o) For every $\mathcal{Y} \in \mathcal{N}$ there exists $T \in \mathcal{G}$ such that $T|_{\mathcal{Y}} \neq 0$.

Note that for every $\mathcal{Y} \in \mathcal{N}$ we have $\mathcal{Y} \in Lat(\mathcal{G})$ hence the set

$$\mathcal{J}(\mathcal{Y}) := \{G \in \mathcal{G} \mid G|_{\mathcal{Y}} = 0\}$$

is a norm closed ideal of $\mathcal{G}$ and $\mathcal{J}(\mathcal{Y}) \neq \mathcal{G}$ by (o). If we can prove that $\mathcal{J}(\mathcal{Y}) \neq \{0\}$ for a certain $\mathcal{Y} \in \mathcal{N}$, then we shall obtain a contradiction with the hypothesis that $\mathcal{G}$ has no closed nontrivial ideals.

To this end, take a non-zero finite-rank operator $A \in \mathcal{G}$. Since $\mathcal{N}$ is a maximal nest in the lattice of all closed subspaces of $\mathcal{X}$, we have

$$\bigcap_{\mathcal{Y} \in \mathcal{N}} \mathcal{Y} = \{0\}.$$

Particularly,

$$\bigcap_{\mathcal{Y} \in \mathcal{N}} (\mathcal{Y} \cap Ran\, A) = \{0\}.$$

But $\{\mathcal{Y} \cap Ran\, A \mid \mathcal{Y} \in \mathcal{N}\}$ is a nest of subspaces of the finite-dimensional vector space $Ran\, A$, hence for a certain $\mathcal{Y}_0 \in \mathcal{N}$ we have $\mathcal{Y}_0 \cap Ran\, A = \{0\}$. Since $A\mathcal{Y}_0 \subseteq \mathcal{Y}_0$ it is easily seen that $A|_{\mathcal{Y}_0} = 0$, so $A \in \mathcal{J}(\mathcal{Y}_0)$. Consequently $\mathcal{J}(\mathcal{Y}_0) \neq \{0\}$. As we have seen, this finishes the proof in the case when the condition (o) holds.

Next assume that the condition (o) is not satisfied. Denote by $\mathcal{Y}_1$ the maximal element of $\mathcal{N}$ with the property $T|_{\mathcal{Y}_1} = 0$ for every $T \in \mathcal{G}$. Moreover we denote by $\rho : \mathcal{G} \to \mathcal{B}(\mathcal{X}/\mathcal{Y}_1)$ the natural representation induced by the factorization of the operators. If we denote by $\pi : \mathcal{X} \to \mathcal{X}/\mathcal{Y}_1$ the canonical projection, then obviously $\{\pi(\mathcal{Y}) \mid \mathcal{Y} \in \mathcal{N}\} \subseteq Lat(\rho(\mathcal{G}))$ and $\{\pi(\mathcal{Y}) \mid \mathcal{Y} \in \mathcal{N}\}$ is a maximal nest in the lattice of all closed subspaces of $\mathcal{X}/\mathcal{Y}_1$. If this nest satisfies a condition of type (o) with respect to the Lie algebra $\rho(\mathcal{G})$, then as above, we obtain a $\mathcal{Y} \in \mathcal{N}$ such that $\{T \in \mathcal{G} \mid \rho(T)|_{\pi(\mathcal{Y})} = 0\}$ is a closed nontrivial ideal of $\mathcal{G}$. This contradicts the hypothesis. On the other hand, if the condition of type (o) is not satisfied, then there exists $\mathcal{Y}_2 \in \mathcal{N}$ such that $\mathcal{Y}_2 \supset \mathcal{Y}_1$, $\mathcal{Y}_2 \neq \mathcal{Y}_1$ and $\rho(T)|_{\pi(\mathcal{Y}_2)} = 0$ for every $T \in \mathcal{G}$. In other words we have $T\mathcal{Y}_2 \subseteq \mathcal{Y}_1$ for every $T \in \mathcal{G}$. Hence $ST = 0$ on $\mathcal{Y}_2$ for every $S, T \in \mathcal{G}$, in view of the choice of $\mathcal{Y}_1$.

Now, consider the representation

$$\tau : \mathcal{G} \to \mathcal{B}(\mathcal{Y}_2),\ T \mapsto T|_{\mathcal{Y}_2}.$$

By Proposition 1 we have $[\mathcal{G}, \mathcal{G}] \neq \{0\}$. But $[\mathcal{G}, \mathcal{G}] \subseteq Ker\,\tau$, since for $S, T \in \mathcal{G}$ we have

$$\tau([S,T]) = ST|_{\mathcal{Y}_2} - TS|_{\mathcal{Y}_2} = 0$$

in view of the previous remarks. On the other hand, $Ker\,\tau$ is a closed nontrivial ideal of $\mathcal{G}$, which contradicts the hypothesis. Consequently $\mathcal{G}$ cannot contain non-zero finite-rank operators. □

Notes

The notion of $\mathcal{A}$-normal operator is introduced here for the first time. The other facts from §29 appeared in M. Şabac [8] and D. Beltiţă [2]. The commutativity of the elements of a complex vector space consisting of normal operators on some Hilbert space was noted by H. Radjavi and P. Rosenthal [1].

Lemmas 2–3 from §30 were proved by J.R. Schue [2]. In the same paper were proved Corollaries 2–3, but only for Hilbert space operators. Corollary 5 was proved by D. Gurarie [1] in another way. All the other facts from §30 are due to D. Beltiţă and appear here for the first time.

Except for Proposition 1, all the other facts from §31 were established by W. Wojtynski [2]. However, our proof of Lemma 1 is different from the original one, inasmuch as we use Theorem 2 from §13. It is worth pointing out that we adopted the term "*ad*-quasinilpotent" used by C.R. Miers [6]. This type of Lie algebras is called "nilpotent" in W. Wojtynski [2], "quasinilpotent" in W. Wojtynski [3] and V.S. Shul'man and Yu.V. Turovskiĭ [1]. In the present book we call "quasinilpotent" another class of Lie algebras (see Definition 5 from §2).

Bibliography

ALAGIA, H.

1. Cartan subalgebras of Banach-Lie algebras of operators. *Pacific J. Math.* 98(1982), 1–15.

ALBRECHT, E.

1. On some classes of generalized spectral operators. *Arch. der Math.* 30(1978), 297–303.

AMBROZIE, C.G.; VASILESCU, F.-H.

1. *Banach Space Complexes.* Kluwer, 1995.

ANDERSON, J.; FOIAŞ, C.

1. Properties which normal operators share with normal derivations and related operators. *Pacific J. Math.* 61(1975), 313–325.

APOSTOL, C.

1. Inner derivations with closed range. *Revue Roum. Math. Pures Appl.* 21 (1976), 249–265.

APOSTOL, C.; STAMPFLI, J.

1. On derivation ranges. *Indiana Univ. Math. J.* 25(1976), 857–869.

AYUPOV, S.A.

1. Skew commutators and Lie isomorphisms in real von Neumann algebras. *J. Funct. Anal.* 138(1996), 170–187.

BARNES, B.A.; KATAVOLOS, A.

1. Properties of quasinilpotents in some operator algebras. *Proc. Roy. Irish Acad. Sect.* A 93(1993), 155–170.
2. Corrigendum to : "Properties of quasinilpotents in some operator algebras". *Proc. Roy. Irish Acad. Sect.* A 95(1995), 249–250.

BARNES, D.W.

1. On the cohomology of soluble Lie algebras. *Math. Z.* 101(1967), 343–349.

BELTIŢĂ, D.

1. Elements of spectral theory for solvable Lie algebras. Master's thesis, Bucharest, 1996 (in Romanian).
2. On certain Lie algebras of normal operators. *Revue Roum. Math. Pures Appl.* 43(1998), 653–658.

3. Spectrum for a solvable Lie algebra of operators. *Studia Math.* 135(1999),163–178.
4. Spectral conditions for the nilpotency of Lie algebras. *J. Operator Theory* (to appear).
5. Analytic joint spectral radius in a solvable Lie algebra of operators. Preprint, 1999.

BELTIŢĂ, D.; ŞABAC, M.

1. An asymptotic formula for the commutators. *J. Funct. Anal.* 153(1998), 262–275.

BERENGUER, M.I.; VILLENA, A.R.

1. Continuity of Lie mappings of skew elements of Banach algebras with involution. *Proc. Amer. Math. Soc.* 126(1998), 2717–2720.

BHATIA, R.; ROSENTHAL, P.

1. How and why to solve the operator equation AX–XB=Y. *Bull. London Math. Soc.* 29(1997), 1–21.

BOASSO, E.

1. Dual properties and joint spectra for solvable Lie algebras of operators. *J. Operator Theory* 33(1995), 105–110.
2. On the joint spectra of the two-dimensional Lie algebra of operators in Hilbert spaces. *Rev. Un. Mat. Argentina* 40(1996), 101–109.
3. On the spectral set of a solvable Lie algebra of operators. *Portugaliae Math.* 54(1997), 85–94.
4. Joint spectra and nilpotent Lie algebras of linear transformations. *Linear Algebra Appl.* 263(1997), 49–62.
5. Tensor products and joint spectra for solvable Lie algebras of operators. *Collect. Math.* 49(1998), 9–16.
6. On the joint spectral radius of a nilpotent Lie algebra of matrices. *Studia Math.* 132(1999), 15–27.

BOASSO, E.; LAROTONDA, A.

1. A spectral theory for solvable Lie algebras of operators. *Pacific J. Math.* 158(1993), 15–22.

BONSALL, F.F.; DUNCAN, J.

1. *Numerical ranges of operators on normed spaces and of elements of normed algebras.* Cambridge Univ. Press, 1971.
2. *Numerical ranges* II. Cambridge Univ. Press, 1973.

CIGLER, G.; DRNOVŠEK, R.; KOKOL-BUKOVŠEK, D.; LAFFEY, T.J.; OMLADIČ, M.; RADJAVI, H.; ROSENTHAL, P.

1. Invariant subspaces for semigroups of algebraic operators. *J. Funct. Anal.* 160(1998), 452–465.

COLOJOARĂ, A.

1. Sur l'algèbre symétrique de certains espaces de Banach. *Revue Roum. Math. Pures Appl.* 18(1973), 1345–1369.

2. Algèbre symétrique du dual d'un espace nucléaire-DF. *Revue Roum. Math. Pures Appl.* 23(1978), 1317–1339.
3. Sur le théorème de Poincaré-Birkhoff-Witt. *Revue Roum. Math. Pures Appl.* 25(1980), 1327–1347.
4. Banach algebras category and Banach Lie algebras category as categories with multiplication. *An. Univ. București Mat.* 39(1990), no. 1–2, 3–5.

COLOJOARĂ, I.; FOIAȘ, C.

1. *Theory of Generalized Spectral Operators.* Gordon and Breach, 1968.

CORTET, J.-C.

1. Sur l'intégrabilité des algèbres de Lie de dimension finie de dérivations d'une C*-algèbre. *C. R. Acad. Sci. Paris, Sér.* A 285(1977), 187–190.

CUENCA MIRA, J.A.; GARCÍA MARTÍN, A.; MARTIN GONZÁLEZ, C.

1. Structure theory for L*-algebras. *Math. Proc. Camb. Phil. Soc.* 107(1990), 361–365.

DIXMIER, J.

1. *Algèbres enveloppantes.* Gauthier-Villars, 1974.

DOSIEV, A.A.

1. Holomorphic functions on the polydisk corresponding to a basis of a nilpotent Lie algebra. *Preprint* 3(1998), *Institute of Mathematics and Mechanics of the Azerbaijan Academy* (in Russian).
2. Słodkowski spectra of representations of a nilpotent Lie algebra. *Preprint* 4(1998), *Institute of Mathematics and Mechanics of the Azerbaijan Academy* (in Russian).

DOWSON, H.R.

1. *Spectral theory of linear operators.* Academic Press, 1978.

DRNOVŠEK, R.; CIGLER, G.; KOKOL-BUKOVŠEK, D.; OMLADIČ, M.; LAFFEY, T.J.; RADJAVI, H.; ROSENTHAL, P. See CIGLER, G.; DRNOVŠEK, R.; KOKOL-BUKOVŠEK, D.; LAFFEY, T.J.; OMLADIČ, M.; RADJAVI, H.; ROSENTHAL, P..

DUNCAN, J.; BONSALL, F.F. See BONSALL, F.F.; DUNCAN, J..

DUNFORD, N.; SCHWARTZ, J.T.

1. *Linear operators. Part* I. Interscience, 1958.
2. *Linear operators. Part* II. *Spectral theory. Self adjoint operators in Hilbert space.* Interscience, 1963.
3. *Linear operators. Part* III. *Spectral operators.* Interscience, 1971.

ESCHMEIER, J.

1. *Analytische Dualität und Tensorprodukte in der mehrdimensionalen Spektraltheorie.* Schriftenreihe des Math. Inst. der Univ. Münster, Heft 42, 2. Serie, 132 pp., 1987.

VAN EST, W.T.; KORTHAGEN, T.J.

1. Non-enlargible Lie algebras. *Nederl. Akad. Wetensch. Proc.* A 26(1964), 15–31.

VAN EST, W.T.; ŚWIERCZKOWSKI, S.

1. The path functor and faithful representability of Banach Lie algebras. *J. Austral. Math. Soc.* 16(1973), 471–482.

FACK, T.; DE LA HARPE, P.

1. Sommes de commutateurs dans les algèbres de von Neumann finies continues. *Ann. Inst. Fourier (Grenoble)* 30(1980), 49–73.

FAINSHTEIN, A.S.

1. Taylor joint spectrum for families of operators generating nilpotent Lie algebras. *J. Operator Theory* 29(1993), 3–27.
2. Fredholm families of operators generating nilpotent Lie algebras. *Obzornik mat. fiz.* 46(1999), no.2, 35–36.
3. On the Fredholm index of the Koszul complex for modules over nilpotent Lie algebras. Preprint, 2000.
4. Corrections to my paper "Taylor joint spectrum for families of operators generating nilpotent Lie algebras". *J. Operator Theory* (to appear).

FARFOROVSKAYA, YU.B.

1. Functions of operators and their commutators in perturbation theory. In: *Functional analysis and operator theory* (pages 147–159). Banach Center Publications No. 30, 1994.

FOIAŞ, C.

1. Invariant paraclosed subspaces. *Indiana Univ. Math. J.* 20(1971), 897–900.

FOIAŞ, C.; ANDERSON, J. See ANDERSON, J.; FOIAŞ, C..

FOIAŞ, C.; COLOJOARĂ, I. See COLOJOARĂ, I.; FOIAŞ, C..

FOIAŞ, C.; ŞABAC, M.

1. A generalization of Lie's theorem IV. *Revue Roum. Math. Pures Appl.* 14 (1974), 605–608.

FONG, C.K.; MIERS, C.R.; SOUROUR, A.R.

1. Lie and Jordan ideals of operators on Hilbert space. *Proc. Amer. Math. Soc.* 84(1982), 516–520.

FONG, C.K.; MURPHY, G.J.

1. Ideals and Lie ideals of operators. *Acta Sci. Math.* 51(1987), 441–456.

FONG, C.K.; RADJAVI, H.

1. On ideals and Lie ideals of compact operators. *Math. Ann.* 262(1983), 23–28.

FÖRSTER, K.-H.; NAGY, B.

1. Lie and Jordan ideals in $\mathcal{B}(c_0)$ and $\mathcal{B}(l^p)$. *Proc. Amer. Math. Soc.* 117(1993), 673–677.

FRUNZĂ, ŞT.

1. The Taylor spectrum and spectral decompositions. *J. Funct. Anal.* 19(1975), 390–421.
2. Generalized weights for operator Lie algebras. In: *Spectral theory* (pages 281–287). Banach Center Publications, No. 8, 1982.
3. Jordan operators on Hilbert spaces. *J. Operator Theory* 18(1987), 201–212.

García Martín, A.; Cuenca Mira, J.A.; Martin González, C. See Cuenca Mira, J.A.; García Martín, A.; Martin González, C..

Grabovskaya, R.Ya.

1. On a formula for the permutation of functions of operators representing a semisimple complex Lie algebra. In *Operator equations in function spaces* (pages 24–27, 131). Voronezh. Gos. Univ., Voronezh, 1986 (in Russian).

Grabovskaya, R.Ya.; Krein, S.G.

1. On the commutation formula for functions of operators representing a Lie algebra. *Funktsional. Anal. i Prilozhen.* 7(1973), 81 (in Russian).

Grosu, C.; Vasilescu, F.-H.

1. The Künneth formula for Hilbert complexes. *Integral Equations Operator Theory* 5(1982), 1–17.

Gurarie, D.

1. Banach uniformly continuous representations of Lie groups and algebras. *J. Funct. Anal.* 36(1980), 401–407.

Gurarie, D.; Lyubich, Yu.I.

1. An infinite-dimensional analogue of Lie's theorem concerning weights. *Funktsional. Anal. i Prilozhen.* 7(1973), 41–44 (in Russian).

Gurarie, D.; Vaksman, L.L.

1. Algebras which contain a compact operator. *Funktsional. Anal. i Prilozhen.* 8(1974), 81–82 (in Russian).

Halmos, P.R.

1. *A Hilbert space problem book.* Springer, 1982.

de la Harpe, P.

1. *Classical Banach-Lie algebras and Banach-Lie groups of operators in Hilbert space.* Lecture Notes in Math., No. 285, Springer, 1972.
2. The algebra of compact operators does not have any finite-codimensional ideal. *Studia Math.* 66(1979), 33–36.
3. Classical groups and classical Lie algebras of operators. *Proc. Symp. Pure Math.* 38(1982), Part I, 477–513.

de la Harpe, P.; Fack, T. See Fack, T.; de la Harpe, P..

Harris, L.A.; Kaup, W.

1. Linear algebraic groups in infinite dimensions. *Illinois J. Math.* 21(1977), 666–674.

Harte, R.

1. Relatively invariant systems and the spectral mapping theorem. *Bull. Amer. Math. Soc.* 79(1973), 138–142.
2. The spectral mapping theorem for quasi-commuting systems. *Proc. Roy. Irish Acad. Sect.* A 73A(1973), 7–18.

HAUSNER, M.; SCHWARTZ, J.T.

1. *Lie Groups; Lie Algebras.* Gordon and Breach, 1968.

HELEMSKIĬ, A. YA.

1. *The homology of Banach and topological algebras.* Kluwer Academic Publishers, 1989.

HOFFMANN, K.H.

1. Hyperplane subalgebras of real Lie algebras. *Geometriae Dedicata* 36(1990), 207–224.

HUDSON, T.D.; MARCOUX, L.W.; SOUROUR, A.R.

1. Lie ideals in triangular operator algebras. *Trans. Amer. Math. Soc.* 350(1998), 3321–3339.

JACOBSON, N.

1. Rational methods in the theory of Lie algebras. *Ann. of Math.* 36(1935), 875–881.
2. *Lie algebras.* Interscience, 1962.

KANTOROVITZ, SH.

1. *Spectral Theory of Banach Space Operators.* Lecture Notes in Math., No. 1012, Springer, 1983.
2. C^n-operational calculus, non-commutative Taylor formula and perturbation of semigroups. *J. Funct. Anal.* 113(1993), 139–152.

KAPLANSKI, I.

1. *Lie algebras and locally compact groups.* The Univ. of Chicago Press, 1971.

KATAVOLOS, A.; BARNES, B.A. See BARNES, B.A.; KATAVOLOS, A..

KATAVOLOS, A.; STAMATOPOULOS, C.

1. Commutators of quasinilpotents and invariant subspaces. *Studia Math.* 128 (1998), 159–169.

KAUP, W.; HARRIS, L.A. See HARRIS, L.A.; KAUP, W..

KISSIN, E.V.

1. On some reflexive algebras of operators and the operator Lie algebras of their derivations. *Proc. London Math. Soc.* 49(1984), 1–35.
2. Invariant subspaces for derivations. *Proc. Amer. Math. Soc.* 102(1988), 95–101.

KOKK, A.

1. *Joint spectral theory and extension of nontrivial multiplicative linear functionals.* Dissertationes Math. Univ. Tartuensis, no. 10, 1995.

KOKOL-BUKOVŠEK, D.; CIGLER, G.; DRNOVŠEK, R.; LAFFEY, T.J.; OMLADIČ, M.; RADJAVI, H.; ROSENTHAL, P. See CIGLER, G.; DRNOVŠEK, R.; KOKOL-BUKOVŠEK, D.; LAFFEY, T.J.; OMLADIČ, M.; RADJAVI, H.; ROSENTHAL, P..

KORTHAGEN, T.J.; VAN EST, W.T. See VAN EST, W.T.; KORTHAGEN, T.J..

KOSZUL, J.L.

1. Homologie et cohomologie des algèbres de Lie. *Bull. Soc. Math. France* 78 (1950), 65–127.

KREIN, S.G.; GRABOVSKAYA, R.YA. See GRABOVSKAYA, R.YA.; KREIN, S.G..

KREIN, S.G.; SHIHVATOV, A.M.

1. Linear differential equations on Lie groups. *Funktsional. Anal. i Prilozhen.* 4(1970), 52–61 (in Russian).

LABROUSSE, J.-PH.

1. Les opérateurs quasi-Fredholm: une généralisation des opérateurs semi Fredholm. *Rendiconti del Circ. Mat. di Palermo, Ser.* II, 29(1980), 161–258.

LAFFEY, T.J.; CIGLER, G.; DRNOVŠEK, R.; KOKOL-BUKOVŠEK, D.; OMLADIČ, M.; RADJAVI, H.; ROSENTHAL, P. See CIGLER, G.; DRNOVŠEK, R.; KOKOL-BUKOVŠEK, D.; LAFFEY, T.J.; OMLADIČ, M.; RADJAVI, H.; ROSENTHAL, P..

LAROTONDA, A.; BOASSO, E. See BOASSO, E.; LAROTONDA, A..

LOMONOSOV, V.I.

1. Invariant subspaces for the family of operators which commute with a completely continuous operator. *Funktsional. Anal. i Prilozhen.* 7(1973), 55–56 (in Russian).

LUMER, G.; ROSENBLUM, M.

1. Linear operator equations. *Proc. Amer. Math. Soc.* 10(1959), 32–41.

LYUBICH, YU.I.

1. *Introduction to the theory of Banach representations of groups.* Birkhäuser, 1988.

LYUBICH, YU.I.; GURARIE, D. See GURARIE, D.; LYUBICH, YU.I..

MAC LANE, S.

1. *Homology.* Springer, 1963.

MARCOUX, L.W.

1. On the closed Lie ideals of certain C^*-algebras. *Integral Equations Operator Theory* 22(1995), 463–475.

MARCOUX, L.W.; HUDSON, T.D.; SOUROUR, A.R. See HUDSON, T.D.; MARCOUX, L.W.; SOUROUR, A.R..

MARTIN GONZÁLEZ, C.; CUENCA MIRA, J.A.; GARCÍA MARTÍN, A. See CUENCA MIRA, J.A.; GARCÍA MARTÍN, A.; MARTIN GONZÁLEZ, C.

MIERS, C.R.

1. Lie isomorphisms in factors. *Trans. Amer. Math. Soc.* 147(1970), 55–63.
2. Lie homomorphisms of operator algebras. *Pacific J. Math.* 38(1971), 717–735.
3. Lie derivations of von Neumann algebras. *Duke Math. J.* 40(1973), 403–410.
4. Derived ring isomorphisms of von Neumann algebras. *Canad. J. Math.* 25 (1973), 1254–1268.
5. Closed Lie ideals in operator algebras. *Canad. J. Math.* 33(1981), 1271–1278.
6. A note on Lie nilpotency in operator algebras. *Studia Math.* 85(1987), 55–59.

MIERS, C.R.; FONG, C.K.; SOUROUR, A.R. See FONG, C.K.; MIERS, C.R.; SOUROUR, A.R..

MIERS, C.R.; PHILLIPS, J.

1. Algebraic inner derivations on operator algebras. *Canad. J. Math.* 35(1983), 710–723.

MURPHY, G.J.

1. Triangularizable algebras of compact operators. *Proc. Amer. Math. Soc.* 84(1982), 354–356.
2. Lie ideals in associative algebras. *Canad. Math. Bull.* 27(1984), 10–15.

MURPHY, G.J.; FONG, C.K. See FONG, C.K.; MURPHY, G.J..

MURPHY, G.J.; RADJAVI, H.

1. Associative and Lie subalgebras of finite codimension. *Studia Math.* 76(1983), 81–85.

NAGY, B; FÖRSTER, K.-H. See FÖRSTER, K.-H.; NAGY, B..

OTT, C.

1. A note on a paper of E. Boasso and A. Larotonda. *Pacific J. Math.* 173(1996), 173–180.
2. The Taylor spectrum for solvable operator Lie algebras. *Preprint*, 1996.
3. *Gemeinsame Spektren auflösbarer Operator-Liealgebren.* Dissertation, Kiel, 1997 (http://analysis.math.uni-kiel.de/wrobel/).

PESTOV, V.

1. Nonstandard hulls of Banach-Lie groups and algebras. *Nova J. Algebra Geom.* 1(1992), 371–381.
2. Free Banach-Lie algebras, couniversal Banach-Lie groups, and more. *Pacific J. Math.* 157(1993), 137–144.
3. Enlargable Banach-Lie algebras and free topological groups. *Bull. Austral. Math. Soc.* 48(1993), 13–22.
4. Regular Lie groups and a theorem of Lie-Palais. *J. Lie Theory* 5(1995), 173–178.
5. Correction to: "Free Banach-Lie algebras, couniversal Banach-Lie groups, and more". *Pacific J. Math.* 171(1995), 585–588.
6. On the centre of free Banach-Lie algebras and a problem studied by van Est and Świerczkowski. *Preprint*, 1994.

PHILIPS, J.; MIERS, R.C. See MIERS, C.R.; PHILLIPS, J..

RADJAVI, H.

1. The Engel-Jacobson theorem revisited. *J. of Algebra* 111(1987), 427–430.

RADJAVI, H.; CIGLER, G.; DRNOVŠEK, R.; KOKOL-BUKOVŠEK, D.; LAFFEY, T.J.; OMLADIČ, M.; ROSENTHAL, P. See CIGLER, G.; DRNOVŠEK, R.; KOKOL-BUKOVŠEK, D.; LAFFEY, T.J.; OMLADIČ, M.; RADJAVI, H.; ROSENTHAL, P..

RADJAVI, H.; FONG, C.K. See FONG, C.K.; RADJAVI, H..

RADJAVI, H.; MURPHY, G.J. See MURPHY, G.J.; RADJAVI, H..

RADJAVI, H.; ROSENTHAL, P.

1. On invariant subspaces and reflexive algebras. *Amer. J. Math.* 91(1969), 683–692.

READ, C.J.

1. Quasinilpotent operators and the invariant subspace problem. *J. London Math. Soc.* 56(1997), 595–606.

ROBART, T.

1. Sur l'intégrabilité des sous-algèbres de Lie en dimension infinie. *Canad. J. Math.* 49(1997), 820–839.

ROSENBLUM, M.

1. On the operator equation BX-XA=Q. *Duke Math. J.* 23(1956), 263–270.

ROSENBLUM, M.; LUMER, G. See LUMER, G.; ROSENBLUM, M..

ROSENTHAL, P.; BHATIA, R. See BHATIA, R.; ROSENTHAL, P..

ROSENTHAL, P.; CIGLER, G.; DRNOVŠEK, R.; KOKOL-BUKOVŠEK, D.; LAFFEY, T.J.; OMLADIČ, M.; RADJAVI, H. See CIGLER, G.; DRNOVŠEK, R.; KOKOL-BUKOVŠEK, D.; LAFFEY, T.J.; OMLADIČ, M.; RADJAVI, H.; ROSENTHAL, P..

ROSENTHAL, P.; RADJAVI, H. See RADJAVI, H.; ROSENTHAL, P..

ŞABAC, M.

1. Une généralisation du théorème de Lie. *Bull. Sci. Math.* 95(1971), 53–57.
2. A generalization of Lie's theorem III. *Revue Roum. Math. Pures Appl.* 19 (1974), 825.
3. Infinite dimensional variants of Lie's theorem for solvable Lie algebras. *St. cerc. şt. mat.* 26(1974), 1241–1278 (in Romanian).
4. A generalization of Lie's theorem II. *Revue Roum. Math. Pures Appl.* 20 (1975), 961–970.
5. Solvable Lie algebras of operators on a Banach space. *Revue Roum. Math. Pures Appl.* 23(1978), 489–493.
6. Scalar generalized operators and irreducible representations of a Lie algebra. *Revue Roum. Math. Pures Appl.* 25(1980), 823–825.
7. Irreducible representations of infinite dimensional Lie algebras. *J. Funct. Anal.* 52(1983), 303–314.
8. Lie subalgebra of normal elements in a Lie algebra with involution. *J. Operator Theory* 31(1994), 319–326.
9. Correction to: "Lie subalgebra of normal elements in a Lie algebra with involution". *J. Operator Theory* 34(1995), 189.
10. Nilpotent elements and solvable actions. *Collect. Math.* 47(1996), 91–104.
11. Some classes of solvable Lie algebras. *An. Univ. Bucureşti Mat.* 46(1997), 89–94.
12. Analytic and polynomial commutativity. In: *Proceedings of the 16th Operator Theory Conference*, THETA, 1997.

13. Taylor type formula for commutators of analytic functional calculus in a complex Banach space. *Revue Roum. Math. Pures Appl.* 43(1998), 231–234.
14. Localization of Wielandt-Wintner Theorem. To appear.

ŞABAC, M.; BELTIŢĂ, D. See BELTIŢĂ, D.; ŞABAC, M..

ŞABAC, M.; FOIAŞ, C. See FOIAŞ; ŞABAC, M..

SAMOILENKO, YU.S.; SHUL'MAN, V.S.

1. On representations of the relations $i[A, B] = f(A) + g(B)$. *Ukrain. Math. Zb.* 43(1991), 110–114 (in Russian).

SCHUE, J.R.

1. Hilbert space methods in the theory of Lie algebras. *Trans. Amer. Math. Soc.* 95(1960), 69–80.
2. Cartan decompositions for L* algebras. *Trans. Amer. Math. Soc.* 98(1961), 334–349.

SCHWARTZ, J.T.; DUNFORD, N. See DUNFORD, N.; SCHWARTZ, J.T..

SCHWARTZ, J.T.; HAUSNER, M. See HAUSNER, M.; SCHWARTZ, J.T..

SEGAL, I.E.

1. Infinite-dimensional irreducible representations of compact semi-simple groups. *Bull. Amer. Math. Soc.* 70(1964), 155–160.

SERRE, J.P.

1. *Lie algebras and Lie groups.* Lectures given at Harward University, Benjamin, 1965.

SHERMAN, T.

1. Representations of Lie algebras by normal operators. *Proc. Amer. Math. Soc.* 16(1965), 1125–1129.

SHIHVATOV, A.M.; KREIN, S.G. See KREIN, S.G.; SHIHVATOV, A.M..

SHUL'MAN, V.S.

1. On transitivity of some operator spaces. *Funktsional. Anal. i Prilozhen.* 16 (1982) (1), 91–92 (in Russian).
2. On invariant subspaces of compact operators. *Funktsional. Anal. i Prilozhen.* 18(1984)(2), 85–86(in Russian).
3. Invariant subspaces and spectral mapping theorems. In: *Functional analysis and operator theory* (pages 313–325). Banach Center Publications, No. 30, 1994.

SHUL'MAN, V.S.; SAMOILENKO, YU.S. See SAMOILENKO, YU.S.; SHUL'MAN, V.S..

SHUL'MAN, V.S.; TUROVSKIĬ, YU.V.

1. Joint spectral radius and invariant subspaces. *Funktsional. Anal. i Prilozhen.* (to appear) (in Russian).
2. Joint spectral radius, operator semigroups and a problem of W. Wojtyński. *J. Funct. Anal.* (to appear).

3. Solvable Lie algebras of compact operators have invariant subspaces. In: *Spectral and evolutionary problems*, Vol. 9 (Sevastopol, 1998), 38–44, Simferopol State Univ., Simferopol, 1999.

SINGER, I.M.

1. Uniformly continuous representations of Lie groups. *Ann. of Math.* 56(1952), 242–247.

SŁODKOWSKI, Z.

1. An infinite family of joint spectra. *Studia Math.* 61(1977), 239–255.

SOUROUR, A.R.; FONG, C.K.; MIERS, C. See FONG, C.K.; MIERS, C.R.; SOUROUR, A.R..

SOUROUR, A.R.; MARCOUX, L.W.; HUDSON, T.D. See MARCOUX, L.W.; HUDSON, T.D.; SOUROUR, A.R..

STAMPFLI, J.; APOSTOL, C. See APOSTOL, C.; STAMPFLI, J..

STEWART, I.

1. *Lie algebras generated by finite dimensional ideals.* Pitman, 1975.

ŚWIERCZKOWSKI, S.; VAN EST, W.T. See VAN EST, W.T.; ŚWIERCZKOWSKI, S..

TAYLOR, J.L.

1. A joint spectrum for several commuting operators. *J. Funct. Anal.* 6(1970), 172–191.
2. The analytic functional calculus for several commuting operators. *Acta Math.* 125(1970), 1–38.
3. Homology and cohomology for topological algebras. *Advances in Math.* 9 (1972), 137–182.
4. A general framework for a multi-operator functional calculus. *Advances in Math.* 9(1972), 183–252.
5. Functions of several noncommuting variables. *Bull. Amer. Math. Soc.* 79 (1973), 1–34.

TUROVSKIĬ, YU.V.

1. On the quasinilpotence of the generalized commutators I. Manuscript in VINITI, No. 1240-82, 1982 (in Russian).
2. Spectral properties of the elements of normed algebras and invariant subspaces. *Funktsional. Anal. i Prilozhen.* 18(1984)(2), 77–78 (in Russian).
3. Polynomial spectral mapping property of Harte spectrum for n-commutative families of elements in Banach algebras. In: *Spectral Theory of Operators and its Applications* (pages 152–177), No. 5, ELM, 1984 (in Russian).
4. On spectral properties of some Lie subalgebras and spectral radius of subsets of a Banach algebra. In: *Spectral Theory of Operators and its Applications* (pages 144–181), No. 6, ELM, 1985 (in Russian).
5. On commutativity modulo the Jacobson radical of the associative envelope of a Lie algebra. In: *Spectral Theory of Operators and its Applications* (pages 199–211), No. 8, ELM, 1987 (in Russian).

6. Volterra semigroups have invariant subspaces. *J. Funct. Anal.* 162(1999), 313–322.

TUROVSKIĬ, YU.I.; SHUL'MAN, V.S. See SHUL'MAN, V.S.; TUROVSKIĬ, YU.I..

UPMEIER, H.

1. *Symmetric Banach manifolds and Jordan C*-algebras.* North Holland, 1985.

VAKSMAN, L.L.; GURARIE, D. See GURARIE, D.; VAKSMAN, L.L..

VASILESCU, F.-H.

1. Radical d'une algèbre de Lie de dimension infinie. *C. R. Acad. Sci. Paris, Sér.* A-B 274(1972), A536–A538.
2. On Lie's theorem in operator algebras. *Trans. Amer. Math. Soc.* 172(1972), 365–372.
3. Normed Lie algebras. *Canad. J. Math.* 24(1972), 580–591.
4. Positive forms on Lie algebras with involution. *Revue Roum. Math. Pures Appl.* 18(1973), 951–957.
5. *Analytic functional calculus and spectral decompositions.* Ed. Academiei and D. Reidel Co., 1982.

VASILESCU, F.-H.; AMBROZIE, C.G. See AMBROZIE, C.G.; VASILESCU, F.-H.

VASILESCU, F.-H.; GROSU, C. See GROSU, C.; VASILESCU, F.-H..

VERONA, A.

1. *Introduction to Lie algebra cohomology.* Ed. Academiei, 1974 (in Romanian).

VILLENA, A.R.; BERENGUER, M.I. See BERENGUER, M.I.; VILLENA, A.R..

WINTER, D.

1. Cartan subalgebras of a Lie algebra and its ideals. *Pacific J. Math.* 33(1970), 263–273.

WOJTYNSKI, W.

1. Engel's theorem for nilpotent Lie algebras of Hilbert-Schmidt operators. *Bull. Acad. Pol. Sci., Ser. Sci. Math. Astronom. Phys.* 24(1976), 797–801.
2. Banach-Lie algebras of compact operators. *Studia Math.* 59(1977), 263–273.
3. Quasinilpotent Banach-Lie algebras are Baker-Campbell-Hausdorff. *J. Funct. Anal.* 153(1998), 405–413.

ŻELAZKO, W.

1. A characterization of Šilov boundary in function algebras. *Ann. Soc. Math. Pol. Ser.* I *Coment. Math. Prace Mat.* 14(1970), 59–64.
2. On a certain class of non-removable ideals in Banach algebras. *Studia Math.* 44(1972), 87–92.

ZELMANOV, E.I.

1. On Engel Lie algebras. *Sibirsk. Mat. Zb.* 29(1988), 112–117 (in Russian).

Index